T0377220

Turbomachinery
Concepts, Applications, and Design

Turbomachinery
Concepts, Applications, and Design

By
V. Dakshina Murty
University of Portland

CRC Press is an imprint of the
Taylor & Francis Group, an **informa** business

CRC Press
Taylor & Francis Group
6000 Broken Sound Parkway NW, Suite 300
Boca Raton, FL 33487-2742

CRC Press is an imprint of Taylor & Francis Group, an Informa business

Printed on acid-free paper

International Standard Book Number-13: 978-1-138-64069-6 (Hardback)

Visit the Taylor & Francis Web site at
http://www.taylorandfrancis.com

and the CRC Press Web site at
http://www.crcpress.com

Cover page photo credit to "Portland District, U.S. Army Corps of Engineers."

In

The Charans of

Huzur Radhasoami Dayal

Contents

Preface

This book is an outgrowth of several years of teaching an undergraduate course on turbomachinery at the University of Portland. It provides an introduction to the subject and is suitable as a single-semester course for seniors and first-year graduate students in disciplines such as mechanical, aerospace, petroleum, and chemical engineering. The prerequisites for the course are one semester in each of thermodynamics and fluid mechanics.

The scope of the book has been defined by two objectives. The first is to provide a unified treatment of the subject of the common forms of a turbomachine while covering the basic fluid mechanics and thermodynamics with a detailed discussion of the governing equations. The second objective is to apply the material to various machines in sufficient detail so that the underlying performance and design factors can be appreciated. With these objectives in mind, topics such as turbine blade cooling, rotor vibrations, computational fluid dynamics, and analysis of the flow, have been omitted as these are best dealt with in specialized courses at graduate level.

The book starts with dimensional analysis for rotating machinery followed by the basic theory of turbomachines. The remainder of the book is divided into two parts. The first part deals with incompressible flow machines such as hydraulic turbines, pumps, fans, and blowers. The second part deals with compressible flow machines such as gas turbines, compressors, and steam turbines. Several worked examples that provide an illustration of the main points have been included in all chapters. In addition, several example problems are included at the end of each chapter. The text uses both SI and English systems of units. Throughout the book, the focus has been on the understanding of the basic principles of various types of turbomachines. With this in mind, only rudimentary design aspects are presented. This is partly due to my belief that design is best taught in an industry setting and any design must be preceded by sound theory and analysis.

I would like to express my sincere gratitude to my mentor, friend, and former colleague Dr. Larry Simmons for a thorough review of the manuscript and critique of various parts of the book, especially on pumps and steam turbines. Over the years, several students took my course ME 415 Turbomachinery offered in the Donald P. Shiley School of Engineering. Their suggestions are appreciated. Thanks are also due to my parents who encouraged me to excel in my profession. I would like to express my sincere appreciation to my children Prem Kiran and Vasant Swarup, who helped in the editing of the book. Finally, I would like to acknowledge my dear wife Lalitha for her constant encouragement and support during this endeavor.

List of Symbols

a	Speed of sound	(m/s; ft./s)
B	Height of wicket gates	(m; ft.)
c_0	Spouting velocity	(m/s;ft./s)
C_D	Drag coefficient	
C_L	Lift coefficient	
C_p	Specific heat at constant pressure	(J/kg·K; ft·lbf/slug·°R)
C_v	Specific heat at constant volume	(J/kg·K; ft·lbf/slug·°R)
C_T	Thrust ratio	
D	Diameter of the rotor (characteristic dimension)	(m; ft.)
D_s	Specific diameter (dimensionless)	
DH	DeHaller number	
E	Energy per unit mass = gH	(m^2/s^2; $ft.^2/s^2$)
f	Friction factor; frequency	
g	Acceleration due to gravity	(m/s^2; $ft./s^2$)
h_f	Head loss due to friction	(m; ft.)
h	Specific enthalpy	(m^2/s^2; $ft.^2/s^2$)
H	Head of machine	(m; ft.)
H_s	Setting of turbine above tails	(m; ft.)
k	Ratio of specific heats	
$\dot{m}$	Mass flow rate	(kg/s; slug/s)
M	Mach number = V/a	
n_B	Number of blades	
N	Rotational speed of the rotor	(rpm; rad/s)
N_s	Dimensionless specific speed	
p	Pressure; number of poles	(Pa; psi)
p_v	Vapor pressure	(Pa; psi)
P	Power	(kW; hp)
q	Axial induction factor	
Q	Volumetric flow rate	(m^3/s; $ft.^3/s$)
Q_F	Quality factor	
r	Radius	(m; ft.)
R	Gas constant; degree of reaction	(J/kg·K; ft·lbf/slug·°R)
R_F	Reheat factor	
Re	Reynolds number = VD/V	
T	Temperature	(K; °R)
u	Blade velocity	(m/s; ft./s)

V	Absolute velocity of fluid	(m/s; ft./s)
v_s	Specific volume	(m^3/kg; $ft.^3$/slug)
w	Relative velocity	(m/s; ft./s)

Greek Symbols

α	Fluid or nozzle angle	(deg; deg)
β	Blade angle	(deg; deg)
ε	Utilization factor	
ϕ	Speed ratio u/V; flow coefficient	
η	Efficiency	
λ	Loss coefficient	
μ_s	Slip coefficient	
π_i	Dimensionless groups	
ρ	Density	(kg/m^3; $slugs/ft.^3$)
σ	Cavitation parameter	
Ψ	Head coefficient, blade loading coefficient	
Ω	Angular velocity = $2\pi N/60$	(rad/s)

Subscripts

0	Stagnation or total condition
1	Inlet condition for pumps or hydraulic turbines; inlet to stator for turbine stage
2	Outlet condition for pumps or hydraulic turbines; inlet to rotor in turbine stage
3	Outlet of rotor for turbine stage
a	Axial direction
f	Friction
i	Inlet
j	Jet conditions
n	Nozzle
o	Outlet
r	Radial direction
s	Constant entropy
tt	Total to total
ts	Total to static
u	Tangential direction
z	Axial direction

1

Introduction and History

The word *turbomachine* is derived from the Latin word *turbo*, which means whirl, or something that spins. The term describes various types of machines involving a pressure head, such as compressors, pumps, blowers, turbines, and so on. The common feature of all these machines is that they have a rotating shaft on which vanes that come into contact with the working fluid are mounted. Such contact produces a change in momentum of the working fluid, and this results in motion of the vanes or increased pressure on the fluid. A turbomachine can be defined as a device in which energy transfer occurs between a fluid in motion and a rotating shaft due to dynamic action that results in changes in pressure and fluid momentum. These are different from positive displacement machines such as reciprocating piston cylinders, because in such machines the work input or output is primarily due to moving boundaries. The development of turbomachines can be classified separately into that of hydraulic machines and gas/steam turbines, including compressors.

From a historical perspective, turbomachines have been around since the time of the Romans, who used paddle-type water wheels for grinding grain around 70 BC. Around the same time, the Chinese used similar machines, also in grinding mills. About a century later, in 62 AD, the Greek engineer Hero built the first steam turbine. Known as the *aeolipile*, it was a spherical vessel containing water and worked on the principle of reaction. As the water was heated, jets of steam issued out of the nozzles on the sides producing torque and thereby motion. Because of the miniscule amount of power it produced, this remained just a toy and its usefulness has essentially remained confined to science projects. However, from a modern turbomachinery point of view, it was a pure reaction machine. An interesting account of the effects of Hero's invention on modern turbomachinery is given by Lyman (2004). Another device that is very similar in principle but works on water jets through the reaction principle is Barker's mill. In this device, water enters the center of a rotor under a static head and emerges tangentially through arms protruding radially, thus providing torque through the reaction of the jet. This is similar in principle to a lawn sprinkler. Although the following centuries saw the invention of several types of water wheels, they were mostly used in grinding mills, water supply, food production, and mining.

The development of modern turbomachinery, as it pertains to hydraulic machines, can be traced to the Swiss mathematician Leonhard Euler. While working at the Berlin Academy of Sciences with his son, Albert, Euler published the now-famous "Euler turbine equation" in 1754. It is based on Newton's second law that torque is proportional to the rate of change of angular momentum of a fluid. This ushered in a more scientific approach to the design and analysis of compressors, turbines, pumps, and other such devices. The term *turbine* itself was not coined until 1822, when a Frenchman named Claude Burdin used the Latin word *turbo, turbinis* to describe "that which spins, as a spinning top." Most of these machines designed during the eighteenth and nineteenth centuries were predominantly waterwheels, either hydraulic turbines or pumps. It was Burdin's student Benoit Fourneyron who improved his teacher's work and is credited with building the first high-efficiency hydraulic turbine in 1824. His turbine achieved an efficiency of close to 85%, which was significantly higher than that of similar devices operating at the time which had efficiencies well below 50%. It is also interesting to note that his turbine was of the radial outflow type. These types of turbines were first introduced into the United States in 1843, at almost the same time as an axial flow turbine called the *Jonval turbine*, another European design, was invented.

The first inward flow turbine can be attributed to Poncelot, who conceptualized it in 1826. The credit for building the first one in 1838 goes to Howd, who also obtained a patent on it. However, it was James B. Francis who extensively analyzed and tested it for several years from 1849. Although the initial designs were purely radial, they were subsequently modified to accommodate mixed flows. Thus, all modern inward mixed flow turbines can be traced back to James Francis, and they are generally called *Francis turbines*. They are highly efficient and account for a major part of the hydroelectric power produced in the world today.

The other type of reaction turbine was conceived by a Czech professor named Victor Kaplan and was not developed until much later. It is named after him, and is an axial flow type. Viktor Kaplan obtained his first patent for an adjustable blade propeller turbine in 1912. Although he was not very successful in the initial designs due to cavitation problems, subsequent design modifications made the turbines suitable for low head applications such as rivers. Today, all axial flow adjustable pitch propeller turbines bear his name. They are highly efficient at both full and partial loads and especially suitable for large flows and low heads.

The third type of hydraulic turbine, which is quite different from the reaction turbines described earlier, is the Pelton turbine. This works on the principle of high-velocity jets impinging on a set of blades. Thus, the turbine extracts energy from the impulse of moving water. The high velocity is obtained by transporting the water from very large heads through pipes called *penstock* and converting it to jets that are then directed, either singly or multiply, onto blades. The entire energy conversion takes place before the water enters the runner and, as such, there is no pressure drop in the Pelton turbine.

Similar to hydraulic turbines, pumps also date back several hundreds of years. The earliest pump concept can be traced to around 2000 BC, when the Egyptians used "shadoofs" to lift water. These devices consisted of long poles mounted on a seesaw with a bucket mounted at one end and a rope attachment on the other. Around 200 BC, the Archimedean screw pump was devised to lift both liquids and mixtures of liquids and solids. Many inventions of the past five to six hundred years, including gear pumps, piston pumps, and plunger pumps belong to the class of positive displacement pumps. In the eighteenth and nineteenth centuries, increased interest and research in moving fluids led to the discovery of various types of pumps and their subsequent design modifications. Although the first patent for a centrifugal pump can be traced to the British engineer John Gwynne in 1851, the first mention of a vaned diffuser, by Sir Osborne Reynolds, did not occur for another quarter of a century, in 1875, in connection with the so-called turbine pump. The following two decades saw several pump manufacturers enter the pump industry. They included Sulzer, Rateau, Byron Jackson, Parsons, Allis Chalmers, and Worthington.

Compared with the invention of hydraulic turbines and pumps, the arrival of compressible flow machines, such as compressors and gas/steam turbines, is more recent. Except for Hero's turbine, which was more of a toy or experiment with no power output, the earliest modern steam turbine can be traced to medieval times, when Taqi-al-Din produced a prime mover for rotating a spit in 1551. Mechanical details of this device are given by Hassan (1976). Almost a century later, John Wilkins and Giovanni Branca produced similar devices, in 1629 and 1648. Although James Watt built the steam engine in 1776, it was based on a reciprocating piston and cylinder, a mechanism that is not a turbine. In 1831, William Avery built the first useful steam turbine and obtained a patent for it. It was an extension of Hero's turbine from almost two millennia earlier! From an engineering perspective, the first steam turbine that had a major impact was built by Sir Charles Parsons in 1884. It was a multistage axial flow reaction turbine that produced 10 hp when spinning at 18,000 rpm. Around the same time, in France, Auguste Rateau developed a pressure-compounded impulse turbine in 1900. In the United States, Charles G. Curtis constructed a velocity-compounded two-stage steam turbine in 1901. As demand for power grew in the twentieth century, interest in steam turbines also continued to grow. Some of the important developments included the harnessing of regenerative feed heating in 1920 and of the reheat cycle in 1925. There was also a steady increase in the inlet pressures over the years. Entry pressure into the turbine, which was about 15 bar at the end of World War I, increased to 30 bar by 1930, and further increased to 100 bar during the 1960s.

Unlike the steam turbine, which forms a part of the Rankine cycle, the development of compressors, which are major part of the Brayton cycle, took place over a much longer period. It needs to be emphasized that the term *steam turbine* here signifies the complete steam plant or cycle, including the boiler, feed pump, and the turbine itself. Compared with the gas turbine cycle, the steam turbine is much easier to design, construct, and operate. This is mainly due to the ease with which water can be made to flow through a boiler and the ease with which the resulting high-pressure steam is directed to the turbine to produce power. Even in the most inefficient scenarios, the turbine produces enough work to drive the feed pump, and thus net power output is always positive from the cycle. This is true for a reciprocating-type expander also; that is, the power produced by the expander is greater than the power consumed by the compressor.

In contrast, the major problem facing gas turbine cycles is the difficulty in obtaining appreciable pressure rises in compressors, especially those of the axial flow type. Since the flow in turbines is from high to low pressure, they are always successful, and, with some care, can also be designed to perform quite efficiently. However, compressors in general, and axial flow compressors in particular, are inherently inefficient due to the adverse pressure gradients. Early designs had such low efficiencies that the power produced by the turbine was not enough to drive the compressor, and, as such, the machines produced by the inventors never ran without external power input. The first US patent for a complete gas turbine was obtained by Charles Curtis in 1895. Following the unsuccessful efforts of Stolze in the early 1900s, Rene Armangand, along with Charles Lemale, produced a gas turbine in 1906 that produced positive net power, albeit at an efficiency of 3%. Other pioneers in the early stages of gas turbine research were Hans Holzwarth, Brown Boveri, and Sanford Moss. The first successful industrial gas turbine engine was created by Aurel Stodola in 1936. He used a twenty-stage axial compressor driven by a five-stage axial turbine. The isentropic efficiency of the compressor was around 85%.

Research on the application of gas turbines to aviation started around the same time as that on commercial applications. Efforts were underway almost simultaneously in the United Kingdom by Frank Whittle and in Germany by Hans von Ohain, Herbert Wagner, and Helmut Schelp. Although Whittle was the earliest of the four to complete the design in 1929, he had to overcome combustion problems with the liquid fuel and also material failure due to high temperature. It was not until 1942, when blades made of nickel–cadmium–cobalt were available, that he could make a successful flight, and his engine was taken to the General Electric laboratories in the United States. Whittle's counterpart in Germany, Hans von Ohain, was more successful, since he avoided the liquid combustion problem by using hydrogen. His first successful flight was in 1939, three years ahead of Whittle's. Herbert Wagner and Helmut Schlep had varying degrees of success in the late 1930s in terms of building a working engine.

Since the end of World War II, gas turbines have been the prime movers of civilian and military aircraft. Several of the smaller companies that existed during the period immediately after the war (1945–1950) either disappeared or consolidated into bigger companies. Today, the three major companies producing the largest engines are General Electric, Pratt and Whitney, and Rolls Royce.

A few closing comments are in order. As opposed to gas and steam turbines, hydraulic turbines exhibit much higher efficiencies. It is not uncommon for hydraulic turbines to have efficiencies as high as 93%–94%. Such high efficiencies coupled with no fuel cost (water power is free!) have resulted in almost no interest in research into hydraulic turbines. A similar argument applies to pumps. Consequently, the major research interest in these two types of machines has been confined to making runner materials more cavitation resistant. In terms of environmental considerations, there is also interest in making hydraulic turbines more "fish friendly." Besides these two aspects, there is little incentive to improve the performance of hydro turbines or pumps. However, for gas turbines, due to their extensive use in aviation, there has been a constant need to maximize the thrust-to-weight ratio, especially for military applications. Hence, there has been consistent support from the military for gas turbine research. This scenario is unlikely to change in the near future.

PROBLEMS

The following problems are of the study and discussion type.

1.1 Perform a literature search and make a timeline for hydraulic machines. Include both hydraulic turbines and pumps.

1.2 Repeat Problem 1 for gas turbines (including compressors) and steam turbines.

2

Dimensional Analysis

The technique of dimensional analysis is most useful in any field of engineering that involves some form of convective transport. These include fluid mechanics, heat transfer, soil mechanics, hydraulics and hydrology, and environmental transport. This technique has been successfully applied in the study of turbomachines, especially hydraulic turbines and pumps. Model testing and accompanying dimensional analysis are imperative in these areas, because the large size of the machines makes full-scale testing impossible.

Upon completion of this chapter, the student will be able to

- Use Buckingham's Pi theorem and the exponent method to identify the relevant dimensionless quantities in turbomachines
- Select turbines and pumps based on specific speed
- Perform preliminary sizing of turbomachines based on specific speed

2.1 Introduction

Unlike problems within the field of solid mechanics, those associated with thermal sciences in mechanical engineering require extensive experimentation. Since most engineering applications are well within the elastic range for stresses and associated deflections, the resulting equations in solid mechanics are linear, and the underlying mathematics is more amenable to exact or at least computer solutions. On the other hand, most problems involving fluid flow or heat transfer are usually unsteady, turbulent, and three-dimensional. These problems are very difficult to formulate mathematically, let alone solve. This being the case, experimentation and some form of empiricism are necessary in their study.

In the field of turbomachinery, the size precludes the possibility of conducting full-scale tests. For instance, the runner diameters of the Francis units at the Grand Coulee dam are 32 feet! To do performance testing on such units, it becomes necessary to use geometrically similar models, much smaller in size. In such tests, the full-scale structure is called the *prototype* and the replica is called the *model*. The performance of the *prototype* can be predicted by testing the *model* using the principle of similitude. In this context, the three types of similitude that are used in model testing are geometric similitude, kinematic similitude, and dynamic similitude. Geometric similitude implies that the ratios of all linear dimensions of the model and prototype are the same, in addition to them having similar shapes. Kinematic similitude implies that kinematic variables, such as blade velocities, relative velocities, flow speeds, and so on, should be in the same proportion for the model and prototype. This is assured when the velocity triangles for the model and prototype are similar. Finally, dynamic similitude implies that the ratios of all forces on the model and prototype are the same. This requires that all the relevant dimensionless numbers involving dynamic quantities are the same for both model and prototype.

2.2 Dimensional Analysis

Since experimentation is inherently an expensive process, it becomes necessary to organize the data in a manner that minimizes the amount of data collection. If this is not done carefully, the number of different experiments, and hence the number of data points, becomes excessive. It is therefore necessary to minimize the variables. Dimensional analysis is a formal procedure of grouping the variables into a smaller number of dimensionless groups. Since the number of such groups will be smaller than the original number of variables, the amount of experimentation will be reduced. Additionally, dimensional analysis provides a convenient method to organize important variables and thus provides better insight into the underlying physics of the problem.

To fix ideas, a hydraulic turbine is considered first. For any such machine, it is of interest to relate the head (specific energy or pressure), flow rate, power, torque, and other features, to size, speed, and fluid properties. These variables can be written as the flow rate Q, specific energy E (this is analogous to head, H, and is often written as gH, potential energy per unit mass), power P, speed N, a characteristic dimension D for size, density ρ, and viscosity μ. The dimensions of the variables are given in Table 2.1.

From this table, it can be seen that all seven variables can be written using three dimensions, namely, mass M, length L, and time T. According to Buckingham's Pi theorem (see Elger et al. (2012)) the maximum number of π groups that can be formed is the difference between the number of variables and the number of dimensions they represent, which in this case would be four (seven minus three). Using the so-called exponent method theorem (see Elger et al. (2012)), these variables can be combined into the following dimensionless groups:

$$\pi_1 = \frac{Q}{ND^3} = \varphi \qquad \pi_2 = \frac{E}{N^2D^2}$$
$$\pi_3 = \frac{P}{\rho N^3 D^5} = C_P \qquad \pi_4 = \frac{\mu}{\rho N D^2} \tag{2.1}$$

The groups π_1 and π_3 are called the *capacity* (or *flow*) *coefficient* and *power coefficient*, respectively. The group π_4 has the form of a reciprocal of Reynolds number and therefore represents the importance of viscous effects of the fluid. If it is recognized that the product ND is twice the peripheral velocity of the rotor u, then π_4 is the reciprocal of Reynolds number.

TABLE 2.1

Variables of Interest in Hydraulic Machines and Their Dimensions

Variable	Symbol	Dimensions
Flow rate	Q	L^3T^{-1}
Specific energy	E $(=gH)$	L^2T^{-2}
Power	P	ML^2T^{-3}
Rotational speed	N	T^{-1}
Size	D	L
Density	ρ	ML^{-3}
Viscosity	μ	$ML^{-1}T^{-1}$

Finally, π_2 is called the *head coefficient*. If E, the specific energy of the fluid, is replaced by its dimensional equivalent gH, then the equation for π_2 can be replaced by

$$\pi_2 = \frac{gH}{N^2D^2} = \psi \tag{2.2}$$

When dealing with hydraulic machines, whether they are power producing or power absorbing, it is common practice to refer to the head, H, on the machine. This is because centrifugal pumps are used to pump fluids between reservoirs located H units above the location of the pump. Similarly, all hydraulic turbines utilize the stored potential energy of the fluid to convert it into mechanical shaft power. Such potential energy is available either in the form of natural or artificially created reservoirs on rivers through dams. Hence, the total specific energy available to the fluid is equivalent to gH (less the losses). Thus, it is quite appropriate to replace E with gH when dealing with hydraulic machines.

The head and power coefficients can be combined by eliminating D from them. This can be achieved in several ways. However, if the combination is done in such a way that a unit integral exponent of the speed, N, is retained, the resulting variable is called the *specific speed*. This can be obtained by combining C_P and Ψ to obtain the following equation:

$$N_s' = \frac{C_P^{1/2}}{\psi^{5/4}} = \frac{N\sqrt{P}}{\sqrt{\rho}(gH)^{5/4}} \tag{2.3}$$

An alternate but perhaps more useful form of Equation 2.3 can be obtained by combining the head and flow coefficients and eliminating D from them (again retaining a unit exponent of N). The result is then

$$N_s = \frac{\varphi^{1/2}}{\psi^{3/4}} = \frac{N\sqrt{Q}}{(gH)^{3/4}} \tag{2.3a}$$

The variables N_s' and N_s described by Equations 2.3 and 2.3a are called the *turbine* and *pump specific speeds*, respectively. This is logical since the variables of most interest in turbines are the power P, speed N, and head H, and for pumps the variables are flow Q, speed N, and head H. A closer look at Equations 2.3 and 2.3a would immediately reveal that they are equivalent, since P, Q, and H are related by the expression

$$P = \rho g Q H \tag{2.4}$$

In writing Equation 2.4, care must be exercised to determine whether H represents the gross head or net head. Here, it is assumed that H is the net head (gross head minus losses). Otherwise, the hydraulic η efficiency needs to be introduced, which for most hydraulic machines operating at the design point is close to 90%. In Equation 2.4, the most basic definition of efficiency has been used, which is the ratio of useful energy out of the machine to the energy into the machine. There are several definitions of efficiency and these will be introduced in later chapters.

The dimensionless variable, specific speed, defined in Equations 2.3 and 2.3a, is one of the most useful parameters in sizing hydraulic machines. For turbines or pumps handling large discharges at low heads, the specific speed would be high; on the other hand, for high heads and low discharges the specific speed would be low. It is interesting to observe that the type (axial, radial, or mixed) of the machine in question is strongly influenced by the specific speed. For high specific speeds, which are due to high discharges (and/or low heads), a large flow area is required to accommodate these high flow rates. An axial flow machine would best accomplish this, since for such machines the flow area is proportional to D^2 (in fact $\sim\pi D^2$ less the area occupied by the hub). Kaplan turbines and axial flow fans fall into this category. These are designed for best efficiency when operating at low heads and high discharges. On the other hand, for low specific speed machines, which require low discharges and correspondingly high heads, centrifugal or radial flow type machines are most suitable. In such machines, since the flow is radial, the flow area ($\sim \pi Dt$, where t is the height or thickness of the blade) is proportional to the circumference, and therefore D, thus producing relatively low discharges. Also, since the flow moves radially, the change in centrifugal head produces high pressure changes. Examples of such machines are the Francis turbine and centrifugal pump. Thus, as the specific speed increases, the shape of the impeller changes from purely centrifugal, to mixed, to purely axial. These are shown pictorially in Figure 2.1.

It is also possible to use the specific speed to select the type of machine appropriate for given requirements. Listed in Table 2.2 are the ranges of specific speed corresponding to each type of hydraulic machine. Thus, the numerical value of the specific speed gives a useful guideline as to the type of hydraulic machine that needs to be used. It also turns out that specific speed is directly related to the most efficient hydraulic machine. The values for the variables in the calculation of specific speed, flow rate Q, head H, and rpm N, are picked at the design point or the best efficiency point.

One final comment needs to be made in connection with specific speeds. The variables N_s and $N_s^{'}$ as defined in Equations 2.3a and 2.3 are dimensionless. However, practicing

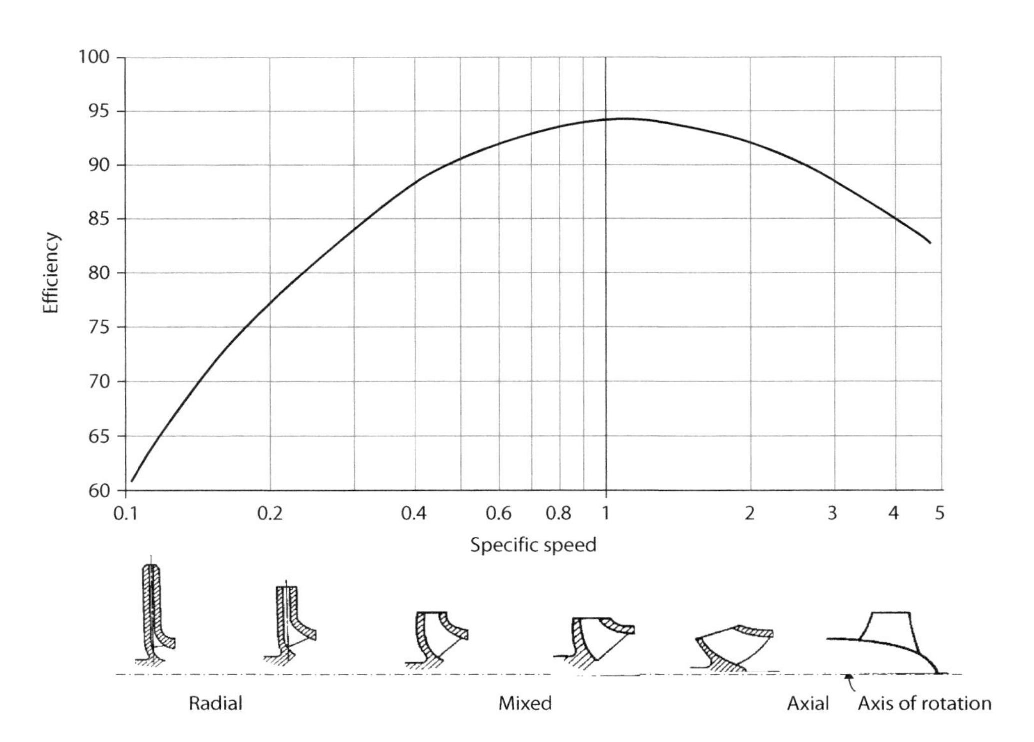

FIGURE 2.1
Variation of hydraulic runner shapes and efficiency with specific speed.

TABLE 2.2

Specific Speed for Various Hydraulic Machines

Turbomachine	Specific Speed[a] Range
Pelton wheel	0.03–0.3
Francis turbine	0.3–2.0
Kaplan turbine	2.0–5.0
Centrifugal pumps	0.2–2.5
Axial flow pumps	2.5–5.5
Radial flow compressors	0.5–2.0
Axial flow compressors, blowers	1.5–20.0
Axial flow steam and gas turbines	0.4–2.0

Source: Logan, Earl Jr., *Turbomachinery: Basic Theory and Applications*, 2nd edn, Marcel Dekker, 1993.
[a] Dimensionless form of specific speed given by Equation 2.3a.

engineers use a slightly different form. Since the values of ρ and g are constant, they are removed from the definitions. Thus, new expressions for specific speed can be written as

$$n_s' = \frac{N\sqrt{P}}{(H)^{5/4}}\ ; \qquad n_s = \frac{N\sqrt{Q}}{(H)^{3/4}} \tag{2.5}$$

where:
- N is in rpm
- Q is in gallons per minute
- H is in feet
- P is in horsepower

Unlike N_s and N_s', the values of n_s and n_s' given by Equation 2.5 are not dimensionless. They are related by the following expressions:

$$N_s = 0.023 n_s' \text{ and } \qquad N_s = \frac{n_s}{2734.7} \tag{2.6}$$

Dimensionless specific speed was obtained by eliminating D from the head and flow coefficients. In a similar manner, by eliminating speed of rotation N such that an integral power (unity) of diameter is retained, the so-called specific diameter D_s is obtained. The resulting expression would be

$$D_s = \frac{\psi^{1/4}}{\varphi^{1/2}} = \frac{D(gH)^{1/4}}{\sqrt{Q}} \tag{2.7}$$

The dimensional form of specific diameter is given by

$$d_s = \frac{D(H)^{1/4}}{\sqrt{Q}} \tag{2.7a}$$

During the 1950s, Cordier performed an extensive experimental study of high-efficiency turbomachines. He found that an interesting relationship existed between specific speed N_s and the specific diameter D_s. If the specific speed and specific diameters are calculated for various classes of turbomachines according to Equations 2.3a and 2.7 and are plotted

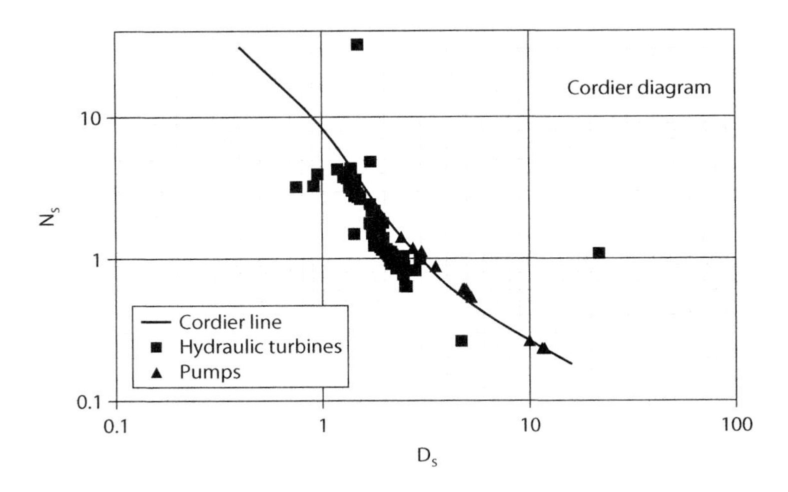

FIGURE 2.2
Cordier diagram.

against each other, the data for various machines fit on a curve with minimal scatter. This curve is called the *Cordier diagram* (see Csanady (1964)) and is shown in Figure 2.2.

The Cordier diagram is a very useful design tool for an engineer. Given the requirements of, say, H and Q, the specific speed can be calculated, since the range of rotational speeds for various machines is usually known from experience. For instance, hydraulic turbines rotate at a few hundred rpm, while pumps rotate at a few thousand rpm. Knowing the specific speed, an approximate diameter of the required machine can be estimated from the Cordier diagram. This gives a good preliminary estimate of the size of the machine for the designer. The diagram can be represented by the following curve-fits (Cordier formulas) well within engineering accuracy:

$$\begin{aligned} N_s &\cong 9.0 D_s^{-2.103} \quad \text{for } D_s \le 2.8 \\ N_s &\cong 3.25 D_s^{-1.126} \quad \text{for } D_s \ge 2.8 \end{aligned} \tag{2.8}$$

The variable π_4 introduced in Equation 2.4 is related to the Reynolds number and hence reflects the viscous effects. Since the term ND in the denominator is the product of angular velocity and diameter, it is a multiple of the blade velocity. Thus, π_4 would be proportional to the reciprocal of Reynolds number. However, incorporating Reynolds number equivalence in model testing is quite difficult. The flow in any turbomachine is three-dimensional, turbulent, and swirling. Hence, it is very difficult to describe the mathematics of the flow, with the result that even numerical solutions based on modern computational fluid dynamics methods are difficult to obtain and are often unreliable. Thus, extensive experimentation is needed to predict the performance of turbomachines. Since the efficiencies of modern hydraulic turbines and pumps are quite high, viscous effects and hence Reynolds number equivalence are largely ignored. This is justifiable for hydraulic turbines, which use low viscosity fluid such as water. However, for pumps that sometimes use heavy fluids, a correction to the efficiency for model and prototype is warranted. Also, due to size differences, the forces on the model and prototype cannot all be scaled exactly. Hence, a correction needs to be applied. For turbines, this correction is given in terms of Moody's (1926) formula. If η_m and η_p are the efficiencies of the model and prototype turbine respectively, then

TABLE 2.3

Conversion Factors for Dimensionless and Dimensional Forms of Specific Speed and Specific Diameter

	Q	H	N	Q	H	N	Q	H	N
Dimensionless	**ft.³/s**	**ft.**	**rpm**	**m³/s**	**m**	**rpm**	**gpm**	**ft.**	**rpm**
$Ns = 1$		Dimensional			Dimensional			Dimensional	
		$ns = 128.8$			$ns = 52.9$			$ns = 2730$	
$Ds = 1$		Dimensional			Dimensional			Dimensional	
		$ds = 0.42$			$ds = 0.565$			$ds = 0.0198$	

$$\frac{1-\eta_p}{1-\eta_m} = \left(\frac{D_m}{D_p}\right)^n \tag{2.9}$$

In Equation 2.9, the term D_m/D_p is the model scale ratio and n is an empirical exponent. Usually, the value of n is 0.2. For pumps, a different formula has been suggested by Wislicenus (1965):

$$\frac{0.95-\eta_p}{0.95-\eta_m} = \left(\frac{\log(32.7Q_p)}{\log(32.7Q_m)}\right)^2 \tag{2.10a}$$

for metric units (flow rates in m³/s) and

$$\frac{0.95-\eta_p}{0.95-\eta_m} = \left(\frac{\log(1154.7Q_p)}{\log(1154.7Q_m)}\right)^2 \tag{2.10b}$$

in imperial units (flow rates in ft.³/s).

The relationship between the dimensionless and dimensional forms of specific speed and specific diameter are given in Table 2.3.

Example 2.1

An interesting ("engineering") proof for Moody's efficiency scaling law can be given by making simplifying assumptions about friction losses in turbomachines. Derive it.

Solution: Efficiency can be written in a simplified form as

$$\eta = \frac{H_{net}}{H} = \frac{H-h_l}{H} = 1-\frac{h_l}{H} \tag{a}$$

or

$$\frac{h_l}{H} = 1-\eta \tag{b}$$

In this equation, H is the gross head and H_{net} is the net head, which is the gross head less the losses. If the friction losses can be reasonably approximated by Darcy's equation, h_l can be written as

$$h_l = f\frac{L}{D}\frac{V^2}{2g} \tag{c}$$

From equation (b),

$$\frac{\left(\dfrac{h_l}{H}\right)_m}{\left(\dfrac{h_l}{H}\right)_p} = \frac{\left(\dfrac{h_{l_m}}{h_{l_p}}\right)}{\left(\dfrac{H_m}{H_p}\right)} = \frac{\left(f\dfrac{L}{D}\dfrac{V^2}{2g}\right)_m}{\left(f\dfrac{L}{D}\dfrac{V^2}{2g}\right)_p}\left(\frac{H_p}{H_m}\right) = \frac{(1-\eta_m)}{(1-\eta_p)} \tag{d}$$

However, by geometric similarity, L/D is the same for the model and prototype. Also, from the application of the Bernoulli equation across the entire unit, $V^2 = 2gH$. Hence, equation (d) becomes

$$\frac{(1-\eta_m)}{(1-\eta_p)} = \frac{f_m}{f_p}\frac{(V^2)_m}{(V^2)_p}\frac{H_p}{H_m} = \frac{f_m}{f_p}\frac{(2gH)_m}{(2gH)_p}\frac{H_p}{H_m} = \frac{f_m}{f_p} \tag{e}$$

The dependence of the friction factor f on the diameter is quite complicated (for laminar flows, it can be expressed simply as $64/Re_D$ where Re is the Reynolds number; however, the flow is never laminar in hydraulic machinery!). It can be approximated as being inversely proportional to $(D)^n$, where n is an empirically determined exponent. Thus,

$$\frac{(1-\eta_m)}{(1-\eta_p)} = \frac{f_m}{f_p} = \frac{(1/D_m)^n}{(1/D_p)^n} = \left(\frac{D_p}{D_m}\right)^n \tag{f}$$

Comment: As noted in the problem statement, the proof is very sketchy, but is quite valid qualitatively.

Example 2.2

A centrifugal pump operating at the best efficiency point produces a head of 26 m and delivers 1 m^3/s of water when rotating at 1500 rpm. Its impeller diameter is 0.5 m. If a geometrically similar pump of impeller diameter 0.8 m is operating at 1200 rpm, calculate the discharge and head.

Solution: Since both the pumps are geometrically similar, *all* the dimensionless variables are expected to be equal:

$$\varphi_1 = \left(\frac{Q}{ND^3}\right)_1 = \left(\frac{Q}{ND^3}\right)_2 = \varphi_2 \tag{a}$$

or

$$Q_2 = Q_1\left(\frac{N_2}{N_1}\right)\left(\frac{D_2}{D_1}\right)^3 = 1\left(\frac{1200}{1500}\right)\left(\frac{0.8}{0.5}\right)^3 = 3.27\ \mathrm{m^3/s} \tag{b}$$

$$\psi_1 = \left(\frac{gH}{N^2D^2}\right)_1 = \left(\frac{gH}{N^2D^2}\right)_2 = \psi_2 \tag{c}$$

or

$$H_2 = H_1\left(\frac{N_2}{N_1}\right)^2\left(\frac{D_2}{D_1}\right)^2 = 26\left(\frac{1200}{1500}\right)^2\left(\frac{0.8}{0.5}\right)^2 = 42.6 \text{ m} \tag{d}$$

For verification purposes, the specific speed and specific diameters, N_s and D_s, for both pumps are computed. The speed of each pump needs to be converted to radians per second. Thus,

$$N_1 = \frac{2\pi(1500)}{60} = 157.1 \text{ rad/s}$$
$$N_2 = \frac{2\pi(1200)}{60} = 125.7 \text{ rad/s} \tag{e}$$

For the first pump,

$$N_s = \frac{N\sqrt{Q}}{(gH)^{3/4}} = \frac{(157.1)\sqrt{1}}{((9.81)(26))^{3/4}} = 2.46 \tag{f}$$

and for second pump

$$N_s = \frac{N\sqrt{Q}}{(gH)^{3/4}} = \frac{(125.7)\sqrt{3.27}}{((9.81)(42.6))^{3/4}} = 2.46$$

The values of specific speeds are within the range for centrifugal pumps, as seen from Table 2.2.

The specific diameters for the first and second pump, respectively, are

$$D_s = \frac{D(gH)^{1/4}}{\sqrt{Q}} = \frac{(0.5)((9.81)(26))^{1/4}}{\sqrt{1}} = 2.0 \tag{g}$$

and

$$D_s = \frac{D(gH)^{1/4}}{\sqrt{Q}} = \frac{(0.8)((9.81)(42.6))^{1/4}}{\sqrt{3.27}} = 2.0$$

Comments: The value of N_s predicted from the Cordier formula is 2.1, which is well within the scatter for the diagram. Thus, the Cordier diagram is a very useful tool in sizing hydraulic machines. An alternate use of the diagram in sizing pumps is presented in the following example.

Example 2.3

As an engineer working in the permits department for the city, you encounter an application wherein a pump is proposed for removing flood water. It runs at 1,800 rpm with a flow rate of 30,000 gpm under a head of 25 ft. What minimum size of pump would be acceptable to you as the engineer issuing permits?

Solution: The specific speed is calculated assuming that the head and flow are at optimum efficiency. There is no need to assume otherwise, since the operating point of pumps is expected to be at the optimum efficiency.

$$N = \frac{2\pi(1800)}{60} = 188.5 \text{ rad/s}$$

$$Q = \frac{30000}{448.83} = 66.8 \text{ ft.}^3\text{/s} \quad \text{(a)}$$

$$N_s = \frac{N\sqrt{Q}}{(gH)^{3/4}} = \frac{(188.5)\sqrt{66.8}}{((32.2)(25))^{3/4}} = 10.2 \quad \text{(b)}$$

From the Cordier diagram, the specific diameter D_s corresponding to this specific speed is 1.0. From the expression for specific diameter,

$$D = \frac{D_s\sqrt{Q}}{(gH)^{1/4}} = \frac{(1)\sqrt{66.8}}{((32.2)(25))^{1/4}} = 1.53 \text{ ft.}$$

The minimum diameter of the runner should be about 18 inches.

Comment: As can be seen from this example, the Cordier diagram gives excellent preliminary estimates of the dimensions of pumps for various applications.

PROBLEMS

In all the following problems, assume that the liquid is water unless specified otherwise.

2.1 The rotor diameter of a pump is 40 cm. It delivers 3 m³/min while rotating at 1800 rpm. A geometrically similar pump has a diameter of 30 cm and rotates at 3600 rpm. What would the flow rate of the second pump be? Also, if the head on the first pump is 15 m, what head would the second pump deliver? What type of pump would this be?

The first pump is tested in a laboratory and it is found that the efficiency is 78%. What is its input power? What would the input power of the second pump be if it is assumed to have the same efficiency?

Ans: $Q = 2.53$, $H = 33.75$, $P = 17.9$

2.2 A pump delivers 2250 gpm against a head of 95 ft. when rotating at 1450 rpm. What would its specific speed be? What type of pump would this be?

If the speed of the pump is changed to 1200 rpm, what would the new head and flow rate be?

Ans: $Q = 1862$ gpm, $H = 65$ ft.

2.3 If a centrifugal pump delivers 650 gpm when operating with a head of 120 ft. and speed of 1800 rpm, what are the corresponding values of flow rate and head when the speed is changed to 1500 rpm?

Ans: $Q = 542$ gpm, $H = 83.3$ ft.

2.4 A pump with impeller diameter of 60 cm is required to deliver 750 L/s at a head of 30 m when the speed is 850 rpm. A model pump with a diameter of 15 cm runs at 1700 rpm. (a) What are the corresponding flow rate and head under dynamically similar conditions? (b) If both the model and prototype have efficiencies of 83%, what power would be required to drive each pump? (c) Calculate the specific speed of each pump and specify the type of pump.

Ans: $Q=23$ L/s, $H=7.5$ m, $N_s=1.03$, $P=2.07$ kW, centrifugal

2.5 A turbine operates under a head of 90 ft. at 550 rpm while producing a power output of 7000 hp. What would the new speed and power be if the head changes to 60 ft.? What type of turbine would this be?

Ans: $N=449$ rpm, $P=3810$ hp

2.6 A turbine with an impeller diameter of 18 in. is tested under a head of 25 ft. while the flow rate is 18 cfs and the speed is 400 rpm. (a) Calculate the specific speed. (b) What type of turbine is this? (c) If a homologous unit with 38 in. runner is used under a head of 120 ft., what would the speed and discharge be? (d) If the original turbine produces 40 hp, what is its efficiency? (e) What is the power produced by the second unit? Assume the same efficiency for both units.

Ans: $N_s=1.18$, $Q=175$ cfs, $H=1874$ hp, $\eta=0.78$

2.7 A turbine operating at a head of 80 ft. and 160 rpm has an efficiency of 85% when the flow rate is 200 cfs. If the same turbine operates at a head of 40 ft., what would the values of speed, flow rate, and brake horsepower be under homologous conditions? Specify the type of turbine

Ans: $Q=141$ cfs, $N=113$ rpm, $P=558$ hp

If another turbine, homologous to the one described, has dimensions exactly half these values, what would the corresponding values of speed, flow rate, and brake horsepower be when the net head is still 80 ft.?

Ans: $N_s=0.55$; Francis turbine

2.8 The power produced by a hydraulic turbine is 25,000 hp when the head is 120 ft. and the rotational speed is 80 rpm. A homologous laboratory model of the turbine produces 54 hp when operating at a head of 24 ft. Calculate the model speed, model flow rate, and the scale ratio. What type of turbine would this be?

Ans: $N=230.2$ rpm, $Q=19.83$ cfs, model ratio$\,=6.43$, $N_s=0.73$; Francis turbine

2.9 The pressure head developed by a fan is 155 mm of water, as measured by a water manometer. For this head, the flow rate is 4.5 m^3/s, while the rotational speed is 1800 rpm. A larger geometrically similar fan runs at 1500 rpm. If the larger fan is to operate at the same head, find its flow rate. Specify the type of fan.

Ans: 6.48 m^3/s, $N_s=1.88$; Axial flow fan

2.10 An axial flow fan operating at 1450 rpm has a diameter of 1.9 m. The average axial velocity is measured to be 12 m/s. A geometrically similar fan which is one-fifth the size is built and tested at 4350 rpm. Assuming dynamically similar conditions, find the flow rate and axial velocity.

Ans: $Q=0.81$ m^3/s; $Va=7.2$ m/s

2.11 A pump rotating at 600 rpm is to be used to pump water between two reservoirs with an elevation difference of 3 m. It is proposed to use a 20 m long pipe of 100 cm diameter. Assume that the friction factor is $f=0.02$ and minor loss coefficient = 0.5. If a flow rate of 0.2 m^3/s is expected, what type of pump would be best suited? Also, what would the head be for the given conditions?

Ans: 3 m, Centrifugal pump

2.12 A centrifugal pump is to be designed to produce a head of 16 m when the rotational speed is 1800 rpm. The shaft power input under these conditions is expected to be 2.8 MW. A model pump is built and tested under dynamically similar conditions with a head of 4 m and a shaft power input of 20 kW. Calculate the model speed and scale ratio. If both the pumps operate at 82% efficiency, estimate the flow rates.

Ans: $Q = 14.6$ m^3/s; $N = 3764$ rpm; scale ratio = 4.18, $Q = 0.42$ m^3/s

2.13 Use a Cordier diagram to estimate the lowest diameter of pump needed to deliver 12 cfs of water when operating at 1800 rpm at a head of 36 ft.

Ans: 1 ft.

2.14 Using the relationship that flow rate Q is the product of velocity V and area A ($Q = VA$), show qualitatively that high specific speed machines tend to be axial while low specific speed tends to be centrifugal.

2.15 As mentioned in the chapter, there are several ways in which the head and flow coefficients can be combined to form other dimensionless numbers (specific speed and specific diameter are one way of combining them). Obtain at least one other combination for N_s and D_s. Comment on the result.

2.16 An inventor claims that he has designed a pump with the following characteristics. While running at 3600 rpm, it produces 200 ft. of head and delivers 4 cfs of water. The rotor diameter is 6 in. Evaluate the inventor's claim. What type of pump is it likely to be?

Ans: Centrifugal pump

2.17 In Example 2.1, Moody's formula for efficiency scaling was derived using the following definition of efficiency:

$$\eta = \frac{H - h_l}{H}$$

where:

H is the gross head

h_l is the head loss

However, the use of an alternative definition for efficiency given as follows, leads to a different form of Moody's formula. Obtain it.

$$\eta = \frac{H_{net}}{H_{net} + h_l}$$

However, if the available head is defined as $H = H_{net} + h_1$, Equation 2.9 is obtained.

2.18 The fluid mechanics laboratory in the Donald P. Shiley School of Engineering at the University of Portland has the following two pieces of equipment for use in several experiments.

a. One of them is a fan. In one of the experiments, the speed was measured as 1200 rpm and the pressure rise across it was 2.5 in. of water. The fan is connected to a duct of diameter 1 ft. and the air speed through the duct was measured at 45 ft./s. Estimate the specific speed of the fan. What type of fan is it likely to be?

Ans: 1.18, centrifugal fan

b. The other is a pump. In the performance testing of the pump, the flow rate and pressure rise were measured as 160 gpm and 3.65 in. of mercury when it was rotating at 1854 rpm. Estimate the specific speed. What type of pump is it likely to be?

Ans: 2.83, Axial flow pump

2.19 Verify the formulas given in Equation 2.6 for various forms of specific speed.

2.20 Sometimes, specific speed is defined in terms of n revolutions per second (rps), instead of revolutions per minute (rpm) as follows:

$$N_1' = \frac{n\sqrt{Q}}{(gH)^{3/4}} \quad \text{and} \quad N_1'' = \frac{N\sqrt{Q'}}{(H)^{3/4}}$$

where:

Q is in gpm
Q' is in ft.^{3}/s
n is in revolutions per second
N is in revolutions per minute
H is in feet

Show that $N_1'' = 17{,}182\ N_1'$.

Similarly, specific speed can also be defined using power in horsepower and n as follows:

$$N_2' = \frac{n\sqrt{P}}{\rho^{1/2}(gH)^{5/4}} \quad \text{and} \quad N_2'' = \frac{N\sqrt{P'}}{(H)^{5/4}}$$

where:

P and P' are in hp
n is in revolutions per second
N is in revolutions per minute
ρ is 1.94 slugs/ft.^{3}
H is in feet

Show that $N_2'' = 273.3\ N_2'$.

2.21 Verify the conversion factors given in Table 2.3

3

Theory of Turbomachines

In the previous chapter, techniques of dimensional analysis were used to obtain various dimensionless quantities relevant to pumps and hydraulic turbines. The concepts of specific speed and specific diameter were introduced and their importance in the selection of various types of machines was discussed. This chapter deals with the theory of fluid flow in turbomachines.

Upon completion of this chapter, the student will be able to

- Use Euler's turbine equation for analyzing turbomachines
- Use velocity triangles to relate kinematical quantities including blade speed and fluid velocity to dynamic quantities such as pressure and enthalpy
- Use the concept of degree of reaction and relate it to velocity triangles

3.1 Euler Turbine Equation

The dimensionless groups derived earlier are helpful in predicting the performance of geometrically similar turbomachines under various conditions. However, they do not by themselves show a way to predict the relevant design variables based on the dynamic conditions that exist in the rotor. For example, if the physical dimensions of the rotor are known, it should be possible to calculate the head or flow rate. For this, the kinematics of fluid flow inside the rotor needs to be related to its dynamics.

In order to analyze the kinematic and dynamic factors in the design of turbomachines, it is necessary to relate the fluid velocities and rotor velocity to the forces of interaction due to the momentum changes of the fluid inside the rotor. Consider the rotor shown in Figure 3.1. A fixed control volume (*cv*) enclosing the fluid in the rotor, with a coordinate system with z-axis aligned with the axis of rotation is chosen. The fluid enters the rotor at a radius r_1 with velocity V_1 and leaves at a radius r_2 with velocity V_2. For steady flow, mass flow rate in should be equal to mass flow rate out of the control volume.

The principle of conservation of linear momentum states that the net force acting on an element of the fluid is the sum of the instantaneous rate of change and convective rate of change of momentum. This can be stated as (details can be found in Elger et al. (2012))

$$\mathbf{F} = \frac{\partial}{\partial t}\int_{cv} \mathbf{V}\rho \, d\forall + \int_{cs} \mathbf{V}(\rho\mathbf{V}\cdot\mathbf{dA}) \tag{3.1}$$

In Equation 3.1, **F** is the total force on the control volume, **V** is the velocity vector, ρ is the density, and $d\forall$ is the volume element; *cv* and *cs* are the control volume and control surface, respectively. The forces on the rotor are caused by the changes in the absolute velocity

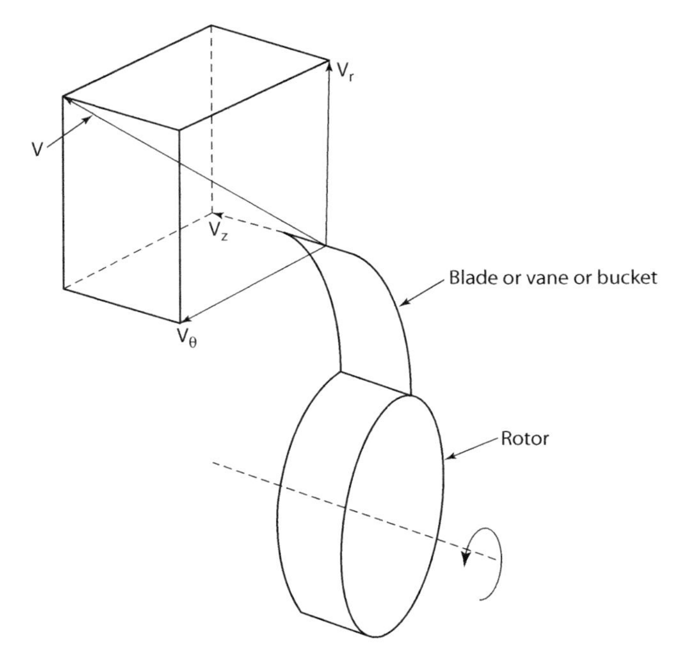

FIGURE 3.1
Velocity components in cylindrical polar system for a rotor.

of the fluid at the rotor inlet and exit. The change in the axial velocity (or axial momentum) causes the axial thrust, while the change in the radial velocity produces bending forces on the rotor. Neither of these forces gives rise to torque to cause rotation of the shaft, and they are therefore of little interest. Thus, the variable of interest is the tangential velocity producing torque.

The moment of momentum equation can be obtained by taking the moment of Equation 3.1 with the position vector, **r**. Then,

$$\mathbf{r}\times\mathbf{F} = \frac{\partial}{\partial t}\int_{cv} \mathbf{r}\times\mathbf{V}\rho d\forall + \int_{cs} \mathbf{r}\times\mathbf{V}(\rho\mathbf{V}\cdot d\mathbf{A}) \tag{3.2}$$

For steady flows, the first term becomes zero. The total force **F** can be written in terms of its components as

$$\mathbf{F} = F_r\,\mathbf{e}_r + F_\theta\,\mathbf{e}_\theta + F_z\,\mathbf{e}_z$$

The moment of **F** is

$$\mathbf{r}\times\mathbf{F} = \mathbf{M} = \begin{vmatrix} \mathbf{e}_r & \mathbf{e}_\theta & \mathbf{e}_z \\ r & 0 & z \\ F_r & F_\theta & F_z \end{vmatrix} = -zF_\theta\,\mathbf{e}_r + (zF_r - rF_z)\,\mathbf{e}_\theta + rF_\theta\,\mathbf{e}_z \tag{3.3}$$

The first two terms of Equation 3.3 are the radial and tangential components of the moment M_r and M_θ. These represent the bending moments about the rotor shaft in two different directions. The only component of relevance would then be the axial or z component, which represents M_z or torque about the axis. By a similar analysis, the cross product of the second term on the right-hand side of Equation 3.2 would be

$$\int_{cs} \mathbf{r} \times \mathbf{V}(\rho \mathbf{V}.\mathbf{dA}) = \int_{cs} \begin{vmatrix} \mathbf{e}_r & \mathbf{e}_\theta & \mathbf{e}_z \\ r & 0 & z \\ V_r & V_\theta & V_z \end{vmatrix} (\rho \mathbf{V}.\mathbf{dA}) \tag{3.4}$$

$$= \int_{cs} \left(-zV_\theta \mathbf{e}_r + (zV_r - rV_z)\mathbf{e}_\theta + rV_\theta \mathbf{e}_z\right)(\rho \mathbf{V}.\mathbf{dA})$$

The first two terms again contribute to the bending moments about two axes of the shaft, and are of little interest. The third term is the torque on the shaft. Assuming that the flow rate is uniform across the inlet and outlet of the rotor, the torque (moment about the z-axis) becomes

$$M_z = \int_{cs} rV_\theta(\rho \mathbf{V} \cdot \mathbf{dA}) = \dot{m}\left[(rV_\theta)_{exit} - (rV_\theta)_{inlet}\right] = \dot{m}\left[(rV_u)_2 - (rV_u)_1\right] \tag{3.5}$$

In Equation 3.5, the tangential component V_θ has been replaced by V_u. It is customary in the study of turbomachines to refer to V_θ and V_z as V_u and V_a, the tangential and axial components of fluid velocity, respectively (see Figure 3.2)

Since the control volume encloses the fluid, the torque calculated by Equation 3.5 is on the fluid. Hence, the torque on the shaft would be its negative. Thus, the torque on the shaft is

$$T = \dot{m}\left[(rV_u)_1 - (rV_u)_2\right] \tag{3.6}$$

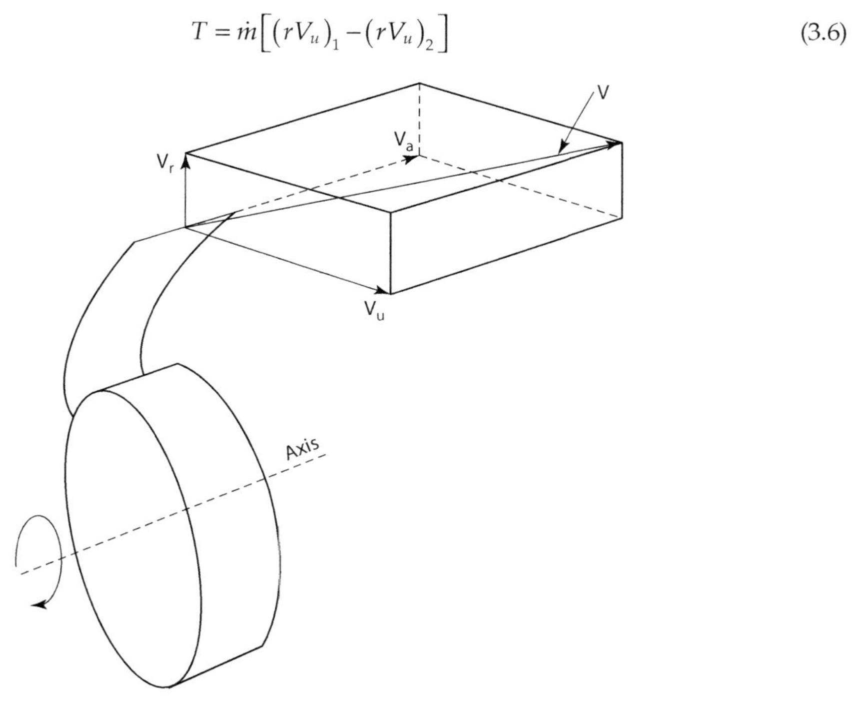

FIGURE 3.2
Velocity components with regard to rotor coordinates.

The power P would be

$$P = T\omega = \dot{m}\omega\left[\left(rV_u\right)_1 - \left(rV_u\right)_2\right] = \dot{m}\left[\left(uV_u\right)_1 - \left(uV_u\right)_2\right] = \dot{m}\left[u_1V_{u1} - u_2V_{u2}\right] \tag{3.7}$$

where the product of radius and the angular velocity has been replaced by the tangential speed of the rotor. There are two other forms in which Equation 3.7 can be expressed. The first is obtained by dividing by $\dot{m}$. Thus,

$$\frac{P}{\dot{m}} = E = u_1V_{u1} - u_2V_{u2} \tag{3.7a}$$

The other form is obtained by dividing Equation 3.7a by g, and thus

$$\frac{E}{g} = H = \frac{u_1V_{u1} - u_2V_{u2}}{g} \tag{3.7b}$$

Equation 3.7 (or alternately Equations 3.7a or 3.7b) are called *Euler's turbine equations* or simply *Euler's equations* and are of the utmost importance in the analysis of all types of turbomachinery.

3.2 Energy Equation

The discussion so far has been on the application of the momentum equation to derive Euler's equation. An alternative approach is the energy equation that can be used to derive an expression equivalent to Equation 3.7a. The energy equation applied to the control volume shown in Figure 3.1 (steady flow, steady state, and single inlet and outlet) becomes

$$\dot{Q} + \dot{m}\left(h_1 + \frac{V_1^2}{2} + gz_1\right) = \dot{m}\left(h_2 + \frac{V_2^2}{2} + gz_2\right) + \dot{W} \tag{3.8}$$

The terms h_1 and h_2 are the static enthalpies at the inlet and the exit, respectively; also, $\dot{Q}$ and $\dot{W}$ are the rate of thermal energy transfer and rate of work for the control volume, respectively. Assuming adiabatic conditions and ignoring potential energy changes, Equation 3.8 becomes

$$\dot{m}\left(h_1 + \frac{V_1^2}{2}\right) = \dot{m}\left(h_2 + \frac{V_2^2}{2}\right) + \dot{W} \tag{3.9}$$

Dividing by $\dot{m}$, and replacing $\frac{\dot{W}}{\dot{m}}$ by E ($= \frac{P}{\dot{m}}$ in Equation 3.7a), the energy becomes

$$h_1 + \frac{V_1^2}{2} = h_2 + \frac{V_2^2}{2} + E \quad \Rightarrow \quad E = h_{01} - h_{02} \tag{3.10}$$

where h_0 is the stagnation enthalpy. If the fluid at a state with velocity V and static enthalpy h is brought to rest isentropically (see Figure 3.3), the static and stagnation states are related by

$$h + \frac{V^2}{2} = h_0 \tag{3.11}$$

By using the equation of state, other state variables such as temperature, pressure, and density can be related between static and stagnation states. The energy equation given by Equation 3.10 is useful in computing the pressure changes in hydraulic machines and pumps since the flow through them will be assumed isothermal. The thermodynamic state variables enthalpy h and internal energy $\hat{u}$ are related by the expression

$$h = \hat{u} + pv_s = \hat{u} + \frac{p}{\rho} \tag{3.11a}$$

The equivalence of Equation 3.7a derived from momentum consideration and Equation 3.10 derived from energy consideration can be shown by considering the velocity triangles.

A slight modification of Equation 3.10 can be made by substituting the Euler's equation, Equation 3.7a, into Equation 3.10. Thus,

$$\begin{aligned} h_1 + \frac{V_1^2}{2} &= h_2 + \frac{V_2^2}{2} + E = h_2 + \frac{V_2^2}{2} + u_1 V_{u1} - u_2 V_{u2} \\ \text{or}\quad h_1 + \frac{V_1^2}{2} - u_1 V_{u1} &= h_2 + \frac{V_2^2}{2} - u_2 V_{u2} \quad or \quad I_1 = I_2 \end{aligned} \tag{3.11b}$$

The quantity $I = h + \frac{V^2}{2} - uV_u$ is of some importance in turbomachinery. It is the rotational stagnation enthalpy, called the *rothalpy*. An examination of Equation 3.11b indicates that the rothalpy is the same at the inlet and the exit of the rotor and hence it is same at all points of the rotor. This form of expression of the energy in the rotor has some relevance for gas turbines and compressors.

The following notation is used throughout this text (see Figure 3.4) unless specified otherwise. Inlet conditions are denoted by subscript 1 and exit conditions by subscript 2. The fluid is always assumed to enter or leave tangentially to the blade. The blade velocity is always tangential to the wheel or the direction of rotation. The angle between the relative velocity w_1 and blade velocity u_1 (wheel tangent) is called the *blade angle;* β_1 for inlet or

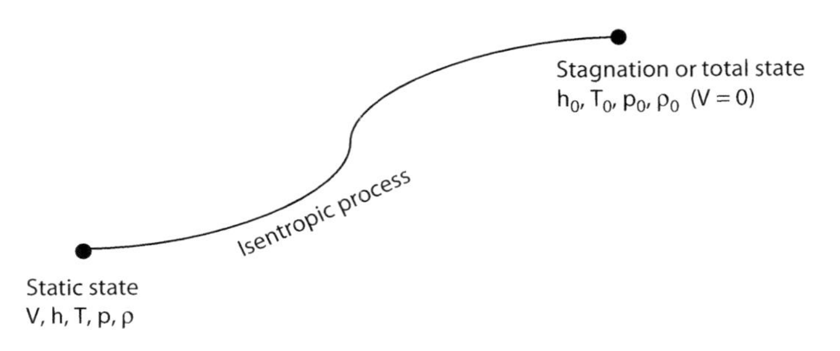

FIGURE 3.3
Static and stagnation states in a fluid.

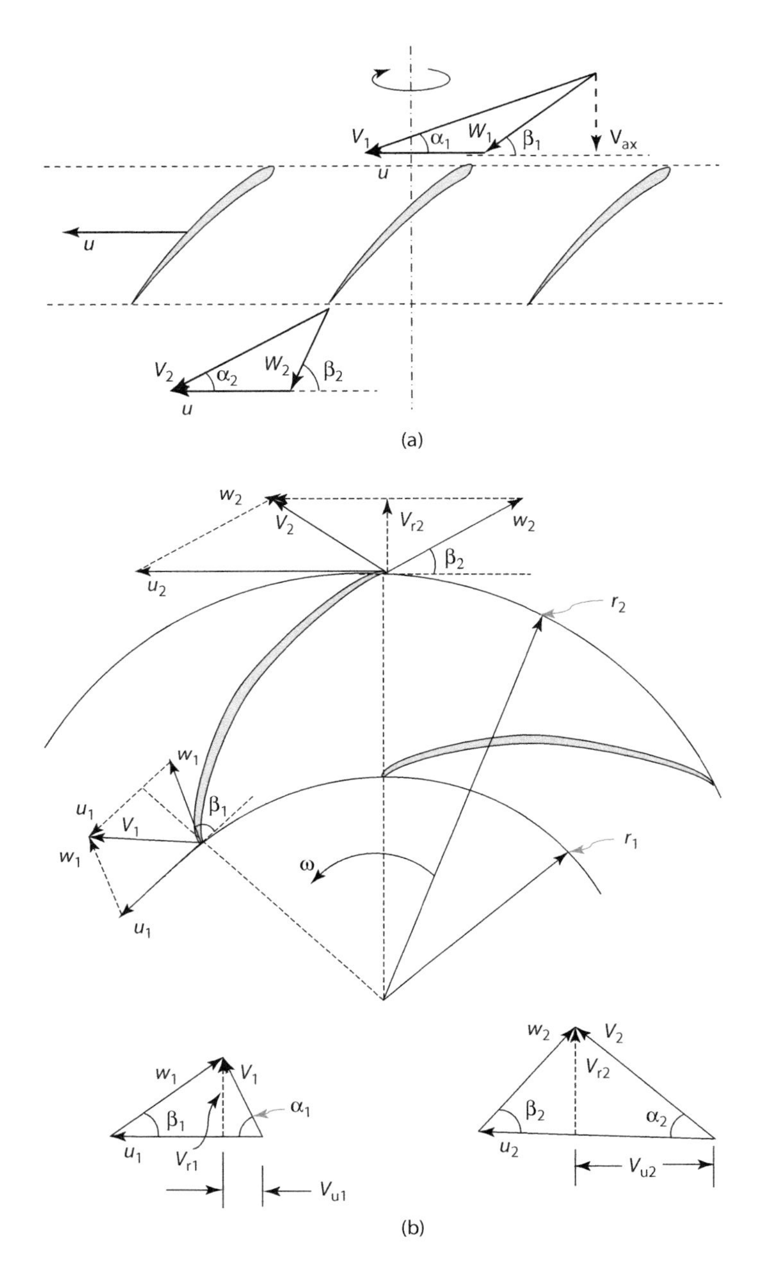

FIGURE 3.4
Velocity triangles and blade angles at inlet and outlet for (a) axial flow and (b) radial flow machines.

β_2 for outlet. Similarly, the angle between the absolute velocity V and blade velocity u is called the *nozzle angle* (or simply the *fluid angle* or *flow angle*); α_1 for inlet or α_2 for outlet. In Figure 3.4, these angles are measured with respect to the blade velocity u. However, these angles could also be measured with respect to the radial velocity V_r (the radial direction) or axial velocity V_a (axial direction). This situation occurs in the case of axial flow machines. However, these will be complements of the angles defined earlier, hence should make very little difference to the analysis. Besides, in most situations the angles will be shown in the velocity diagram to remove any ambiguity. It is customary to measure the angles with respect to axial or radial directions for gas turbines and compressors, while for hydraulic turbines and pumps the angles are measured with respect to the wheel tangent or blade velocity.

The sign convention adopted in this text is as follows: The blade angles β and the fluid angle α will be considered positive if the component of w and V point in the direction of the blade velocity u and will be negative if they are in opposite direction to u. *All angles will be reported as acute angles only.* All the angles shown in Figure 3.4a are positive.

From the velocity triangles shown in Figure 3.4b (using the law of cosines),

$$w_1^2 = u_1^2 + V_1^2 - 2u_1V_1\cos\alpha_1 = u_1^2 + V_1^2 - 2u_1V_{u1}$$

$$w_2^2 = u_2^2 + V_2^2 - 2u_2V_2\cos\alpha_2 = u_2^2 + V_2^2 - 2u_2V_{u2}$$

Subtracting and simplifying, the result would be

$$E = u_1V_{u1} - u_2V_{u2} = \frac{V_1^2 - V_2^2}{2} + \frac{u_1^2 - u_2^2}{2} - \frac{w_1^2 - w_2^2}{2}$$

$$= \frac{V_1^2 - V_2^2}{2} + \frac{u_1^2 - u_2^2}{2} + \frac{w_2^2 - w_1^2}{2} \tag{3.12}$$

$$\text{or} \quad H = \frac{E}{g} = \frac{V_1^2 - V_2^2}{2g} + \frac{u_1^2 - u_2^2}{2g} + \frac{w_2^2 - w_1^2}{2g}$$

The first term in Equation 3.12 represents the change in *dynamic* head across the rotor. Its value could be considerable, as in the case of impulse turbines wherein most of the energy change is due to the absolute kinetic energy change, in the form of impulse action on the blades. Well-designed power-producing machines would have minimal exiting kinetic energy head. In power-consuming devices, the increase in dynamic head would be the result of the power input into the rotor. If the exit kinetic energy head is too large, then a diffuser is attached to the end of the rotor to convert it partially into pressure. This is particularly meaningful in pumps and blowers, since their primary purpose is to produce pressurized fluid. However, this energy conversion occurring outside the rotor would be an energy transformation rather than a transfer.

The second term in Equation 3.12 represents the change in the energy of the fluid due to movement along the radius. In a sense, it represents the centripetal effect on the fluid and this is used in centrifugal pumps and centrifugal compressors to produce changes in pressure. Its relationship to pressure changes in the fluid can be demonstrated from Figure 3.5.

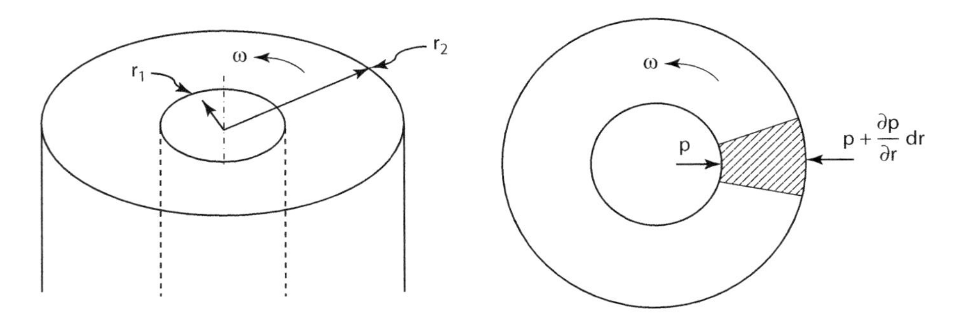

FIGURE 3.5
Fluid mass in rotation.

The equation of equilibrium for a non-viscous incompressible fluid in steady motion is given by

$$\frac{dp}{dr} = \rho\omega^2 r \quad \Rightarrow \quad p_2 - p_1 = \rho\omega^2(r_2^2 - r_1^2)/2 = \rho(u_2^2 - u_1^2)/2$$

$$\text{or} \quad \frac{p_2 - p_1}{\rho} = \frac{(u_2^2 - u_1^2)}{2}$$

Thus, as the fluid moves out along the radius from r_1 to r_2, the *static* pressure increases due to the centrifugal effect.

It is difficult to give a physical interpretation to the third term of Equation 3.12. It is simply the change in kinetic energy due to change in relative velocity, which is measured with respect to the rotor. The algebraic sum of the three changes in kinetic energy gives the net head on the rotor as shown. The relative contributions of the dynamic portion of the head (the first term) and the static portion of the head depend significantly on the design, and vary considerably from one type of machine to another. This relative proportion is described by the concept of *degree of reaction*, which is discussed in the following section.

3.3 Impulse, Reaction, and Utilization Factor

Among the several methods used to classify turbomachines, an important one is by the action of the fluid on the vanes. Machines wherein power is produced predominantly due to the impingement of the fluid jet on the blades are called *impulse turbines*. On the other hand, when the power is produced by a combination of the action of the jet and the pressure difference on the blade surfaces, these machines are called *reaction machines*. The variable that quantifies this concept is called the *degree of reaction* or simply *reaction*. It is defined as the ratio of static enthalpy change across the rotor to the total (or stagnation) enthalpy change across the rotor. In other words,

$$R \equiv \frac{h_1 - h_2}{h_{01} - h_{02}} = \frac{h_1 - h_2}{E} = \frac{h_1 - h_2}{u_1 V_{u1} - u_2 V_{u2}} \tag{3.13}$$

It can be seen that Equations 3.10 and 3.12 relate the static and stagnation enthalpy changes to the energy transfer E. By combining them, other forms of degree of reaction can be obtained as shown:

$$h_1 + \frac{V_1^2}{2} = h_2 + \frac{V_2^2}{2} + E = h_2 + \frac{V_2^2}{2} + \frac{V_1^2 - V_2^2}{2} + \frac{u_1^2 - u_2^2}{2} - \frac{w_1^2 - w_2^2}{2}$$

$$= h_2 + \frac{u_1^2 - u_2^2}{2} - \frac{w_1^2 - w_2^2}{2}$$

$$\text{or } h_1 - h_2 = \frac{u_1^2 - u_2^2}{2} - \frac{w_1^2 - w_2^2}{2} = \frac{u_1^2 - u_2^2}{2} + \frac{w_2^2 - w_1^2}{2}$$

Thus,

$$R = \frac{(u_1^2 - u_2^2) + (w_2^2 - w_1^2)}{(V_1^2 - V_2^2) + (u_1^2 - u_2^2) + (w_2^2 - w_1^2)} \tag{3.14}$$

Several simplifications are possible especially in the context of axial flow machines. But before these are studied in detail, it is useful to introduce the concept of the *utilization factor*, which is discussed next.

The efficiency of power-generating devices depends on two factors. The first is how effectively the available energy of the fluid is utilized and converted into torque. The second is the fraction of energy lost to friction, turbulence, eddies, and so on. From a theoretical perspective, all the kinetic energy available for the fluid before it enters the rotor should be utilized. However, this is not feasible, since, even for ideal situations, not all energy supplied is converted to useful work, because the fluid has to have a finite residual kinetic energy in order to leave the rotor. Otherwise, the fluid would enter the rotor and not leave, a physical impossibility. The ratio of the energy produced under ideal conditions to available energy at the entrance is called the *utilization factor* or *diagram efficiency* (see Shepherd (1956)), denoted by the symbol ε. Since E is the energy produced under ideal circumstances, the available energy at the entrance would be the sum of E and exit kinetic energy. Thus,

$$\varepsilon = \frac{E}{E + \frac{V_2^2}{2}} \tag{3.15}$$

Using the definition of E from Equation 3.12, the utilization factor can be written as

$$\varepsilon = \frac{(V_1^2 - V_2^2) + (u_1^2 - u_2^2) + (w_2^2 - w_1^2)}{V_1^2 + (u_1^2 - u_2^2) + (w_2^2 - w_1^2)} \tag{3.16}$$

Other forms of the utilization factor can be obtained by using different expressions for E. A well-designed turbine would have the highest value of ε, which would make the exit kinetic energy as low as possible. Theoretically, the highest value of ε would be 100%, but this would make the exit velocity zero. This is impossible, since the fluid would then enter the rotor but would not leave it. Hence, the maximum utilization factor is always less than 100%, and occurs when the *kinetic energy of the exiting fluid is a minimum.*

A detailed study of the utilization factor ε, degree of reaction R, and their variations in the context of axial flow turbines provides useful insights into their design, especially in the number and features of stages. Commonly, multiple stages are used in axial flow turbines (and compressors), since the pressure drops (and rises) are considerably smaller than their counterparts, namely, radial flow machines. Besides, axial flow turbines represent the largest class of turbines using compressible fluids (e.g. steam and gas turbines). In the following analysis, the simplifying assumptions of frictionless flow and constant axial velocity are made. Since the fluid inlet and exit occurs at the same radius in axial flow turbines, $u_1 = u_2$. Hence,

$$R = \frac{(w_2^2 - w_1^2)}{(V_1^2 - V_2^2) + (w_2^2 - w_1^2)}$$
$$\text{or } (w_2^2 - w_1^2) = \frac{R}{1-R}(V_1^2 - V_2^2) \tag{3.17}$$

Similarly,

$$\varepsilon = \frac{(V_1^2 - V_2^2) + (w_2^2 - w_1^2)}{V_1^2 + (w_2^2 - w_1^2)} \tag{3.18}$$

By combining Equations 3.17 and 3.18, a useful relationship between ε and R is obtained.

$$\varepsilon = \frac{V_1^2 - V_2^2}{V_1^2 - RV_2^2} \tag{3.19}$$

Although Equation 3.19 has been derived for axial flow turbines, it is valid for radial turbines also (see Problem 3.2). Typical velocity triangles for an axial flow device are shown in Figure 3.6a. The angle between the inlet velocity V_1 and blade velocity u is the nozzle angle α_1, which can be measured with respect to the axial (radial) or tangential direction. For maximum utilization, V_2 should be a minimum. Since it cannot be equal to zero, V_2 would be a minimum when it is purely in the axial direction, as shown in Figure 3.6b. Then, $V_2 = V_1 \text{Sin}\,\alpha_1$, and the maximum utilization factor ε_m would be (from Equation 3.19)

$$\varepsilon_m = \frac{V_1^2 - V_1^2 \sin^2\alpha}{V_1^2 - RV_1^2 \sin^2\alpha} = \frac{V_1^2\left(1 - \sin^2\alpha\right)}{V_1^2\left(1 - R\sin^2\alpha\right)} = \frac{\cos^2\alpha_1}{1 - R\sin^2\alpha_1} \tag{3.20}$$

Equation 3.20 can be obtained from the velocity triangles shown in Figure 3.6b (see Problem 3.5). The variation of maximum utilization with nozzle angle for various values of degree of reaction R can be seen in Figure 3.6c.

It can be seen from Figure 3.6c that maximum utilization is very high for low values of α_1 (less than 20°) and falls off quite rapidly thereafter. This factor has an important bearing on the inlet nozzle angle for impulse turbines, especially Pelton wheels, and will be discussed in Chapter 4.

3.4 Zero and 50% Reaction Machines

The flow situation pertaining to three values of R, namely $R = 0$, $R = 0.5$, and $R = 1$ will be discussed now. The first two, $R = 0$ and $R = 0.5$, are of significant practical importance since many axial flow turbine stages belong to one of these categories. The third value, namely

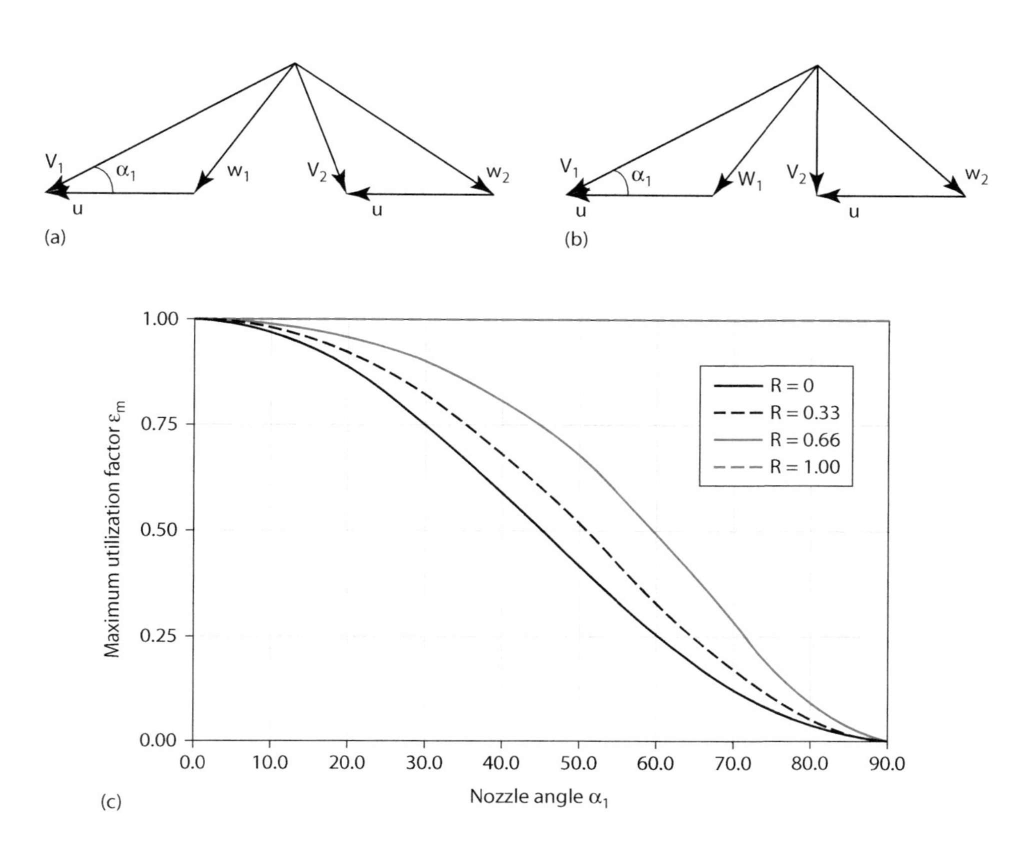

FIGURE 3.6
(a, b) Velocity triangles for axial flow impulse turbines, (c) variation of maximum utilization with nozzle angle.

$R=1$, is more of theoretical interest. The following discussion is restricted to axial flow machines; hence $u_1=u_2=u$.

When $R=0$, the turbine is called a *pure impulse* or *impulse turbine*, or *impulse turbine stage*. The most common example of an impulse turbine is the Pelton wheel, a hydraulic turbine useful for very high heads. Detailed discussion of the features of Pelton wheel is deferred for now and can be found in the chapter on hydraulic turbines. Since $R=0$, it follows from Equation 3.17 that $w_1=w_2$. The velocity triangles for this situation are shown in Figure 3.7a. For maximum utilization, V_2 should be as small as possible; that is, when it is in the axial direction, as shown in Figure 3.7b. The triangle ACE is isosceles, and hence $CD=ED=u$. Hence, $BC = CD=ED=u$. Thus, $V_2=V_{ax}=V_1 \text{ Sin } \alpha_1$, and $V_{u1}=2u$ and $V_{u2}=0$. The energy produced and maximum utilization factor would be

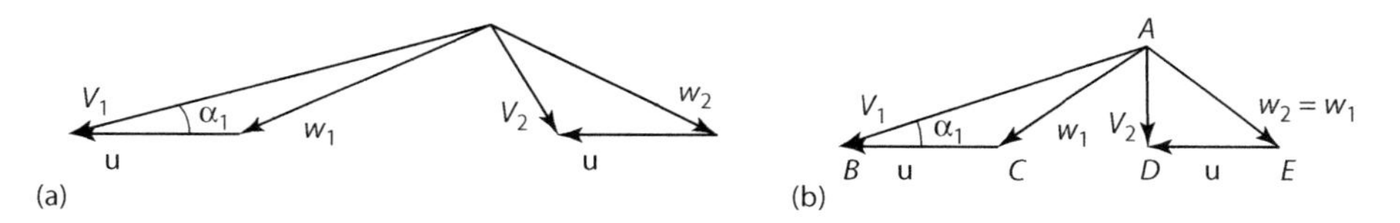

FIGURE 3.7
Velocity triangles when $R=0$ ($w_1=w_2$): (a) typical value of ε and (b) maximum ε.

$$E = u_1 V_{u1} - u_2 V_{u2} = 2u^2$$
$$\varepsilon_m = \cos^2 \alpha_1 \tag{3.21}$$

The second case of practical interest is when $R = 0.5$, as in what are also called *50% reaction turbines*. From the definition of degree of reaction,

$$R = \frac{1}{2} = \frac{(w_2^2 - w_1^2)}{(V_1^2 - V_2^2) + (w_2^2 - w_1^2)} \tag{3.21a}$$
$$\text{or} \quad (V_1^2 - V_2^2) = (w_2^2 - w_1^2)$$

This is nonlinear and hence has multiple solutions. One possible solution is $V_1 = w_2$ and $V_2 = w_1$. The velocity triangles would then be symmetric, and are shown in Figure 3.8a. For maximum utilization, V_2 again should be axial. This situation is shown in Figure 3.8b. Thus, $V_2 = w_1 = V_{ax} = V_1$ Sin α_1, and $V_{u1} = u$ and $V_{u2} = 0$. The energy produced and maximum utilization factor would be

$$E = u_1 V_{u1} - u_2 V_{u2} = u^2$$
$$\varepsilon_m = \frac{\cos^2 \alpha_1}{1 - 0.5 \sin^2 \alpha_1} \tag{3.22}$$

A comparison of Equations 3.21 and 3.22 indicates that an impulse turbine produces twice the amount of power as a reaction turbine for the same blade speed and nozzle angle, but less efficiently, since the utilization factor is lower. This has important implications on the number and types of stages in gas and steam turbines, but this discussion is deferred until later chapters. A compromise between higher power production and higher utilization is to use impulse stages for higher pressures and temperatures (entry to gas turbines) and use 50% reaction for later stages.

One variable that determines the shape of impulse blades is the speed ratio $\phi = u/V_1$, which is also known as the *peripheral velocity factor*. The correct definition of the velocity factor is $\phi = \frac{u}{\sqrt{2gH}}$. However, for impulse turbines that operate under high heads, the inlet

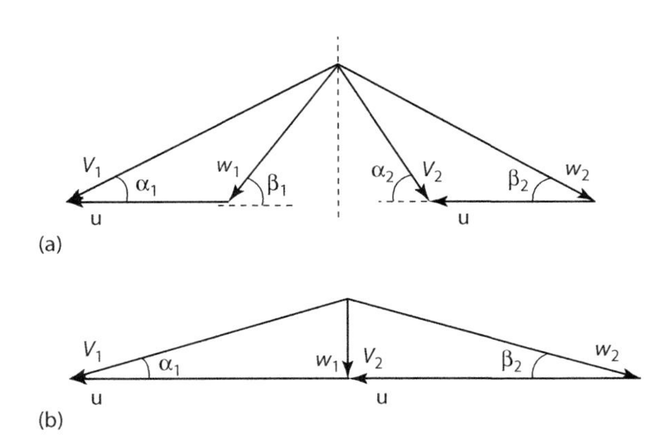

FIGURE 3.8
Velocity triangles for 50% reaction turbine ($w_1 = V_2$; $w_2 = V_1$), (a) for typical value of ε and (b) for maximum ε.

velocity (jet velocity) is almost equal to $\sqrt{2gH}$. For impulse and 50% reaction cases, the values of ϕ are given by the following expressions:

$$\phi = \frac{\cos\alpha_1}{2} \text{ for } R = 0$$
$$\text{and } \phi = \cos\alpha_1 \text{ for } R = 0.5 \tag{3.22a}$$

These can be seen from Figures 3.7 and 3.8. A comparison of different variables for impulse and 50% reaction turbines is given in Table 3.1.

The situation when $R=1$ would result in V_1 and V_2 being equal (this can be concluded from Equation 3.14). The velocity triangles for this condition are shown in Figures 3.9a and 3.9b. The former corresponds to the scenario when the axial velocities are equal at the inlet and exit, while the latter corresponds to when they are different. The case when $R>1$ is relevant in discussion of blowers and centrifugal fans and will be discussed in Chapter 6.

The preceding discussion also has an effect on the shape of impulse and reaction blading. Since impulse stages handle higher pressures (correspondingly higher densities) and there is relatively less pressure drop, the blades are shorter with high curvature, as the fluid turns in the rotor. The reaction stages tend to be longer and the blade curvature is also much less. The shape of blades in various turbine stages is shown in Figure 3.10.

TABLE 3.1

Comparison between Impulse and Reaction Stages for Axial Flow Turbines ($V_{ax1} = V_{ax2}$)

Impulse ($R=0$)	**50% Reaction ($R=0.5$)**
$E=2u^2$	$E=u^2$
$\varepsilon_m=\cos^2\alpha_1$	$\varepsilon_m=\cos^2\alpha_1/(1-0.5\sin^2\alpha_1)$
$\phi=\cos\alpha_1/2$	$\phi=\cos\alpha_1$

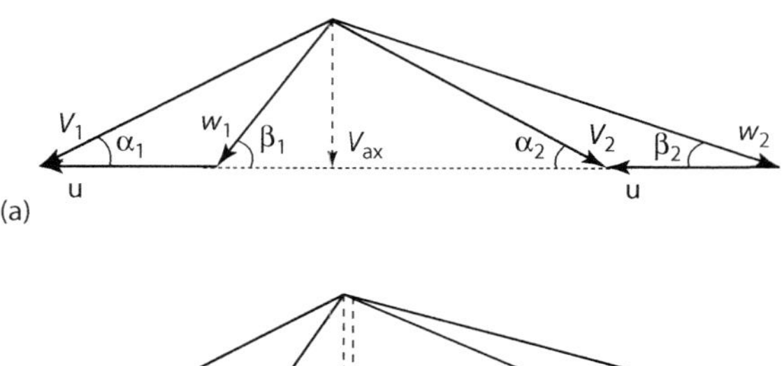

FIGURE 3.9
Velocity triangles when $R=1$; (a) axial velocity same at inlet and exit, (b) axial velocity different at inlet and outlet.

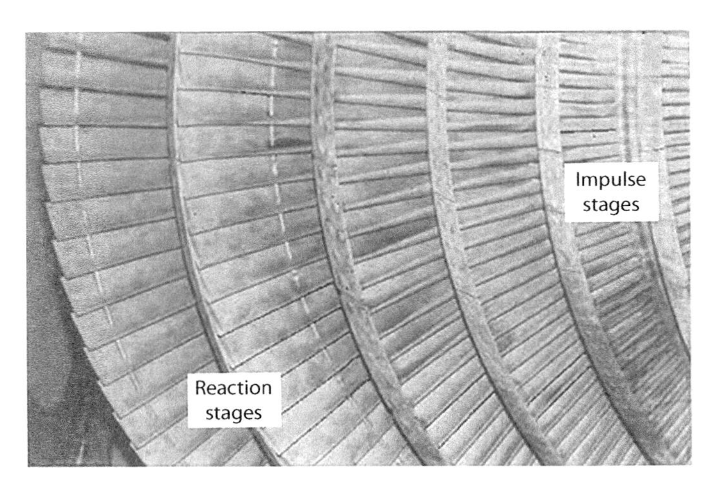

FIGURE 3.10
Shapes of impulse and reaction stage blading.

Example 3.1

Consider an axial flow turbine stage with 50% reaction. The blade speed is 200 m/s. Air leaves through a nozzle making an angle of 30° with the wheel tangent. The axial component of the inlet fluid speed is 200 m/s. (a) Calculate the blade angles, energy per unit mass, and utilization factor. (b) It is proposed to maximize the utilization for the same blade speed, axial velocity, and nozzle angle. What would be the degree of reaction, utilization factor, and energy produced?

Solution:
Using trigonometry from the velocity triangles,

$$V_1 = \frac{V_{ax}}{\sin 30} = \frac{200}{\sin 30} = 400\,\text{m/s} = w_2;\; u_1 = u_2 = u = 200\,\text{m/s} \tag{a}$$

$$w_1 = \sqrt{V_1^2 + u^2 - 2uV_1\cos 30} = 247.9\,\text{m/s} = V_2$$

The blade angles can be calculated as

$$\alpha_1 = |\beta_2| = 30\,\text{deg (given)}$$

$$\beta_1 = \arctan\left(\frac{V_{ax}}{V_1\cos\alpha_1 - u}\right) = \arctan\left(\frac{200}{400\cos 30 - 200}\right) = 53.8\,\text{deg} = \alpha_2 \tag{b}$$

Energy per unit mass is given by

$$E = \frac{\left(V_1^2 - V_2^2\right)}{2} + \frac{\left(u_1^2 - u_2^2\right)}{2} + \frac{\left(w_2^2 - w_1^2\right)}{2} = V_1^2 - V_2^2 = 98546\frac{\text{m}^2}{\text{s}^2}\left(= \frac{\text{J}}{\text{kg}}\right)$$

$$\varepsilon = \frac{E}{E + \frac{V_2^2}{2}} = \frac{98546}{98546 + \frac{247.9^2}{2}} = 0.762 \tag{c}$$

(since $V_1 = w_2$ and $V_2 = w_1$ and $u_1 = u_2$)

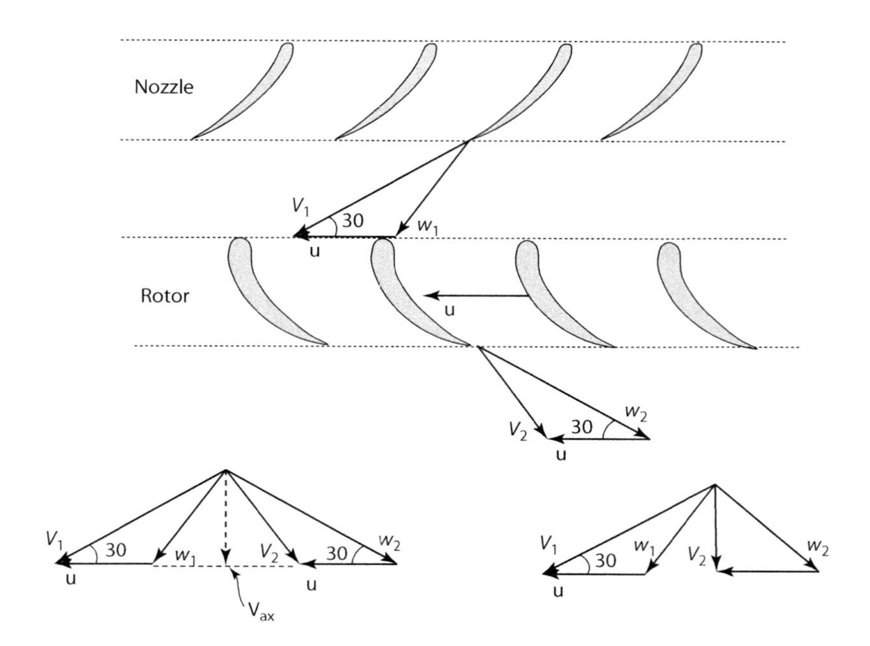

The velocity triangle at the inlet would be the same. However, the exit velocity triangle would be a right triangle, since for maximum utilization, exit velocity needs to be a minimum or axially directed. The corresponding velocities are then

$$V_1 = \frac{V_{ax}}{\sin 30} = \frac{200}{\sin 30} = 400\,\text{m/s};\; u_1 = u_1 = u = 200\,\text{m/s}$$

$$w_1 = \sqrt{V_1^2 + u^2 - 2uV_1 \cos 30} = 247.9\,\text{m/s} \tag{d}$$

$$V_2 = V_{ax} = 200\,\text{m/s};\; w_2 = \sqrt{V_2^2 + u^2} = 282.8\,\text{m/s}$$

Then,

$$R = \frac{\left(w_2^2 - w_1^2\right)}{\left(V_1^2 - V_2^2\right) + \left(w_2^2 - w_1^2\right)} = \frac{\left(282.8^2 - 247.9^2\right)}{\left(400^2 - 200^2\right) + \left(282.8^2 - 247.9^2\right)} = 0.133$$

$$E = \frac{\left(V_1^2 - V_2^2\right)}{2} + \frac{\left(u_1^2 - u_2^2\right)}{2} + \frac{\left(w_2^2 - w_1^2\right)}{2} = \frac{\left(V_1^2 - V_2^2\right)}{2} + \frac{\left(w_2^2 - w_1^2\right)}{2}$$

$$= \frac{\left(400^2 - 200^2\right)}{2} + \frac{\left(282.8^2 - 247.9^2\right)}{2} = 69260\,\frac{\text{m}^2}{\text{s}^2}\left(= \frac{\text{J}}{\text{kg}}\right)$$

$$\varepsilon = \frac{E}{E + \frac{V_2^2}{2}} = \frac{69260}{69260 + \frac{200^2}{2}} = 0.775$$

Note: Utilization factor and energy produced can also be calculated using alternative definitions as shown by the following:

$$\varepsilon = \frac{(V_1^2 - V_2^2)}{(V_1^2 - RV_2^2)} = \frac{(400^2 - 247.9^2)}{(400^2 - (0.5)(247.9^2))} = 0.762$$

$$\varepsilon = \frac{(V_1^2 - V_2^2)}{(V_1^2 - RV_2^2)} = \frac{(400^2 - 200^2)}{(400^2 - (0.133)(200^2))} = 0.775 \tag{e}$$

for the utilization factors and

$$E = u_1 V_{u1} - u_2 V_{u2} = u(V_{u1} - V_{u2}) = (200)(400\cos 30 - (-247.9\cos 53.8))$$

$$= 98564 \frac{\text{m}^2}{\text{s}^2} \tag{f}$$

$$E = u_1 V_{u1} - u_2 V_{u2} = uV_1\cos\alpha_1 = (200)(400\cos 30) = 69282 \frac{\text{m}^2}{\text{s}^2}$$

Comment: These are the same answers that were obtained earlier. Thus, many of the relationships between degree of reaction, utilization factor, and velocities in the earlier discussion are relevant.

Example 3.2

A centrifugal pump produces 20 L/s while rotating at 1500 rpm. The vanes have a tip width of 1.25 cm and a blade angle of −30°. The vanes are backward-curved, and the radial velocity is 3 m/s. It is proposed use this pump against a head of 15 m. What should the minimum speed that will deliver this head be? Assume that water enters the pump with no whirl.

Solution: The velocity triangles will look as follows:

$$H = \frac{E}{g} = \frac{u_1 V_{u1} - u_2 V_{u2}}{g} = -\frac{u_2 V_{u2}}{g} \tag{a}$$

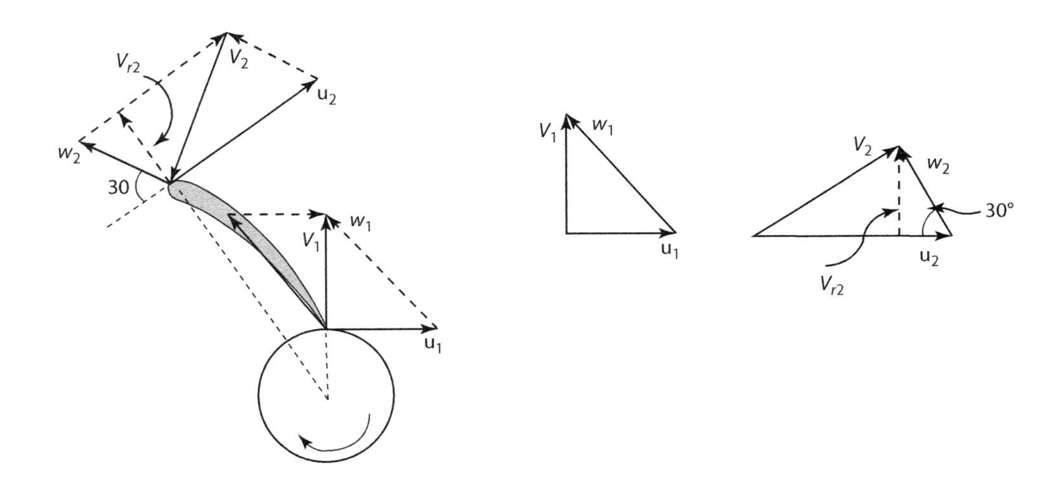

$$Q = 2\pi r_2 b_2 V_{r2} \Rightarrow r_2 = \frac{Q}{2\pi b_2 V_{r2}} = \frac{0.02}{2\pi(0.0125)(3)} = 0.0848\,\text{m}$$

$$\omega = \frac{2\pi N}{60} = \frac{2\pi(1500)}{60} = 157.1\ \text{rad/s} \qquad \text{(b)}$$

$$u_2 = \omega r_2 = (157.1)(0.0848) = 13.33\,\text{m/s}$$

From the exit velocity triangle,

$$V_{u2} = u_2 - w_{u2} = u_2 - \frac{V_{r2}}{\tan 30} = 13.33 - \frac{3}{\tan 30} = 8.137\,\text{m/s}$$

Thus, the head delivered at 1500 rpm would be

$$|H| = \frac{u_2 V_{u2}}{g} = \frac{(13.33)(8.137)}{9.81} = 11.06\,\text{m/s}$$

To calculate the speed required to deliver a head of 15 m, the pump affinity laws are used:

$$\left(\frac{gH}{N^2D^2}\right)_1 = \left(\frac{gH}{N^2D^2}\right)_2 \Rightarrow \left(\frac{H}{N^2}\right)_1 = \left(\frac{H}{N^2}\right)_2 \Rightarrow N_2 = N_1\left(\frac{H_2}{H_1}\right)^{0.5}$$

or

$$N_2 = 1500\sqrt{\left(\frac{15}{11.06}\right)} = 1746\,\text{rpm}$$

Thus, the pump needs to rotate at a rate higher than 1746 rpm to deliver a head of 15 m. The flow rate at this new rpm can also be calculated using the equality of the flow coefficient. This is left as an exercise.

Comments: For the purpose of verification, the specific speed of the pump is calculated.

$$N_s = \frac{N\sqrt{Q}}{(gH)^{0.75}} = \frac{(157.1)\sqrt{0.02}}{((9.81)(11.1))^{0.75}} = 0.66$$

This is well within the range of centrifugal pumps.

Example 3.3

The following data refers to a hydraulic reaction turbine. The inner and outer diameters are 6 and 12 in., respectively. The blade speed at the inlet is 50 ft./s. Assuming that the radial velocity of flow is constant at 10 ft./s and there is no whirl velocity at the exit, calculate the blade angles at the inlet and the exit. The degree of reaction is 0.6. Also, calculate the head and static pressure drop in the turbine.

Solution:

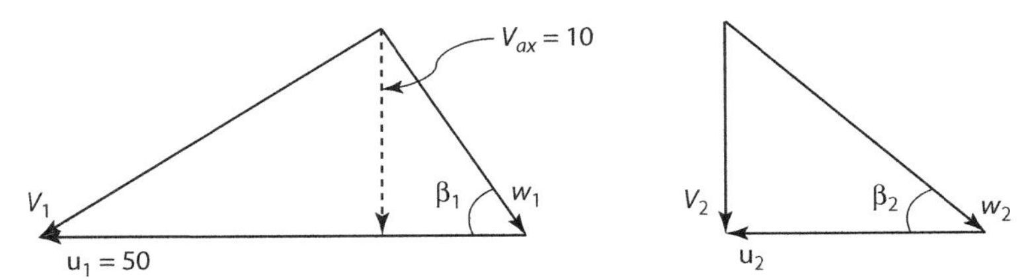

Since the inlet diameter is half the exit diameter, $u_2 = \dfrac{u_1}{2} = 25\text{ft./s}$
Since radial velocity is a constant, $V_2 = V_{ax} = 10\text{ft./s}$
From the exit velocity triangle, $w_2 = \sqrt{V_2^2 + u_2^2} = \sqrt{10^2 + 25^2} = 26.92$ ft./s
Blade velocity at the exit is $\beta_2 = \arctan\left(\dfrac{10}{25}\right) = -21.8^\circ$
From the definition of degree of reaction,

$$0.6 = R = \frac{\left(u_1^2 - u_2^2\right) + \left(w_2^2 - w_1^2\right)}{2\left(u_1 V_{u1} - u_2 V_{u2}\right)} = \frac{\left(u_1^2 - u_2^2\right) + \left(w_2^2 - w_1^2\right)}{2u_1 V_{u1}} \tag{a}$$

However, from the inlet velocity triangle,

$$\begin{aligned} V_{u1} &= u_1 - 10\cot\beta_1 = 50 - 10\cot\beta_1 \\ w_1^2 &= 10^2 + (10\cot\beta_1)^2 = 100 + 100\cot^2\beta_1 \end{aligned} \tag{b}$$

Substituting the values in Equation (a),

$$0.6 = \frac{\left(50^2 - 25^2\right) + \left(26.92^2 - 100\cot^2\beta_1\right)}{(2)(50)(50 - 10\cot\beta_1)} = \frac{2600 - 100\cot^2\beta_1}{(2)(50)(50 - 10\cot\beta_1)} \tag{c}$$

Equation (c) can be solved to get $\beta_1 = -45^\circ$
To calculate the static pressure drop, the energy transfer E needs to be calculated from Euler's equation:

$$E = u_1 V_{u1} - u_2 V_{u2} = u_1 V_{u1} = (50)(50 - 10\cot\beta_1) = (50)(40) = 2000\frac{\text{ft}^2}{\text{s}^2}\left(= \frac{\text{ft} - \text{lbf}}{\text{slug}}\right)$$

$$H = \frac{E}{g} = \frac{2000}{32.2} = 62.11\text{ft.}$$

The static pressure drop can be calculated from the definition of degree of reaction as being the ratio of the static and stagnation enthalpy changes. Also, the energy transfer in the rotor can be written in terms of enthalpy. Furthermore, since the fluid is incompressible, enthalpy changes can be written as

$$E = \left(h_1 + \frac{V_1^2}{2}\right) - \left(h_2 + \frac{V_2^2}{2}\right) = h_{01} - h_{02}$$

$$h_1 - h_2 = \left(\hat{u} + \frac{p}{\rho}\right)_1 - \left(\hat{u} + \frac{p}{\rho}\right)_2 = \frac{p_1}{\rho} - \frac{p_2}{\rho}$$

Hence, the enthalpy changes can be written in terms of pressure as

$$0.6 = R = \frac{(h_1 - h_2)}{(h_{01} - h_{02})} = \frac{1}{E}\left(\frac{p_1}{\rho} - \frac{p_2}{\rho}\right) = \frac{1}{(2000)(1.94)}(p_1 - p_2)$$

$$p_1 - p_2 = (0.6)(2000)(1.94) = 2328\,\text{psf} = 16.16\,\text{psi}$$

Note: An equally effective procedure would be to calculate all the velocities for the inlet and exit triangles and use the definitions of E and Δp in terms of the velocities. This procedure is sometimes more convenient since Euler's equation, Equation 3.7, is derived from the momentum principle and is a vector equation, whereas Equation 3.10 is derived from the energy equation, which is a scalar equation. This procedure is also illustrated in the following.

For the inlet velocity triangle,

$$u_1 = 50 \text{ ft./s}; \quad w_1 = \frac{10}{\sin 45} = 14.14 \text{ ft./s}; \quad V_{u1} = u_1 - 10 = 40 \text{ ft./s}$$

$$V_1 = \sqrt{V_{u1}^2 + 10^2} = 41.23 \text{ ft./s}$$

$$u_2 = 25 \text{ ft./s}; \quad w_2 = 26.92 \text{ ft./s}; \quad V_2 = 10 \text{ ft./s}$$

$$E = \frac{(V_1^2 - V_2^2)}{2} + \frac{(u_1^2 - u_2^2)}{2} + \frac{(w_2^2 - w_1^2)}{2} = 2000\frac{\text{ft}\cdot\text{lbf}}{\text{slug}}$$

The energy equation across the rotor, Equation 3.10, becomes

$$h_1 + \frac{V_1^2}{2} = h_2 + \frac{V_2^2}{2} + E \quad \Rightarrow h_1 - h_2 = E + \frac{V_2^2}{2} - \frac{V_1^2}{2}$$

$$\Rightarrow \quad \frac{p_1}{\rho} - \frac{p_2}{\rho} = E + \frac{V_2^2}{2} - \frac{V_1^2}{2} = 2000 + \frac{10^2}{2} - \frac{41.23^2}{2} = 1200\frac{\text{ft}\cdot\text{lbf}}{\text{slug}}$$

$$\text{or} \quad p_1 - p_2 = \Delta p = (1200)(1.94) = 2328 \text{ psf} = \frac{2328}{144} = 16.17 \text{ psi}$$

Comments: As expected, the pressure drop through the turbine is not zero since the turbine is a reaction turbine. The utilization factor can now be calculated: this is left as an exercise.

Example 3.4

In the design of turbines, an important variable is the ratio of the blade speed to the inlet fluid speed, u/V_1, called the *peripheral velocity factor* or *speed ratio*, ϕ. Consider an axial flow impulse turbine with constant axial velocity. Obtain an expression for the utilization factor in terms of peripheral speed factor ϕ and the nozzle angle α_1.

Solution:

Since the flow is axially directed, $u_1 = u_2 = u$.

$$R = \frac{(w_2^2 - w_1^2)}{(V_1^2 - V_2^2) + (w_2^2 - w_1^2)} = 0 \quad \Rightarrow \quad w_1 = w_2 \tag{a}$$

The velocity triangles are shown.

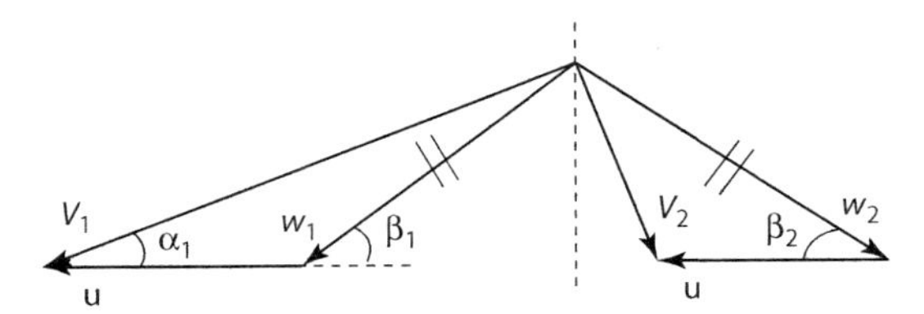

From the triangles,

$$\begin{aligned} V_2^2 &= w_2^2 + u^2 - 2uw_2\cos\beta_2 \\ V_1^2 &= w_1^2 + u^2 + 2uw_1\cos\beta_1 \\ \text{or } V_1^2 - V_2^2 &= 4uw_1\cos\beta_1 = 4u(V_1\cos\alpha_1 - u) \end{aligned} \tag{a}$$

Using Equation 3.19, an expression for utilization factor can be obtained. Thus,

$$\begin{aligned} \varepsilon &= \frac{\left(V_1^2 - V_2^2\right)}{\left(V_1^2 - RV_2^2\right)} = \frac{V_1^2 - V_2^2}{V_1^2} = \frac{4u(V_1\cos\alpha_1 - u)}{V_1^2} \\ &= 4\frac{u}{V_1}\left(\cos\alpha_1 - \frac{u}{V_1}\right) = 4\phi(\cos\alpha_1 - \phi) \end{aligned} \tag{b}$$

For maximum utilization, the derivative $d\varepsilon/d\phi = 0$. Thus,

$$\frac{d\varepsilon}{d\phi} = 0 \Rightarrow 4\cos\alpha_1 - 8\phi = 0 \Rightarrow \phi = \frac{\cos\alpha_1}{2} \tag{c}$$

A plot of Equation (b) is shown in the following figure.

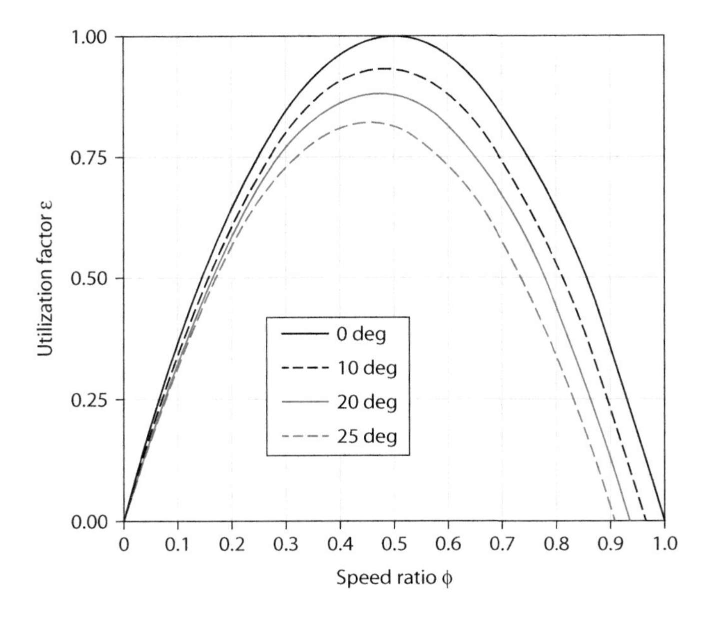

As seen from the figure, the maximum occurs around a speed ratio of 0.5 (for so-called zero angle turbines, the maximum occurs at exactly 0.5).

Comments: The theoretical maximum value of utilization factor as shown in this problem is never achieved in practice. This aspect of impulse turbines will be discussed quite extensively in the chapter on hydraulic turbines.

Example 3.5

An axial flow reaction turbine has a degree of reaction $R = 1/3$. If the utilization factor for such a turbine is to be maximum, show that the relationship between blade velocity and inlet fluid velocity is given by the relationship

$$\frac{u}{V_1} = \frac{3}{4}\cos\alpha_1$$

Solution:

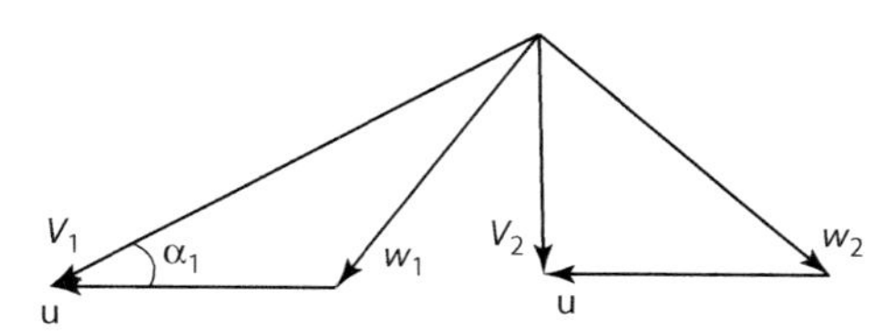

For axial flow turbines, since $u_1 = u_2$, the expression for degree of reaction becomes

$$R = \frac{\left(w_2^2 - w_1^2\right)}{\left(V_1^2 - V_2^2\right) + \left(w_2^2 - w_1^2\right)} = \frac{\left(w_2^2 - w_1^2\right)}{2\left(u_1 V_{u1} - u_1 V_{u2}\right)} = \frac{\left(w_2^2 - w_1^2\right)}{2u_1 V_{u1}} \tag{a}$$

From the velocity triangles,

$$w_2^2 = u^2 + V_2^2 = u^2 + V_1^2 \sin^2\alpha_1$$

$$w_1^2 = u^2 + V_1^2 - 2uV_1\cos\alpha_1$$

Thus,

$$\begin{aligned}\left(w_2^2 - w_1^2\right) &= 2uV_1\cos\alpha_1 - V_1^2\left(1 - \sin^2\alpha_1\right) = 2uV_1\cos\alpha_1 - V_1^2\cos^2\alpha_1 \\ &= \left(2u - V_1\cos\alpha_1\right)V_1\cos\alpha_1\end{aligned} \tag{b}$$

Similarly,

$$u_1 V_{u1} = uV_1\cos\alpha_1 \tag{c}$$

Hence, the degree of reaction from Equation (a) can be written as

$$\frac{1}{3} = R = \frac{\left(w_2^2 - w_1^2\right)}{2u_1 V_{u1}} = \frac{\left(2u - V_1\cos\alpha_1\right)V_1\cos\alpha_1}{2uV_1\cos\alpha_1} = 1 - \frac{V_1}{2u}\cos\alpha_1 \tag{d}$$

After simplifying, the result is

$$\frac{u}{V_1} = \frac{3}{4}\cos\alpha_1$$

Comment: A more generalized form of the speed ratio can be obtained for other values of R (see Problem 3.28).

Example 3.6

As the final example, let us consider the degree of reaction and utilization factor for a lawn sprinkler. The arm of a lawn sprinkler is 4 in. and it rotates at 100 rpm. The fluid exits with an absolute velocity of 6 ft./s in the tangential direction. Calculate the degree of reaction and utilization factor.

Solution: At the inlet, the velocity is assumed to be zero. Since the blade velocity at the inlet will be zero, the velocity triangle at the inlet will degenerate to a point. At the exit, the velocity triangle is shown as follows.

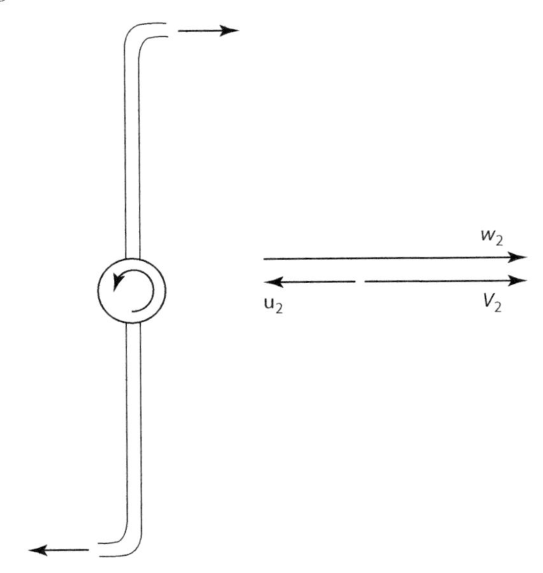

$$N = 120\,\text{rpm} = \frac{2\pi N}{60} = \frac{2\pi(120)}{60} = 10.47\,\text{rad/s} \tag{a}$$

$$u_2 = (10.47\,\text{rad/s})\frac{4}{12}\,\text{ft.} = 3.46\,\text{ft./s}$$

From the velocity triangle at the exit,

$$V_2 = u_2 + w_2 \quad \Rightarrow \; 6\,\text{ft./s} = -3.46\text{ ft./s} + w_2 \;\Rightarrow w_2 = 9.46\text{ ft./s} \tag{b}$$

$$V_{u2} = V_2 = 6\text{ ft./s}$$

Hence,

$$E = \left|u_1 V_{u1} - u_2 V_{u2}\right| = (3.46\text{ ft./s})(6\text{ ft./s}) = 20.73\frac{\text{ft}\cdot\text{lbf}}{\text{slug}} \tag{c}$$

$$\varepsilon = \frac{E}{E + \dfrac{V_2^2}{2}} = \frac{20.44}{20.43 + \dfrac{6^2}{2}} = 0.535$$

Also,

$$R = \frac{(u_1^2 - u_2^2) + (w_2^2 - w_1^2)}{(V_1^2 - V_2^2) + (u_1^2 - u_2^2) + (w_2^2 - w_1^2)} = \frac{(w_2^2 - u_2^2)}{2E} \tag{d}$$

$$= \frac{(9.45^2 - 3.46^2)}{2(20.73)} = 1.87$$

Comments: The degree of reaction is greater than one. The reason for this is that all the energy transfer is a result of the increase of static head due to centrifugal action and changes in relative velocity in the rotor. Since the inlet kinetic energy has been assumed to be zero, the exit kinetic energy makes the change in stagnation enthalpy less than the change in static enthalpy. The relationship between the utilization factor and degree of reaction given by Equation 3.19 can be verified.

PROBLEMS

In all the problems of this chapter, use water as the working fluid unless specified otherwise.

3.1 In a certain hydraulic turbomachine, the following data apply:

Inlet blade velocity is 20 m/s. $V_{u1} = 21$ m/s; $V_{r1} = 10$ m/s.

At the exit, V_2 is purely axial and has a value of 14 m/s, and blade speed is 9 m/s. Is this power absorbing or power producing? Find E and the changes in static and stagnation pressures. Would this machine be classified as a purely radial, purely axial, or mixed flow type?

Ans: $E = 420$ J/kg; $\Delta p_{stag} = 420$ kPa; $\Delta p_{stat} = 247.4$ kPa

3.2 In the text, the relationship for utilization ε and degree of reaction R was derived for axial flow turbines. Show that the same expression is valid for radial or mixed flow turbines also; that is, show that

$$\varepsilon = \frac{V_1^2 - V_2^2}{V_1^2 - RV_2^2}$$

3.3 Show that for axial flow turbines, maximum utilization occurs when V_2 is in the axial direction.

3.4 Show that for a 50% reaction axial flow turbine, the velocity triangles are symmetric; that is, $V_1 = w_2$ and $V_2 = w_1$.

3.5 Show that for axial flow turbines, maximum utilization factor is given by the following expression (α_1 is the nozzle angle):

$$\varepsilon_{max} = \frac{\cos^2 \alpha_1}{1 - R\sin^2 \alpha_1}$$

In the equation given, α_1 is the nozzle (measured with respect to the wheel tangent) angle at the inlet.

3.6 The blade speed of an axial flow turbine using air is 250 ft./s. The tangential components of absolute velocities at the inlet and outlet are 900 ft./s (in the direction of rotation) and 300 ft./s (opposite to the direction of rotation). Calculate (a) power output if mass flow rate is 20 lbm/s, (b) change in total enthalpy, and (c) change in total temperature across the rotor.

Ans: $P = 338.8$ hp; $\Delta h_{stag} = 11.97$ BTU/lbm; $\Delta T_{stag} = 49.89$ R

3.7 For a certain turbomachine, the inlet and outlet blade speeds are 100 ft./s and 25 ft./s, respectively. At the point of entry, the radial velocity component of the absolute velocity is 30 ft./s. It also has a tangential component of 114 ft./s in the direction of blade speed. If absolute velocity at the exit is directed radially and has a magnitude of 50 ft./s, calculate (a) power, (b) degree of reaction, (c) utilization factor and (d) the pressure change of the fluid if it is water and the flow rate is 23 cfs.

Ans: $P = 923.8$ hp; $R = 0.5$; $\varepsilon = 0.90$

3.8 The following data apply to a radial flow turbine:

Degree of reaction = 0.6

Rotor angular velocity = 300 rpm

Rotor inlet diameter = 4 ft.

Rotor exit diameter = 2 ft.

If the radial velocity at the inlet is the same as at the exit and has a value of 10 ft./s, find the blade angles at the inlet and outlet. Assume no whirl velocity at the exit.

Ans: $\beta_1 = 38.6°$ or $141.4°$; $\beta_2 = 17.6°$

3.9 For a certain stage of an axial flow turbine, the nozzle makes an angle of 28° with the wheel tangent. The axial velocity component at the inlet is 500 ft./s. (a) If the degree of reaction is 50% and the blade speed is 550 ft./s, find the blade angles. (b) If nozzle angle, blade speed, and axial velocity remain unchanged, what should the value of degree of reaction be for maximum utilization?

Ans: $\beta_1 = \alpha_2 = 52°$; $\beta_2 = \alpha_1 = 28°$; $R = 0.145$

3.10 While rotating at 900 rpm, a centrifugal pump delivers 2500 gpm of water at a head of 75 ft. If the impeller diameter is 1.5 ft. and the blade width is 1.75 in., find the (a) blade angle β_2, (b) radial velocity V_{r2}, and (c) blade speed u_2. Assume no whirl at the inlet.

Ans: $\beta_2 = 12.5°$; $V_{r2} = 8.1$ ft./s; $u_2 = 70.7$ ft./s

3.11 The following data refer to the impulse stage of an axial flow turbine:

Nozzle angle = 20°

Mean blade speed = 700 ft./s

Axial velocity = 350 ft./s and is constant through the stage

Calculate (a) energy transfer E, (b) change in total (stagnation) enthalpy, (c) change in static enthalpy, (d) utilization factor, (e) blade angles β_1 and β_2. Is this the maximum utilization?

Ans: $E = \Delta h_0 = 366000$ ft-lb/slug; $\varepsilon = 0.7$; $\beta_1 = \beta_2 = 53.22°$; Not maximum utilization

3.12 Solve the Problem 11 for a 50% reaction turbine.

Ans: $E = 856000$ ft-lb/slug; $\varepsilon = 0.9$; $\beta_1 = 53.22$; $\beta_2 = 20$

3.13 The hub radius of a mixed-flow-type pump is 5 cm and the impeller tip radius is 15 cm. At the inlet, the absolute velocity is axial and the outlet relative velocity is purely radial. Also, inlet axial velocity is equal to outlet radial velocity and the magnitude of exit relative velocity is equal to the blade speed at the inlet. Under these conditions, find the energy input, and all the relevant angles if its rotational speed is 2800 rpm.

Ans: $E = -1934$ J/kg; $\beta_1 = 45°$; $\alpha_1 = \beta_2 = 90°$; $\alpha_2 = 18.4°$

3.14 The following data refer to an axial flow blower with a single stage:

Mass flow rate = 0.75 kg/s

Rotor tip and hub diameters = 24 cm and 16 cm

Rotational speed = 3500 rpm

As the air flows through the rotor, it turns through an angle of 22° away from the axial direction. Assuming standard atmospheric conditions and that the axial velocity is a constant, compute the degree of reaction and power input to the blower. Also, find the utilization factor. Inlet absolute velocity is purely axial.

Ans: $P = -252.8$ kW; $R = 0.875$

3.15 The inlet and outlet velocity triangles of a particular stage of a turbomachine are given. Answer the following:

a. What is the degree of reaction?
b. What is the change in stagnation enthalpy across the stage?
c. Is this a power-generating or power-absorbing machine?
d. What is the energy exchange per unit mass of the machine?

Ans: $R = 0.5$; $E = 60000$ ft-lb/slug; power-generating

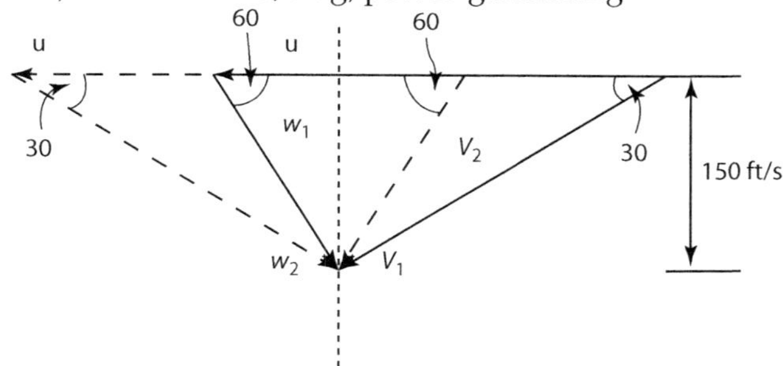

3.16 For an axial stage of a turbomachine, the nozzle at the entry makes an angle of 30° while the axial component of the absolute velocity is 75 m/s. Given that the blade speed is 75 m/s and the degree of reaction is 0.5, find the blade angles at the inlet and outlet.

Ans: $\beta_1 = 53.8°$

For the same nozzle angle, blade speed, and axial velocity, if exit absolute velocity is equal to inlet axial velocity, what would the degree of reaction be?

Ans: $R = 0.134$

3.17 The rotational speed of an impulse turbine is 3000 rpm and the mean diameter of the rotor is 90 cm. Given that the utilization factor is 0.9, find the inlet blade angle and power output if the mass flow rate is 2.5 kg/s. The inlet to the turbine is through a nozzle that makes an angle of 22° with the blade tangent. The inlet fluid velocity is 250 m/s.

Ans: $\beta_1 = 46.0°$; $P = 70.3$ kW

3.18 The blade speed of a 50% reaction axial flow steam turbine is 400 ft./s. The nozzle through which steam enters the turbine makes an angle of 28° with the wheel tangent. The axial component of the inlet velocity is equal to the blade speed. Find the rotor blade angles and utilization factor.

Ans: $\alpha_2 = \beta_1 = 48.6°$; $E = 441000$ ft-lb/slug; $\varepsilon = 0.76$

It is required that the utilization factor be made a maximum for the same axial velocity, blade speed, and nozzle angle. What would the degree of reaction be under such conditions?

3.19 For a reaction turbine, the utilization factor and degree of reaction are 0.93 and 0.85, respectively. The fluid leaves the rotor in a purely radial direction. The flow rate of the water is 15 cfs, and the radial velocity is 12 ft./s and is constant throughout the rotor. If the blade velocity at the inlet is 56.44 ft./s, find the power developed by the turbine.

Ans: $P = 50.53$ hp

3.20 Inward flow reaction turbines are called *Francis turbines*, and are not open to the atmosphere. A particular Francis unit rotates at 380 rpm. Water enters at a radius of 0.4 m and leaves at a radius of 0.24 m. The gates are set so that water enters the rotor at a speed of 17 m/s ($=V_1$) and at an angle of 10°. Find the inlet and outlet blade angles, utilization factor, degree of reaction, and the pressure drop through the rotor. Water exits in the radial direction with an absolute velocity of 3 m/s.

Ans: $\beta_1 = 74.4°$; $\beta_2 = 17.4°$; $\varepsilon = 0.98$; $R = 0.47$

3.21 Consider an axial flow turbine with degree of reaction R. Show that the relationship between blade velocity u and inlet velocity V_1 for maximum utilization is given by the following expression:

$$\frac{u}{V_1} = \frac{\cos\alpha}{2(1-R)}$$

where α is the nozzle angle.

3.22 For a radial inward flow turbine, the nozzle angle is measured with respect to the wheel tangent and is 45°. The radial velocity is constant throughout the rotor. If the inlet blade angle is 90°, show that the maximum utilization factor should be 0.666.

Hint: Inlet triangle is an isosceles right triangle

3.23 Devices called *pump turbines* can operate either in the pump mode (while absorbing power) or in the turbine mode (while producing power). For a certain axial flow machine, the mean radius is 1.0 m. The axial velocity is constant at 10 m/s and the pump rotates at 450 rpm. Under ideal conditions, that is, no inlet whirl in pump mode and no exit whirl in turbine mode, draw the velocity triangles for each mode and calculate the inlet and exit blade angles if $E = 110$ J/kg.

Ans: $\beta_1 = 11.98°$; $\beta_2 = 12.58°$

3.24 Consider an axial flow impulse turbine stage. For such a stage, the degree of reaction would be zero and the blade velocities at the inlet and exit would be the same. Obtain an expression for the variation of utilization factor in terms of the inlet blade angle α and speed ratio ϕ $(= u/V_1)$.

3.25 The rotational speed of a reaction turbine is 200 rpm and the inlet and outlet diameters are 10 ft. and 3 ft., respectively. The inlet absolute velocity has a radial component of 25 ft./s and tangential component of 45 ft./s. At the outlet, the fluid is directed radially and has a magnitude of 50 ft./s. Calculate (a) degree of reaction, (b) utilization factor. If the exit velocity is 51.4 ft./s, calculate (c) the degree of reaction.

Ans: $R = 98.4\%$; $\varepsilon = 0.792$; $R = 100\%$ (almost)

3.26 The nozzle angle for an axial flow impulse turbine is 15°. The inlet absolute velocity is 800 ft./s and the blade speed is 550 ft./s. Axial speed at the inlet and outlet is assumed to be a constant. Calculate (a) energy transfer E, (b) change in total (stagnation) and static enthalpies, (c) utilization factor.

Ans: $E = \Delta h_0 = 245{,}038$ ft-lbf/slug; $\Delta h = 0$; $\varepsilon = 0.77$

3.27 Obtain expressions for degree of reaction and utilization factor for a lawn sprinkler.

Ans: $R = \dfrac{w_2 + u_2}{2u_2} = \dfrac{1}{\varepsilon}$

3.28 For special cases of axial flow reaction turbines with degree of reaction in the form $R = 1/(k+1)$, where k is an integer, a special relationship exists between the blade velocity u and fluid inlet velocity for maximum utilization. Show that this relationship is given by

$$\frac{u}{V_1} = \frac{k+1}{2k}\cos\alpha$$

Verify this relationship when $k = 1$ and 3. See Example 3.5 and Problem 3.21 of the text.

3.29 Calculate the flow rate for the new speed in Example 3.2.

3.30 Calculate the utilization factor in Example 3.3. Is this the maximum?

Ans: $\varepsilon = 0.975$; Yes

4

Hydraulic Turbines

Up to this point, the emphasis has been on the theoretical aspects of turbomachines with limited application to specific machines. In this chapter, the analysis learned previously will be used in studying the performance characteristics of hydraulic turbines, both impulse and reaction types. An example of the former is the Pelton wheel, while Francis, Kaplan, and bulb turbines fall under the reaction category.

Upon completion of this chapter, the student will be able to

- Select turbines based on specific speeds
- Learn the features of different types of hydraulic turbines
- Analyze and perform preliminary design of hydraulic turbines
- Understand cavitation in the context of hydraulic turbines

4.1 Introduction

The energy stored by either flowing water or water stored in reservoirs at suitable heights can be converted into mechanical power through the use of suitable machinery. In the early days, such machines were called *waterwheels* and were used mainly for household tasks such as grinding corn, lifting water, and sawing wood. In modern times, such machines are used almost exclusively for driving electric generators. These machines are called *hydraulic turbines* and the electric power generated by such machines is called *hydroelectric power*. Hydroelectric power has the advantages of being inexpensive to produce and environmentally friendly by virtue of not emitting pollutants. It is, however, limited in terms of choice of location; consequently, the transportation costs of the generated electric power are usually higher. A schematic diagram of a typical hydroelectric power plant is given in Figure 4.1. Although the turbine shown is a Kaplan turbine, the setting for other vertical axis turbines would be similar. Water enters the turbine through the scroll case (1) and is directed onto the runner (6) through wicket gates (3). The generator stator (8) and rotor (9) are mounted on the same main shaft (7) as the turbine itself. Other significant parts of the unit are the thrust bearing (12), air cooler (13), and oil head (10). Although not present in Pelton turbines, water leaving the turbine is typically directed into the downstream tails through a draft tube in Francis and Kaplan turbines.

Hydraulic turbines are classified broadly into impulse and reaction turbines. Pelton wheel is the most common form of impulse type turbine while Francis turbine and Kaplan turbine are reaction type. A guide to the selection of turbines based on the head and power requirement is shown in Figure 4.2. Pelton wheels are most suitable for high heads and low flow rates. Kaplan turbines (somewhat the reverse of an exhaust fan!) are suited to high flows and low heads. This is related to the specific speeds, as discussed in Chapter 2, wherein it was shown that higher heads and lower flows give rise to lower specific speeds (impulse turbines) and lower heads with higher flows imply higher specific speeds (reaction turbines). A detailed discussion on turbine selection, plant capacity determination, and economic and hydrologic analysis can be found in Warnick et al. (1984).

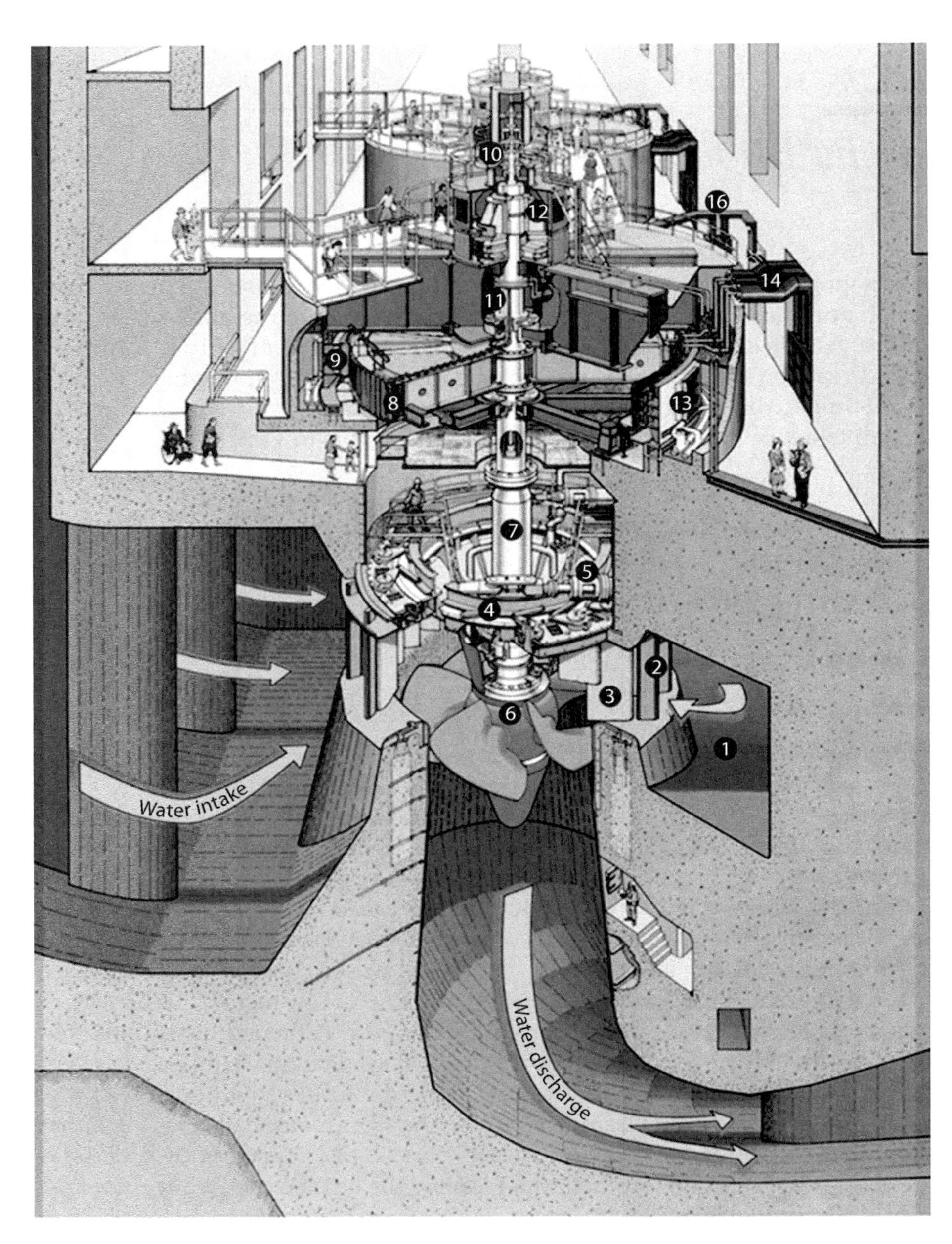

FIGURE 4.1
Components of a hydroelectric power plant. (Photo credit to Portland District, U.S. Army Corps of Engineers.)

4.2 Impulse Turbines

An impulse turbine (in the context of hadraulic turbines) is one in which the total pressure drop of the fluid occurs inside one or more nozzles and there is no drop in pressure as the fluid flows through the rotor. Such turbines are most useful when water is available at

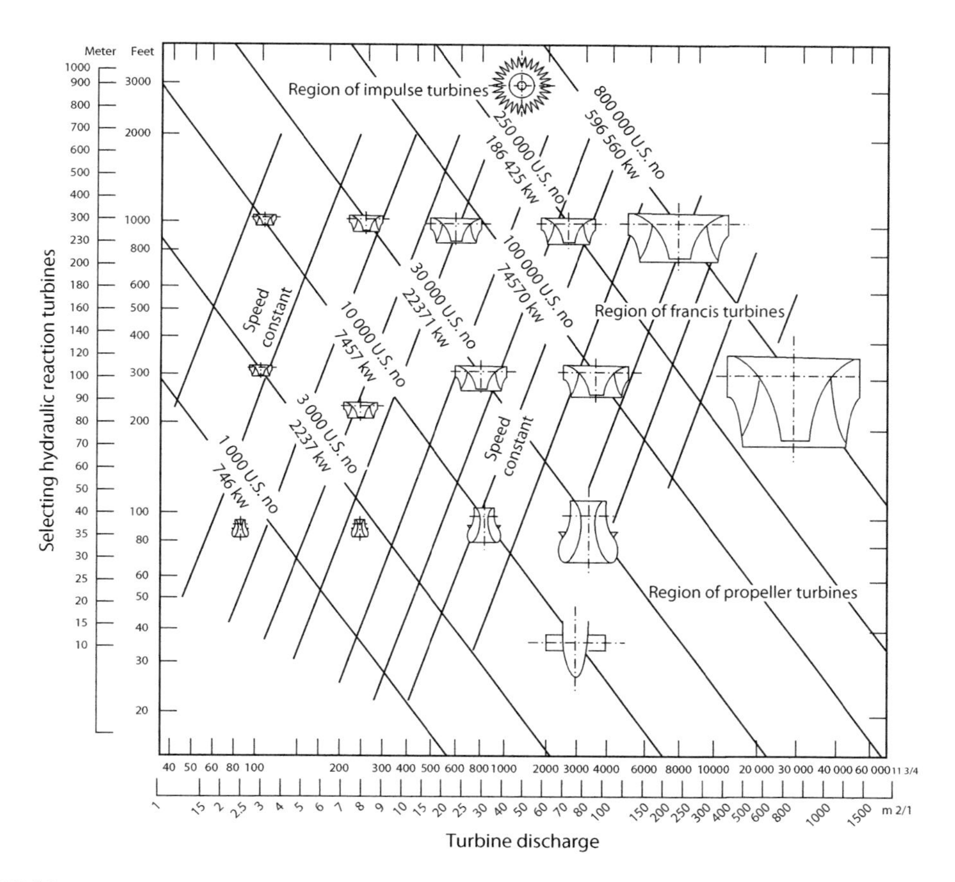

FIGURE 4.2
Guide to selecting hydraulic turbines. (From Walters, R. N. and Bates, C. G., (1976) *Selecting Hydraulic Reaction Turbines*, Engineering Monograph No. 20, Bureau of Reclamation, U.S. Department of the Interior, Washington, D.C.)

very high heads, that is, with large potential energy. When such large heads are available, the potential energy is first converted into kinetic energy by letting the fluid flow through large pipes called *penstock*, and the high kinetic energy fluid then impinges on vanes or buckets. Thus, kinetic energy is converged into force, which results in torque on the shaft. A small part of the kinetic energy is lost in fluid friction on the wheels, and another small part is lost as kinetic energy of the fluid leaving the vanes.

4.2.1 Pelton Wheel

A hydraulic impulse turbine is called a *Pelton wheel*, which is by far the most important machine among the hydraulic turbines. It has evolved from the development of the earliest hydraulic turbines, frequently called *waterwheels*. It is named after Lester A. Pelton (1829–1908), who patented it in 1890.

The locations of some Pelton turbine installations are shown in Table 4.1. A brief perusal of the table shows that the range of specific speeds is well within the accepted limits for impulse turbines indicated in Table 2.2.

The basic features of a Pelton wheel are shown in Figure 4.3.

It consists of a rotor on which ellipsoidal or hemispherical buckets are mounted along the circumference, one or several nozzles directing the high kinetic energy fluid onto the

TABLE 4.1
Pelton Turbine Installations around the World

	Location	H (ft.)	Power (hp)	N (rpm)	N_s
1	Reissek, Austria	5800	31,000	750	0.060
2	Dixence, Switzerland	5735	50,000	500	0.051
3	Pragnieres, France	3920	100,000	428	0.100
4	Kitimat, British Columbia, Canada	2500	140,000	328	0.160
5	Upper Molina, Colorado, USA	2490	12,000	600	0.086
6	Big Creek, California, USA	2418	61,000	300	0.100
7	Bucks Creek, California, USA	2350	35,000	450	0.118
8	Kings River, California, USA	2243	40,000	360	0.107
9	Santos, Brazil	2231	91,800	360	0.163
10	Cementos El Cairo, Colombia	1900	3000	900	0.090
11	Green Springs, Oregon, USA	1800	23,500	600	0.180
12	Paucartambo, Peru	1580	28,000	450	0.174
13	Middle Fork, California, USA	1325	91,500	240	0.209
14	San Bartolo, Mexico	1233	39,000	428	0.266
15	Thorpe, North Carolina, USA	1188	30,000	258	0.147
16	Feather River, California, USA	1110	76,000	240	0.237
17	Wahiawa, Hawaii, USA	700	1800	360	0.097

buckets, and a flow-controlling mechanism in the nozzles. This mechanism is usually in the form of a spear, the movement of which is controlled hydraulically. The spear controls the amount of flow and at the same time provides smooth flow with minimal losses. An important feature of Pelton wheels is the conversion of the available potential energy of the fluid into kinetic energy *outside* the rotor (in this case inside the penstock and nozzles). The Pelton runner typically operates in an atmospheric chamber and the energy transfer occurs purely through impulse action. The rotor is not enclosed, and the water leaving the turbine goes to the tail race. A picture of a rotor with bolted buckets is shown in Figure 4.4a. The buckets can be either bolted as shown in the figure or they can be cast and made an integral part of the rotor.

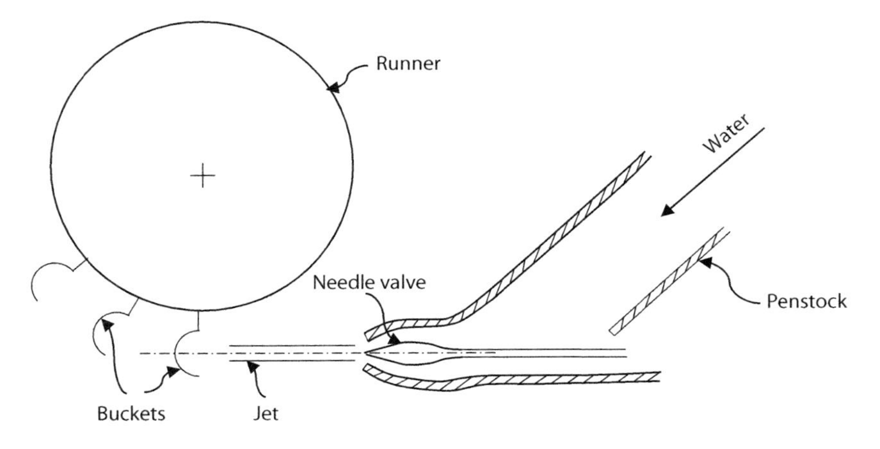

FIGURE 4.3
Schematic diagram of a Pelton wheel.

FIGURE 4.4
Typical runners (a) Pelton turbine, (b) Francis turbine, and (c) Kaplan turbine. (Courtesy of Voith Hydro, Inc.)

The velocity diagram for a Pelton wheel is shown in Figure 4.5b. At the inlet, the jet is almost aligned with the blade and the inlet velocity degenerates to a straight line. At the outlet, the fluid leaves with a relative velocity w_2 making an angle of β_2 with the blade speed. The force and power are given by the following expressions:

$$F = \rho Q(V_1 - u)(1 + \cos\beta_2) \tag{4.1}$$

$$P = Fu = \rho Qu(V_1 - u)(1 + \cos\beta_2) \tag{4.2}$$

In Equations 4.1 and 4.2, u is the blade speed, V_1 is the jet speed, and β_2 is the blade angle at the exit. The maximum power is obtained when β_2 is equal to zero (the fluid turns through an angle of 180°, which from our convention for blade angles would be 0°). The velocity triangle at the exit would degenerate into a straight line. This is called the *zero angle blade* (Figure 4.5a) and would represent the ideal case. However, when the blade angle

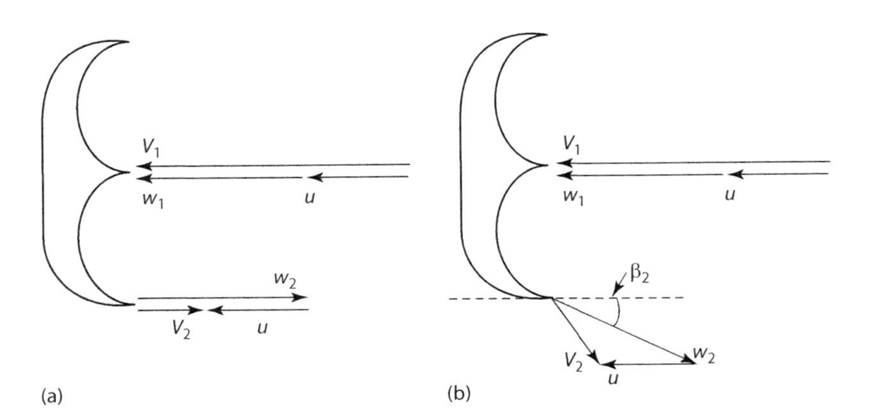

FIGURE 4.5
Velocity triangles for Pelton wheel.

is zero, water exiting from one bucket would strike the back of the succeeding bucket, with the result that the overall efficiency of the turbine would be reduced. Hence, the value of β_2 is normally set between 10° and 15°.

The bucket of a typical Pelton wheel is shown in Figure 4.6. For reasons of symmetry, the jet emerging from the nozzle is split into two parts and the fluid is equally distributed in the hemispherical buckets. Theoretically, the centerline of the bucket, called the *splitter* should be shaped like a cusp, with the two hemispheres meeting at the center, making the angle of contact of the two surfaces zero. However, this is not possible for two reasons: first, because of the difficulty in manufacturing such an intersection, and second because of the entrained sand and other abrasive particles in the jet that invariably erode the splitter surface. Typical relative dimensions of the bucket shown in Figure 4.6 are given as Equation 4.3 (see Kadambi and Prasad (1977) and Shepherd (1956)).

$$B/d = 3;\ e/d = 3.5;\ T/d = 0.8 - 0.95;\ L/d = 2.5 - 2.8;\ E/d = 0/85 \tag{4.3}$$

The mechanical shaft power generated by the Pelton wheel is converted into electrical power by means of the electric generator, which is mounted on the same shaft as the turbine. The generator must run at a constant speed since the electric power generated must have a constant frequency. Hence, the turbine must operate at a constant speed. This is possible when the water to the turbine is supplied at a constant rate and under a constant

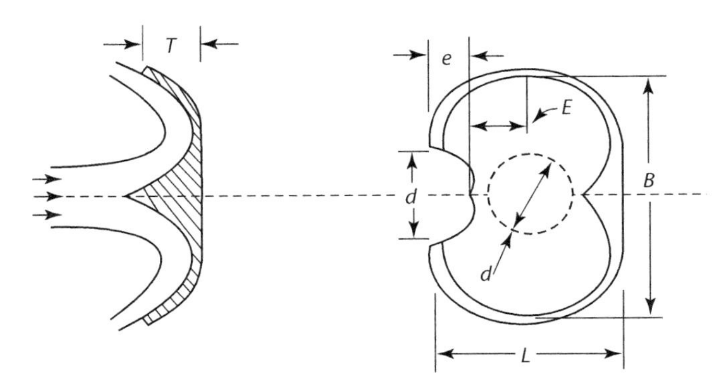

FIGURE 4.6
Typical Pelton wheel dimensions.

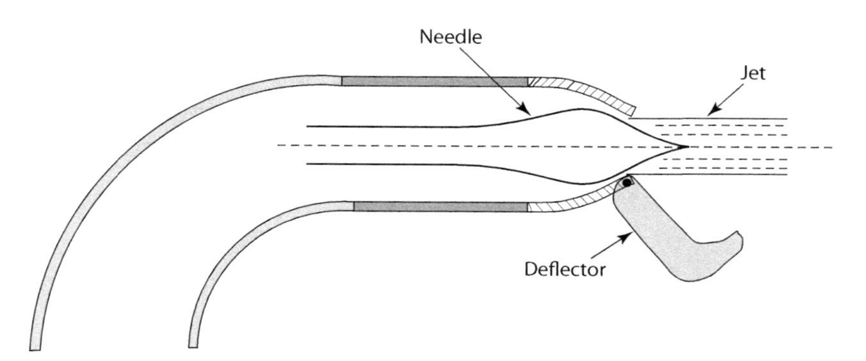

FIGURE 4.7
Needle assembly for flow control in a Pelton wheel.

head. Any changes in the electrical power requirements would result in corresponding changes in the load on the turbine and the turbine would have to modify its power output accordingly. The power output from the Pelton wheel can be altered by changing the head on the turbine or the flow rate through it. It is generally not feasible to change the head. Hence, it is necessary to regulate the water supply to the turbine. This is accomplished by equipping the nozzle with a moveable, conical-spear-shaped needle as indicated in Figure 4.7. The longitudinal motion of the spear controls the annular area available for flow and thus controls the flow rate. The needle itself is carefully designed so as to allow a uniform jet of water at all times to impinge the buckets, thereby minimizing losses.

The principal advantage of flow control using a needle in the nozzle is that the velocity of the jet is nearly constant irrespective of the position of the needle. However, the needle mechanism has a disadvantage that becomes apparent when the load on the turbine is suddenly reduced. Pelton wheels operate under high heads. This necessitates a long penstock, and a sudden closing of the nozzle in response to an abrupt drop in demand which results in water hammer in the penstock. Thus, water control using needle movement is suitable only when the penstock is relatively short. Otherwise, suitable precautions must be taken to avoid water hammer. This is accomplished using the jet deflector, as shown in Figure 4.7. If the load on the turbine is reduced suddenly, the jet deflector is first moved into position to deflect a portion of the water from impinging the turbine until the needle can be moved slowly to the proper position. It is reported that the movement of the deflector can be accomplished in as little as 3, while the movement of the needle may take a few minutes (see Brown and Whippen (1976)). Nozzles have been in use for several years, and over time their design has been perfected. The typical proportions of a nozzle are shown in Figure 4.8. The dimensions as recommended by Shepherd (1956) are summarized here:

$$d_1/d = 1.2 - 1.4;\ d_2/d = 3 - 4;\ d_3/d = 1.25 - 1.5;\ \alpha = 60^\circ - 90^\circ;\ \beta = 40^\circ - 60^\circ \tag{4.4}$$

4.2.2 Design Aspects of Pelton Turbines

In most applications involving the design of a Pelton turbine, the power required (or expected) and available head are usually known. From these, the flow rate can be calculated by making a suitable assumption of the efficiency ($P = \eta\gamma QH$), which for hydraulic turbines is quite high, usually close to 90%. Only hydraulic design aspects are considered here; mechanical design aspects such as strength, materials, and vibration are beyond the scope of this text and are not addressed. Some of the variables that need to be decided

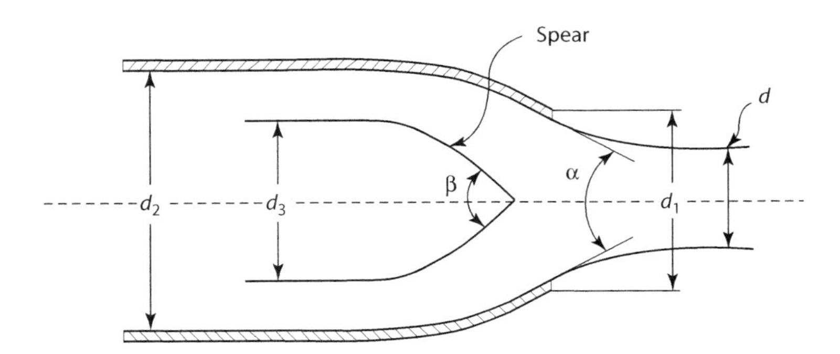

FIGURE 4.8
Typical geometry of Pelton wheel nozzle.

when designing the turbine are jet speed, blade velocity, and pitch diameter. These are discussed in the following paragraphs.

Jet Velocity: By assuming that water is supplied from a reservoir, the jet velocity V_j is given by

$$V_1 = V_j = C_v\sqrt{2gH} \tag{4.5}$$

where C_v is the nozzle velocity coefficient or simply the nozzle coefficient. Nozzles are inherently very efficient devices since they convert pressure head to velocity, and this process is quite efficient. Diffusers, on the other hand, are less efficient since they convert velocity to pressure. The value of C_v is between 0.94 and 0.98.

Speed of the Buckets: Theoretically, power from Pelton wheels (and hence utilization factor) would be a maximum when the bucket or blade speed is half of the jet speed. This can be conveniently expressed in terms of the speed ratio ϕ, which is defined as the ratio of the blade speed to inlet or jet speed. Thus, the ideal value of ϕ ($=u/V_1$) would be 0.5 (see Example 3.4). This analysis ignores energy losses due to friction. Thus, it is found from empirical considerations that for the best operating conditions, the value of ϕ is taken to be 0.43–0.48. In general, the speed ratio ϕ is dependent on specific speed. However, the range of specific speeds for Pelton wheels is quite narrow, and hence any value picked from the range described should give satisfactory values of the blade speed for design considerations.

Speed of Rotation of Runner: If the designer is free to select the speed of rotation N, the usual procedure is to select a suitable specific speed N_s for the turbine. For Pelton wheels, the range is between 0.03 and 0.3 (see Table 2.2). Knowing the specific speed, along with head and flow rate, the rotational speed of the runner can then be calculated. However, there is an additional factor that needs to be considered when selecting N. The runner rotation is related to the synchronous speed of the generator. If it is mounted on the same shaft, then both speeds would be the same. The speed of the generator is dependent on the number of poles p and frequency f of alternating current being produced. The equation for rotation then is

$$N = \frac{120f}{p} = \frac{(120)(60)}{p} = \frac{7200}{p} \tag{4.6}$$

The number of poles p has to be an even number. For the United States, the frequency of AC current f is 60 Hz, while in Europe it is 50 Hz. At this stage, the specific speed N_s should be calculated to be sure that it is in the acceptable range for Pelton wheels.

Pitch Diameter of Runner: The pitch diameter D of the runner is related to the blade speed and rotation by the expression $u = \omega r = \pi ND/60$.

Diameter of the Nozzle: The flow rate is related to the jet velocity and nozzle diameter by the expression of mass flow rate given by

$$Q = n_j V_j \frac{\pi d_j^2}{4} \tag{4.7}$$

where:
n_j is the number of nozzles
d_j is the jet diameter (indicated by d in Figure 4.8)

Jet diameter is typically restricted to 6–10 in. If larger nozzles are required, multiple nozzles are used. Generally, for a given jet diameter, the runner diameter can be as large as desired. However, it should not be less than nine times the jet diameter, (Brown and Whippen (1976)) that is,

$$D/d_j > 9 \tag{4.8}$$

Number of Buckets: The number of buckets, and, correspondingly, the distance between them are governed by two competing considerations. The buckets should be close enough that all the water in the jet is utilized. At the same time, water leaving one bucket should not strike the back of the adjacent bucket. Using these two criteria, Tygun (Kadambi and Prasad (1977)) developed an empirical formula that relates the number of buckets n_B and the wheel diameter to jet diameter ratio m ($=D/d_j$). This is given by Tygun's formula as follows:

$$n_B = \frac{m}{2} + 15 \tag{4.9}$$

Similar data have been given by Nechleba (1957). A comparison of both data is provided in Table 4.2. As can be seen, the predicted number of buckets is quite close.

The steps in the design process for Pelton turbines can be summarized as follows. Usually, the available head H and desired power P are known. Often, the expected speed of the runner is also specified since it is constrained by the speed of the generator.

TABLE 4.2

Variation of Number of Buckets for Pelton Wheels

m ($=D/d_j$)	6	8	10	15	20	25
n_B (Tygun)	18	19	20	23	25	28
n_B (Nechelaba)	17–21	18–22	19–24	22–27	24–30	26–33

1. Calculate the specific speed either using the power P or discharge Q; a reasonable value of efficiency (88%–93%) can be assumed to calculate Q.
2. Calculate jet velocity V_j from Equation 4.5.
3. Calculate speed of buckets u by assuming a value of ϕ $(=u/V_j)$ between 0.43 and 0.48. From this, calculate the pitch diameter of the runner.
4. Calculate the number of nozzles and their diameters using Equations 4.7 and 4.8.
5. Estimate the number of buckets using either Tygun's or Necheleba's recommendations as in Table 4.2.

The procedure is illustrated in Example 4.1.

Example 4.1

Consider the design of the Pelton wheel at Donnell power plant in Oakdale, California. The specifications are: $H=1325$ ft., $P=91{,}000$ hp, and $N=240$ rpm. Estimate some of the features of the turbine.

Solution: The flow rate can be calculated from the power and head assuming a suitable value for the efficiency. Thus (assuming the efficiency is 90%),

$$Q=\frac{P}{\eta\gamma H}=\frac{(91000)(550)}{(0.9)(62.4)(1325)}=672.6\,\text{ft.}^3/\text{s}$$

$$N=240\,\text{rpm}=25.1\,\text{rad/s} \tag{a}$$

$$N_s=\frac{N\sqrt{Q}}{(gH)^{0.75}}=0.218 \text{ which is well within the range for Pelton wheels}$$

The inlet fluid speed and blade speed can be calculated from (assuming $C_v=0.98$)

$$V_1=V_j=C_v\sqrt{2gH}=0.98\sqrt{2gH}=286.2\,\text{ft./s}$$

$$\frac{u}{V_1}=0.46\Rightarrow u=(0.46)(286.2)=131.7\,\text{ft./s} \tag{b}$$

The wheel and jet diameters can be calculated as

$$u=\frac{ND}{2}\Rightarrow D=\frac{2u}{N}=\frac{2(131.7)}{25.1}=10.49\,\text{ft.}=126''$$

$$Q=n_jV_j\frac{\pi d_j^2}{4}\Rightarrow d_j=\sqrt{\frac{4Q}{\pi n_jV_j}}=\sqrt{\frac{4(672.6)}{\pi(6)(286.2)}}=0.71\,\text{ft.}=8.5'' \tag{c}$$

where it is assumed that six nozzles are used. The number of buckets can be evaluated from Tygun's formula as

$$n_B=\frac{m}{2}+15=\frac{1}{2}\frac{D}{d_j}+15=\frac{1}{2}\left(\frac{126}{8.5}\right)+15=22.4 \text{ or } 22 \text{ buckets}$$

Nechleba's formula will also give a similar number of buckets. If the generator is mounted on the same shaft, the number of poles can be calculated as

$$p=\frac{7200}{N}=\frac{7200}{240}=30 \text{ poles, an even number as required}$$

Comments: The actual diameter of the rotor is 124 in. and the power plant uses six jets. Thus, the rudimentary calculations give quite accurate results for the final design.

Example 4.2

Show that for Pelton wheels operating under ideal conditions, the power is maximized when the blade speed is half the jet speed.

Solution: The force acting on the bucket from the linear momentum equation is

$$F = \rho Q(V_1 - u)(1 + \cos\beta_2) \tag{a}$$

Power would be the product of mass flow rate and the force.

$$P = uF = \rho Qu(V_1 - u)(1 + \cos\beta_2) \tag{b}$$

Maximum power would be obtained by setting the derivative dP/du to zero. Thus,

$$\frac{dP}{du} = \rho Q(1 + \cos\beta_2)[V_1 - 2u] = 0 \Rightarrow u = \frac{V_1}{2} \Rightarrow \phi \equiv \frac{u}{V_1} = 0.5 \tag{c}$$

where ϕ is the speed ratio or the peripheral velocity factor.

Comments:

1. For a single bucket, the force would be $\rho A_j(V_1 - u)(V_1 - u)(1 + \cos\beta_2)$, since not all the water that leaves the jet reaches the bucket. However, with several buckets, all the water is captured for momentum transfer and the force would then be $\rho A_j V_1(V_1 - u)(1 + \cos\beta_2)$. Replacing Q with $A_j V_1$, we get Equation (a).
2. By calculating the efficiency from Equation (b) and introducing the velocity coefficient C_v, a similar but more realistic result would be obtained. (See problem 4.14).
3. In the problem, it was assumed that the buckets are shaped such that the inlet blade angle $\beta_1 = 0$, that is, the fluid enters the bucket tangentially. In reality, the water enters at a small angle α_1 (usually $\leq 10°$). A small error, usually less than two percent, is introduced by neglecting this. (See Problem 4.15).

4.3 Reaction Turbines

A reaction turbine is one in which a significant portion of the pressure drop takes place in the rotor. Hence, these are designed so that the fluid fills the rotor completely. Also, any such machine must have the rotor enclosed so that the decrease in pressure of the fluid takes place in the runner passages, and not freely in all directions. The water exiting the turbine has a considerable amount of kinetic energy and also pressure. Hence, it is essential that this type of turbine is completely enclosed. The most common example of a reaction turbine is the lawn sprinkler. Reaction turbines are classified according to the direction of flow of water in the runner. If the direction of flow is predominantly radial, then the turbine is called a *radial flow turbine*; if the direction is mainly axial, then it is called an *axial turbine*. Two of the most common examples of these turbines are the Francis and Kaplan turbines respectively, which are shown in Figure 4.4 b and c.

4.3.1 Francis Turbine

The first well-designed Francis turbine was built in 1849 by a hydraulic engineer named James B. Francis. As originally conceived, it was a purely radial machine. Generally, all inward flow reaction hydraulic turbines are called Francis turbines. Modern designs of Francis turbines have some mixed flow characteristics. Some locations of Francis turbines around the world are shown in Table 4.3. The specific speeds are well within the limits mentioned in Table 2.2.

A schematic diagram of a Francis turbine is shown in Figure 4.9. It consists of a spiral casing, a runner, a set of guide vanes, and a draft tube. The inlet of the spiral casing is connected to the penstock. Water is uniformly distributed around the circumference of the runner by using the spiral casing. In order to distribute the water uniformly around the circumference, it is necessary to reduce the area available for flow. Thus, the casing takes the shape of a spiral, which gives the maximum area at the entrance and almost zero at the tip; hence the name spiral casing. Water from the spiral casing is guided onto the runner by a set of circumferentially placed guide vanes, which are sometimes called *wicket gates*. These can be set to any desired angle, and their shape is designed so that the water passes over them with minimum friction losses. The water then passes through the runner and discharges into a draft tube.

As has been mentioned already, the flow in the runner of modern Francis turbines is not purely radial, but a combination of radial and axial flows. The width of the runner depends on the specific speed. The higher the specific speed, the higher the width of the runner, since higher specific speeds necessarily mean larger flows (correspondingly larger flow areas) and/or higher rotational speeds. The runners can typically be classified as slow, medium or normal, and fast runners. Typical shapes of such runners are shown in Figure 4.10.

The manufacturers of hydraulic turbines have, over the years, developed several techniques for estimating the various parameters and dimensions of different parts of turbines.

TABLE 4.3

Francis Turbine Installations around the World

	Location	H (ft.)	Power (hp)	N (rpm)	N_s
1	Rovina-Piastra, Italy	1962	179,254	600	0.444
2	Fionnay, Switzerland	1817	742,914	600	1.002
3	Tafjord, Norway	1490	63,200	750	0.468
4	Telemark, Norway	1378	93,870	500	0.419
5	Churchill Falls, Canada	1237	148,785	375	0.453
6	Oak Grove, Oregon, USA	1005	618,000	200	0.639
7	Santa Barbara, Mexico	850	35,000	514	0.481
8	Bogota, Colombia	808	36,000	500	0.506
9	Kerala, India	921	30,100	720	0.566
10	Bhakra Nangal, India	590	13,500	600	0.550
11	Grand Coulee, Washington, USA	328	150,000	166.7	1.063
12	Kootnay Canal, BC, Canada	285	960,000	85.7	1.648
13	Del Rio Lampa, San Salvador	268	196,000	128.6	1.207
14	Brewster, Washington, USA	187	93,000	150	1.520
15	Udhampur, India	165	100000	100	1.229
16	Conowingo, Maryland, USA	104	36,000	150	1.978

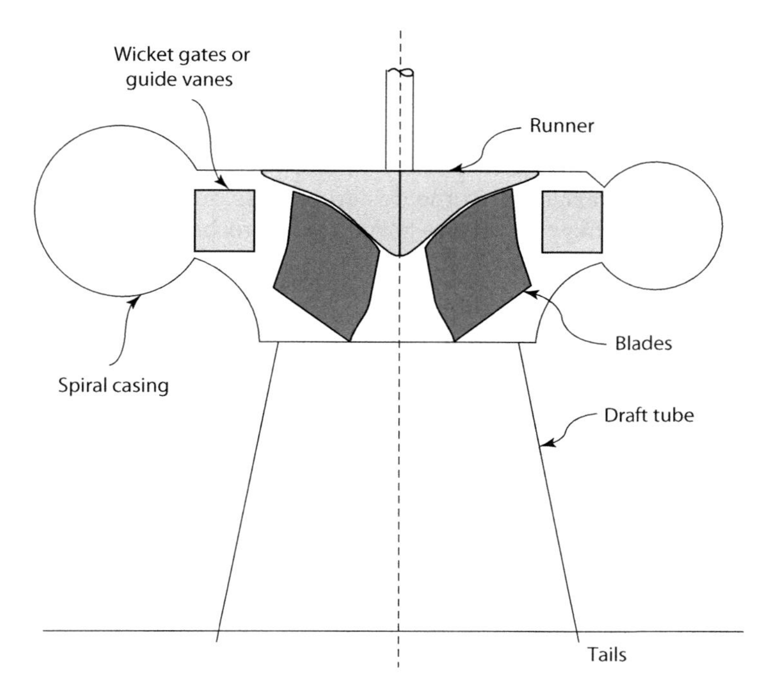

FIGURE 4.9
Schematic diagram of a Francis turbine.

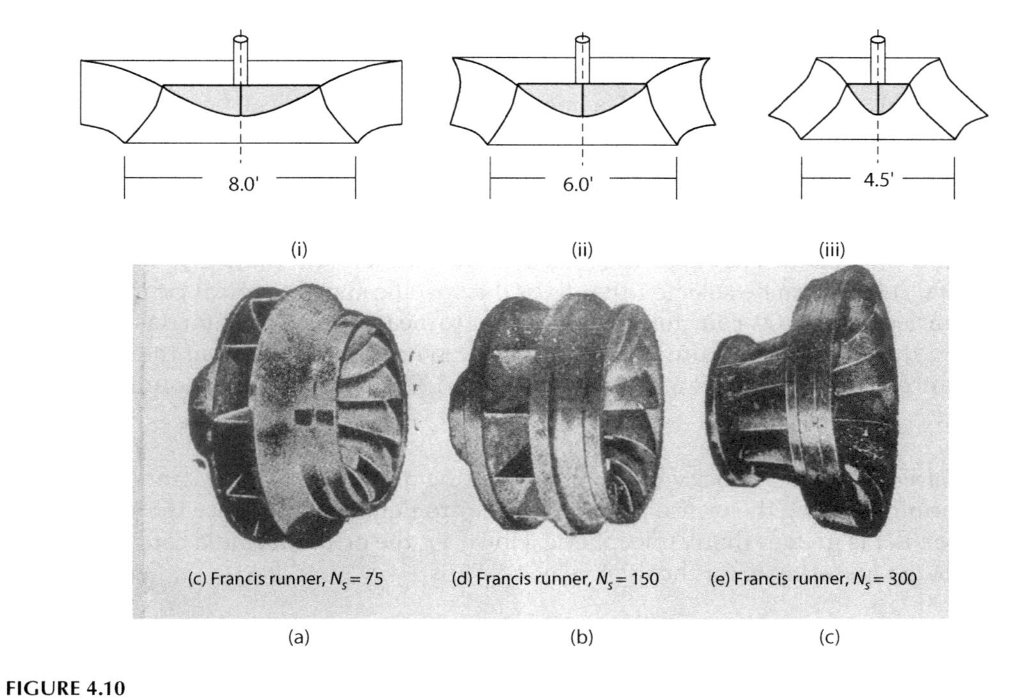

FIGURE 4.10
Schematic and actual pictures of Francis turbine runners: i, (a) slow, ii, (b) normal, iii, (c) fast. N_s is the dimensional specific speed as defined in Equation 2.5. (Adapted from Lal, J., *Hydraulic Turbines*, 6th edn, Metropolitan Book, New Delhi, 1986.)

These techniques are based on experience and a thorough knowledge of the theory of turbines. Since the variables are many, and design practices are not all standardized, it is quite likely an engineer will experience differences in the design of turbines from different companies. It is therefore highly desirable for the engineer to compare several designs and combine them with his or her experience before the final selection is made. Thus, the reader will find large variations in the design process depending on the source. The procedures mentioned in this chapter are based on the recommendations of Brown and Whippen (1976).

4.3.2 Design Aspects of Francis Turbines

In a typical design of the Francis turbine, the input variables are the head under which the turbine will operate, the required speed of the runner, and the required power output. Often, for preliminary design, only the available head and power would be known. The output variables that need to be determined are then the diameter of the runner, the number and angle of wicket gates, the number of buckets on the runner, and the height of the runner.

A reasonable estimate of the flow rate through the turbine can be made from the expression for power as follows:

$$P = \eta \gamma Q H \tag{4.10}$$

where:

η is the hydraulic efficiency
γ is the specific weight

At the design point, the efficiency of hydraulic turbines is between 0.88 and 0.94 (see Shepherd (1956)). Once the flow rate Q is known, the specific speed can be computed. This fixes the most suitable type of turbine that can be selected.

Sometimes, only the head and power are known. After calculating the flow rate from Equation 4.10, one of two procedures can be followed. From the turbine selection diagram given in Figure 4.2, the type of turbine is fixed. From this, a suitable value of specific speed N_s is picked. This fixes the speed of the runner. Alternatively, a reasonable speed of the runner can be selected that fixes the specific speed. Typical runner speeds range from less than 100 rpm for slow runners to nearly 1000 rpm for fast runners. Sometimes, the speed of the runner is fixed by the synchronous speed of the generator, in accordance with Equation 4.6, if the turbine and the generator are mounted on the same shaft.

Inlet Fluid Velocity: The velocity of the fluid entering a reaction turbine cannot be accurately determined using the orifice formula (similar to Equation 4.5), since the pressure at the turbine inlet is greater than atmospheric. However, the orifice formula can be suitably modified to get an estimate for the inlet velocity. Thus,

$$V_1 = C\sqrt{2gH} \tag{4.11}$$

where C is a coefficient that depends on the specific speed. The variation is given in Table 4.4.

TABLE 4.4

Typical Variation of Design Variables of Francis Turbines with Specific Speed

$N_s\left(=\frac{N\sqrt{Q}}{(gh)^{3/4}}\right)$	0.5	2
C	0.8	0.6
α_1	15°	35°
ϕ	0.7	0.8

Number and Angle of Wicket Gates: The water from the wicket gates makes an angle α_1 with the circumference of the rotor. This angle also depends on the specific speed, and its variation is shown in Table 4.4. The number of wicket gates and the number of stay vanes for reaction turbines are equal and are selected somewhat arbitrarily. These are usually multiples of four and range from 12 to 28 depending on the size of the runner, with the highest number used for large diameters.

Blade Velocity: The blade velocity u_1 is determined by using the speed ratio ϕ, which is defined as the ratio of blade velocity to free jet velocity under head H; that is,

$$\phi = \frac{u_1}{\sqrt{2gH}} \tag{4.12}$$

with the dependence of ϕ on specific speed given in Table 4.4. The knowledge of the blade velocity can be used in computing the runner diameter.

Gate Dimensions: The next step in the design process is to size the gates and the runner. The most important dimension in this regard is the height of the gates, which in a sense also determines the height of the runner. This can be easily done since the flow rate, velocity of water at the entrance, and the angle α_1 are known. Thus, the height of the gates B is given by

$$B = Q/\left(V_1 \pi C' D \sin\alpha_1\right) \tag{4.13}$$

In Equation 4.13, C' is a coefficient that allows for the thickness of the gates. The value of C' is usually about 0.95. The relative velocity can now be computed using the law of cosines as follows (see Figure 4.11):

$$w_1 = \left(u_1^2 + V_1^2 - 2u_1V_1\cos\alpha_1\right)^{1/2} \tag{4.14}$$

The direction of the relative velocity can be calculated from the relationship

$$w_1 \sin\beta_1 = V_1 \sin\alpha_1 \tag{4.15}$$

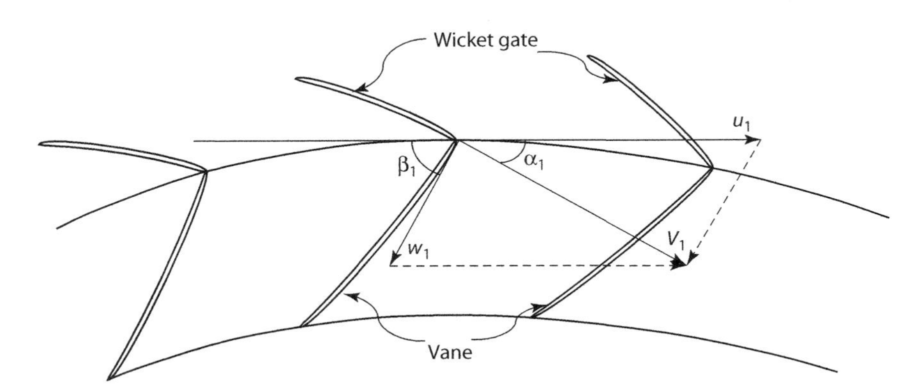

FIGURE 4.11
Velocity triangle at entrance of a Francis runner.

The blade angle $|\beta_1|$ is typically less than 45°.

In most theoretical analysis, it is assumed that the water leaving the turbine runner is purely in the axial direction. This condition results in the greatest efficiency for the Francis turbine, since all the inlet momentum of the fluid would have been converted into torque. However, this is seldom achieved in practice since the water exiting the turbine will still have a non-zero, but small, radial velocity. The common practice is to ignore this component, and assume that the exit velocity from the runner is the same as the inlet velocity of the draft tube. In order that the water from the runner enters the draft tube properly, the blades should be nearly tangential to the circle through the exit edges of the buckets. The blade angle at the exit of the buckets is typically between 15° and 30°.

Number and Shape of Buckets (Vanes or Blades): The final design variable is the number and shape of the buckets. The standard practice in the turbine industry is to use 15 buckets. Some modern turbines have 13, 17, or 19 buckets, with a larger number of buckets being used when the specific speed is low. The blade angles at the inlet and outlet are selected within the broad guidelines mentioned earlier. The contour of the intermediate portion of the bucket is formed so that there is a gradual change in the direction of the relative velocity and the area available for flow.

The design procedure can be summarized as follows:

1. From the head H, power P, and speed N, calculate the specific speed N_s. If the speed is not given, an appropriate value (between 100 and 1000 rpm) can be assumed. If needed, flow rate can be calculated from the relationship $P = \gamma QH/\eta$, with an efficiency of about 90%.
2. Calculate the inlet fluid velocity V_1 and inlet blade velocity u_1 from Equations 4.11 and 4.12, respectively.
3. From the blade speed u_1 and the rotational speed N, calculate the runner diameter D_1.
4. Pick an appropriate value for α_1 from Table 4.4 and calculate the blade height using Equation 4.13.

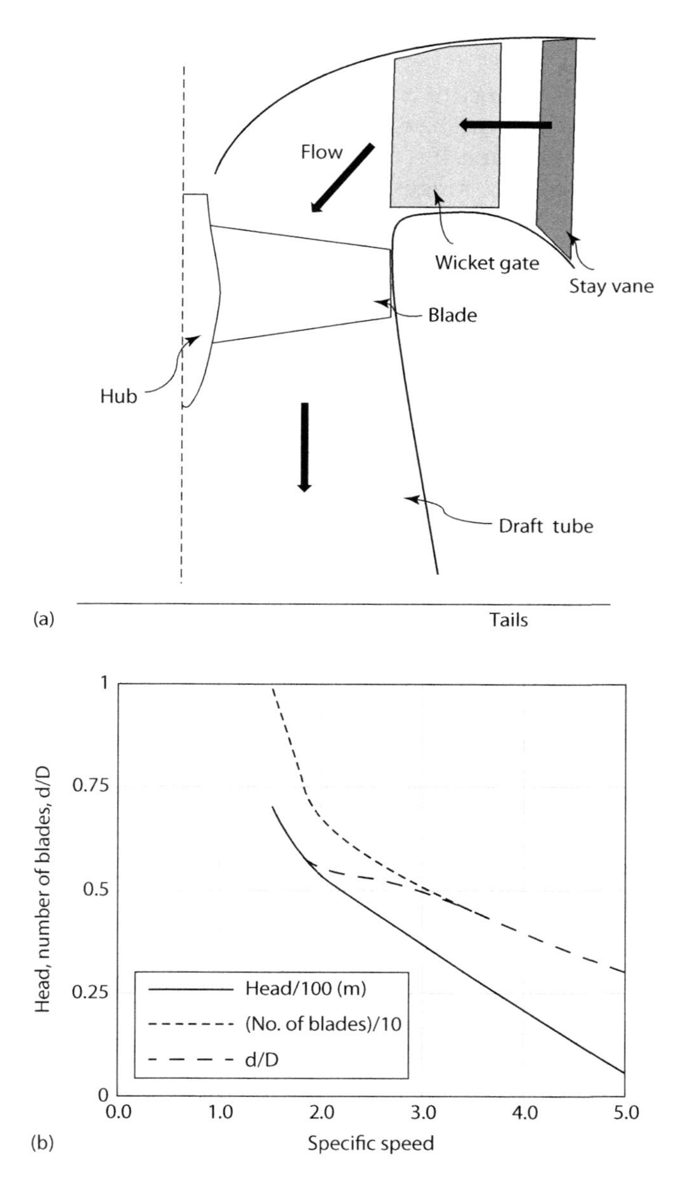

FIGURE 4.12
(a) Schematic diagram of a Kaplan turbine and (b) variation of some quantities for Kaplan turbines.

5. Choose the inlet blade angle $|\beta_1|$ between 0° and 45°; exit blade angle β_2 is between 15° and 30°.
6. Choose the number of buckets to be any odd number between 13 and 19 (15 is the most common number); the number of wicket gates is equal the number of stay vanes (see Figure 4.12) and it is chosen as a multiple of four between 12 and 28.
7. Check if the required flow rate requirement is met; if not, adjust the blade width B.

4.3.3 Kaplan Turbine

In situations where large volumes of water are available with low heads, it is essential to provide large flow areas, which can be easily accomplished using axial flow machines. A propeller turbine is such a machine. It is a reaction turbine that operates under heads of 15 to 250 feet when specific speed N_s ranges from 2.0 to 5.0. It is primarily an axial flow type of device. A propeller turbine with adjustable blades is called a *Kaplan turbine*, named after the German Professor Victor Kaplan who developed such a machine between 1910 and 1924. Some Kaplan turbine installations around the world are shown in Table 4.5. Again, the values of specific speed are within the limits indicated in Table 2.2.

A schematic diagram for a Kaplan turbine is shown in Figure 4.12.

Water passes over stay vanes and passes through wicket gates before coming into contact with the blades. The space between the wicket gates and the blades is used to turn the direction of flow from radial to nearly axial. After passing over the blades, the water flows through a draft tube. Typical efficiencies of Kaplan turbines are in the range 90%–93% at the design point.

4.3.4 Design Aspects of Kaplan Turbines

The design methodology for propeller turbines proceeds along the same lines as a Francis turbine. There are, however, some important differences, the most important being the value of ϕ, the speed ratio. While this value was low for Francis turbines, it can be as high as 1.4–2.0 for propeller turbines. The ratio of hub diameter to blade diameter, d/D, is a very important parameter in propeller turbines. Some variations of head, specific speeds, d/D, and number of blades for axial flow turbines is shown in Figure 4.13.

The volumetric flow of water is given by the expression

$$Q = V_{ax}\frac{\pi\left(D^2 - d^2\right)}{4} = \zeta\sqrt{2gH}\,\frac{\pi\left(D^2 - d^2\right)}{4} \tag{4.15a}$$

TABLE 4.5

Kaplan Turbine Installations around the World

	Location	H (ft.)	Power (hp)	*N* (rpm)	Ns
1	Jactaquac, Canada	111	142,000	112.5	2.718
2	Porsi, Sweden	108	121,800	115	2.642
3	Bhakra Nangal, India	98	34,000	166.7	2.298
4	Camargos, Brazil	89	35,700	150	2.396
5	Jupia, Brazil	80	140,000	90	3.249
6	Cochrane, Montna, USA	80	42,000	128.6	2.531
7	Volgograd, Russia	74	171,400	62.2	2.736
8	Tungabhadra Dam, India	66	13,800	214	3.094
9	Kuibishev, Russia	62	146,000	62.2	3.119
10	Saf Harbor, Maryland, USA	53	42,500	109.1	3.614
11	Bonneville, Oregon, USA	50	66,000	75	3.330
12	Wheeler Dam, Alabama, USA	48	45,000	85.7	3.307
13	Ozark, Arkansas, USA	38	26,809	62	2.473
14	Webbers Falls, Oklahoma, USA	25	26,809	62	4.173

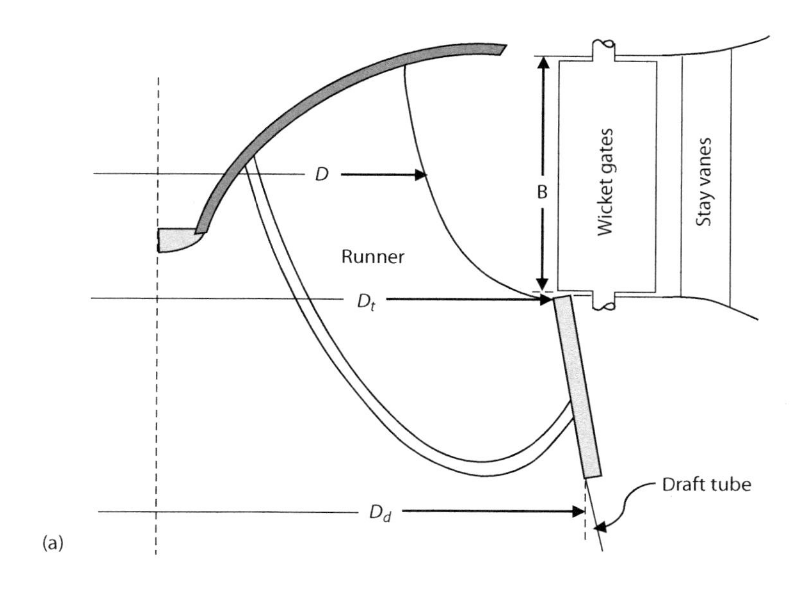

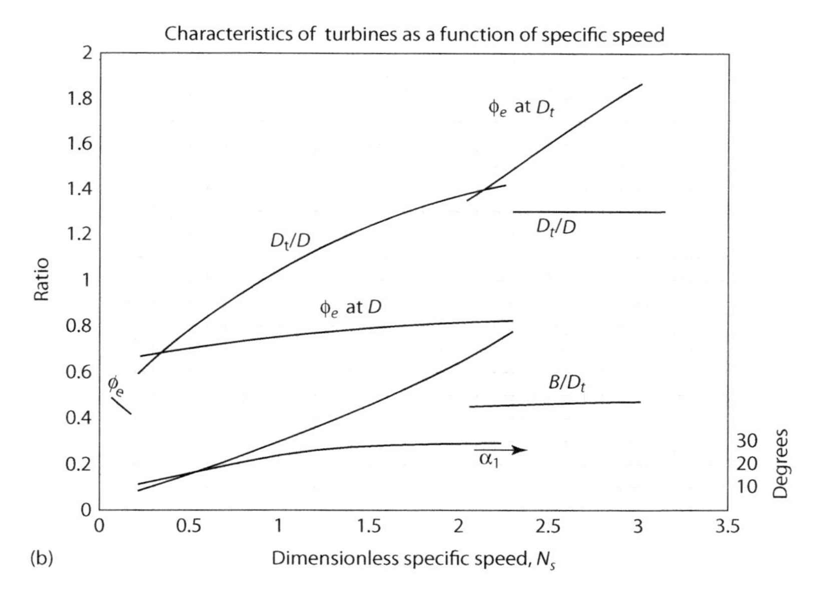

FIGURE 4.13
(a) Schematic diagram of Francis turbine and (b) approximate proportions for reaction turbines. (Modified from Dougherty et al., *Fluid Mechanics with Engineering Applications*, 8th edn, McGraw-Hill, New York, 1985.)

In this equation, the factor ζ lies in the range 0.35–0.75.

An important feature of axial flow turbines is that the blade speed varies along the blade, and hence the blade angle β also varies from point to point. However, the blade speeds at the inlet and outlet are equal. If the diameters at the tip and hub are known, blade speeds can be determined, and vice versa. By assuming that the axial velocity is constant, the inlet and exit velocity triangles can be drawn at the hub and tip. From these, the blade angles β_1 and β_2 can be determined. The typical number of blades in Kaplan turbines is between five and eight. The highest number of blades is ten.

An alternative design procedure given by Dougherty et al. (1985) is based on the specific speeds from which basic design parameters such as runner diameter and blade height are calculated. The basic features of reaction turbines are shown in Figure 4.13. Typical parameters that are known in the design of turbines are the head H, power P, and often the speed N. From these, the specific speed can be calculated, based on which the other design parameters can be picked. The variation of these design parameters with specific speed is shown in Figure 4.13.

4.4 Draft Tubes

As pointed in the previous sections, a Pelton wheel operates at high heads (and low specific speeds). By their very design, such turbines are open to the atmosphere. Also, the kinetic energy of the fluid leaving the runner is a very small fraction of the total available head (typically a few percent). This is in direct contrast with reaction turbines, in which a considerable portion of kinetic energy is available at the exit of the runner that would be wasted if the turbine were placed above the tails, and the outgoing water just exhausted to the atmosphere. Since reaction turbines operate under small heads, the exit kinetic energy constitutes a significant portion of the total head. It was reported by Jagdish Lal (1986) that the exit kinetic energy could range from 15% to 45% of the net head for Francis and Kaplan turbines, respectively. Hence, any efforts to recover this energy would greatly improve the overall efficiency. A draft tube primarily serves this purpose. If the turbine is placed at some distance above the tails and leading the outgoing water to the tails through a tube, two things happen. First, the pressure immediately outside the runner, that is, the inlet of the draft tube, becomes less than atmospheric and the water reaches atmospheric pressure only at the tail race. This increases the net head and hence improves both the output and the efficiency of the turbine. Second, by letting the tube from the turbine to the tails be slightly divergent, a part of the exit kinetic energy can be converted into pressure. This improves the turbine output further. Such a tube is called a *draft tube*. It should be pointed out that the addition of a draft tube is neither feasible nor desirable in the case of an impulse turbine such as the Pelton wheel, since, as already mentioned, the exit kinetic energy is a very small fraction of the overall head and Pelton wheel runners are not confined as reaction turbines. A schematic diagram of the draft tube is shown in Figure 4.14.

An approximate analysis of the draft tube can be performed as follows. If the subscripts e and t represent the conditions at the exit of the turbine (nearly the same as inlet of the draft tube) and tails, the pressures and velocities are related by the following expression (subscript a refers to atmospheric conditions):

$$\frac{p_e}{\gamma}+\frac{V_e^2}{2g}+H_s=\frac{p_a}{\gamma}+\frac{V_t^2}{2g}+h_f \tag{4.16}$$

where:
H_s is the height of the exit of the turbine above the tail
h_f is the losses

It has been assumed that the pressure at the exit of the draft tube (inside) is nearly atmospheric. After rearrangement, Equation 4.16 becomes

$$\frac{p_e}{\gamma}=\frac{p_a}{\gamma}-\frac{\left(V_e^2-V_t^2\right)}{2g}-\left(H_s-h_f\right) \tag{4.17}$$

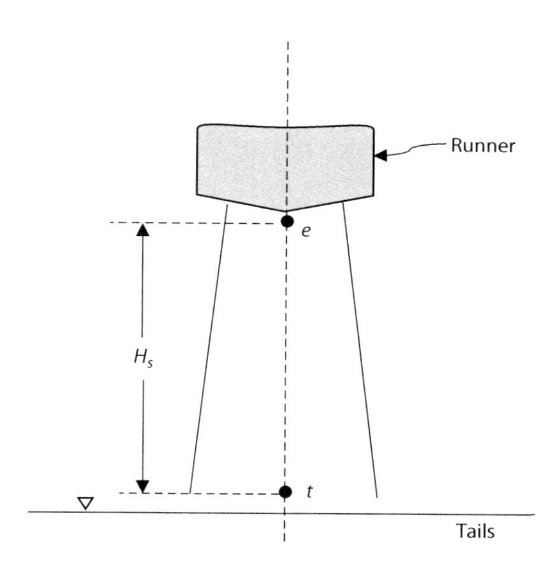

FIGURE 4.14
Schematic diagram of a draft tube.

Notice that even in the absence of diffusion effects, that is, if $V_e = V_f$, the value of p_e is lower than p_a. Thus, the addition of a draft tube creates a partial vacuum at the turbine exit, thereby increasing the net head under which the turbine operates. Notice also that the second term on the right hand side of Equation 4.17 is the gain in the velocity head due only to diffusion effects, which will be denoted by h_d. Thus, the efficiency of the draft tube can be defined as the ratio of real diffusion effects to ideal diffusion effects, that is,

$$\eta_d = \frac{(h_d - h_f)}{h_d} = 1 - \frac{h_f}{\left[\left(V_e^2 - V_t^2\right)/2g\right]} \tag{4.18}$$

Draft tubes are broadly classified into four categories. These are shown in Figure 4.15. The first type is shaped like the frustum of a cone and is simply called the *straight divergent tube*. It has an efficiency of about 85% and is useful for low specific speed vertical shaft Francis turbines.

The second type is called the *Moody's spreading tube* or *hydracone*. This is also a straight tube, but is bell-shaped, with a solid conical core in the central portion of the tube. This type has the advantage that it can allow flow with a whirl component of velocity with minimal losses. This type of draft tube also has an efficiency of about 85% (see Lal (1986)). The third type is called the *simple elbow tube*, which can be used when there are space limitations. The efficiency is usually much lower, of the order of about 60%. The last type is similar to the third except that the exit from the draft tube is rectangular in shape.

The geometrical parameters of draft tubes such as the length and included angle are governed by two factors. First, the length of the tube should not be so long as to cause cavitation at the inlet of the draft tube. Secondly, the diffuser angle should be small enough so that separation does not occur. The latter is achieved by keeping the diffuser angle below 10°.

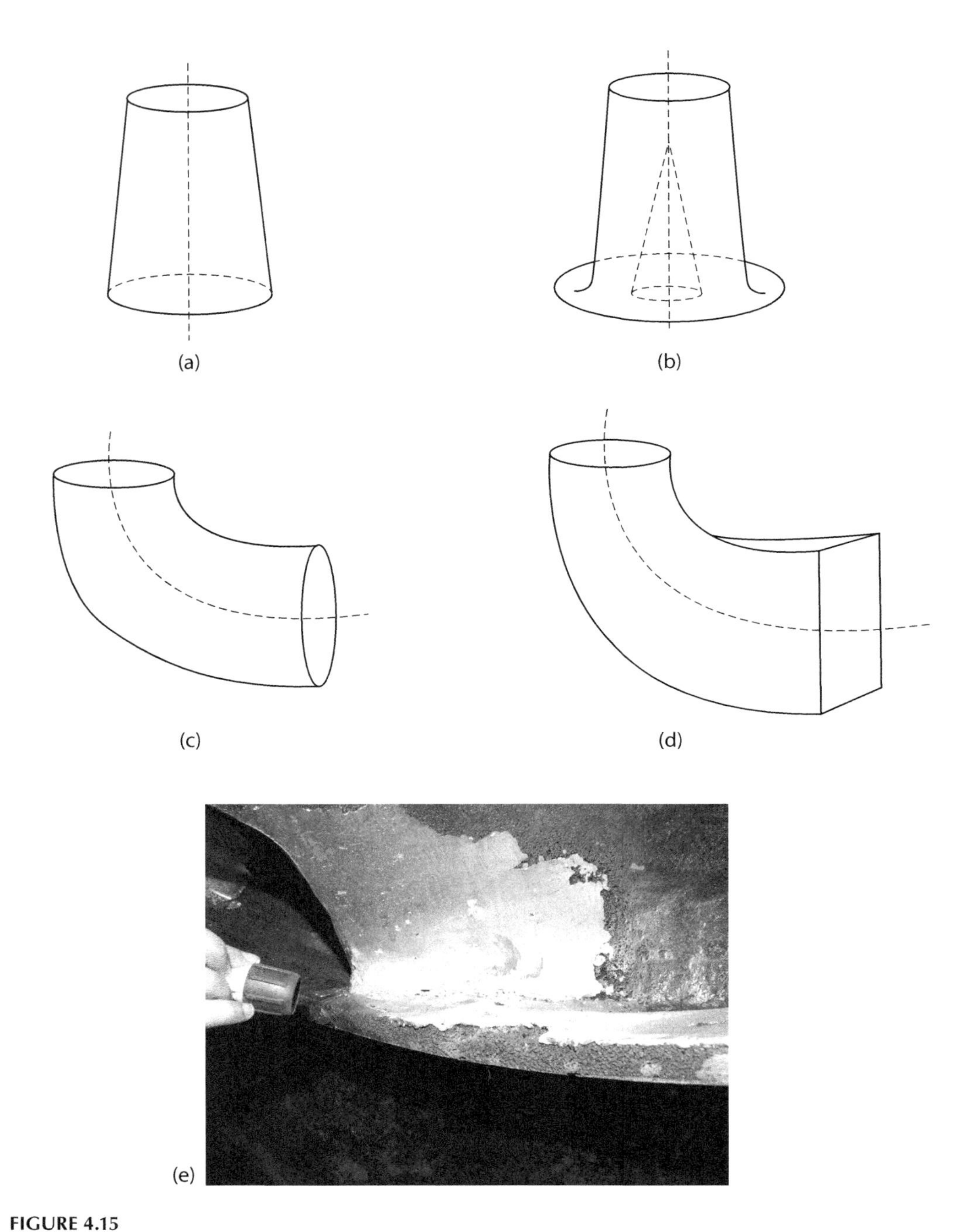

FIGURE 4.15
Types of draft tubes: (a) straight divergent type, (b) hydrocone, (c,d) elbow type, (e) Cavitation damage in turbine runner. (Courtesy of Voith Hydro, Inc.)

4.5 Cavitation in Hydraulic Turbines

According to Bernoulli's theorem (see Wilson (1984)), an increase in velocity in a fluid is accompanied by a decrease in pressure. For any liquid, there is a lower limit beyond which the absolute pressure cannot decrease. This value is called the *vapor pressure*, and it depends on the nature of the liquid and local temperature. If at any point, the liquid flows into a region where the pressure is reduced to vapor pressure, the liquid boils locally and vapor pockets are formed. This process is called *cavitation*. The vapor bubbles are carried along with the liquid, and when these reach regions of higher pressure, they suddenly collapse. The collapsing action produces large pressure pulses, often of the order of gigapascals. When such collapse takes place adjacent to solid walls continually and at high frequency, the material in that zone can be damaged, resulting in pitting of solid surfaces. Cavitation is also accompanied by noise, vibration, and can, depending on the severity, result in a noticeable drop in efficiency. It can be expected to occur in draft tubes, trailing edges of turbine blades, and so on. A picture of a cavitated turbine with pitting damage is shown in Figure 4.15e.

In order to avoid cavitation, it is necessary that the absolute pressure at every point in the flow field is above vapor pressure. A simplified analysis of cavitation can be performed as follows. Consider the draft tube shown in Figure 4.16. By applying the energy equation between the exit of the turbine and the tails, the following equation is obtained. The velocity at the exit has been ignored (compare with Equation 4.16):

$$\frac{p_e}{\gamma}+\frac{V_e^2}{2g}+H_s=\frac{p_a}{\gamma}+h_f \tag{4.19}$$

where the subscript e refers to the exit of the turbine, which is nearly the same as the inlet of the draft tube. Thus,

$$\frac{V_e^2}{2g}=\frac{(p_a-p_e)}{\gamma}-(H_s-h_f) \tag{4.20}$$

Cavitation occurs when the value of p_e becomes equal to the vapor pressure of the fluid at the local temperature. A parameter σ, called the *cavitation parameter*, can be defined as the ratio of exit velocity head and head on the machine.

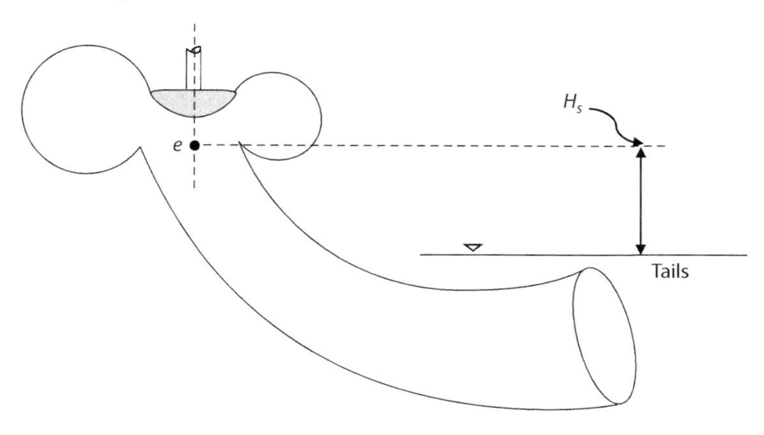

FIGURE 4.16
Draft tube setting.

Thus,

$$\sigma = \frac{\left(V_e^2/2g\right)}{H} = \frac{V_e^2}{2gH} = \frac{\left(p_a - p_v\right)/\gamma - \left(H_s - h_f\right)}{H} \tag{4.21}$$

where p_e has been replaced by the vapor pressure p_v. Usually, the head loss h_f is small and is dropped from the equation. Thus, a number called the *Thoma cavitation coefficient*, named after the German engineer Dietrick Thoma (1881–1943) can be defined as

$$\sigma = \frac{\left(p_a - p_v\right)/\gamma - H_s}{H} \tag{4.22}$$

In order to avoid cavitation, it is necessary that $\sigma > \sigma_c$, where σ_c is called the *critical cavitation coefficient*. The values of σ_c are determined experimentally by plotting σ versus efficiency. It is seen that for large values of σ, efficiency is nearly constant. When σ falls below σ_c, however, the efficiency rapidly drops off. This is also accompanied by increased noise and vibration. Typical values of σ_c for reaction turbines are given in Figure 4.17.

An alternative method of designing turbine settings against cavitation is provided by Wislicenus (1965) using the concept of the suction specific speed as follows:

$$N_{ss} = \frac{N\sqrt{Q}}{(H')^{3/4}} \geq 54.3$$
$$\text{where } H' = \left(\frac{p_i}{\gamma} + \frac{V_i^2}{2g}\right) - \frac{p_v}{\gamma} \tag{4.22a}$$

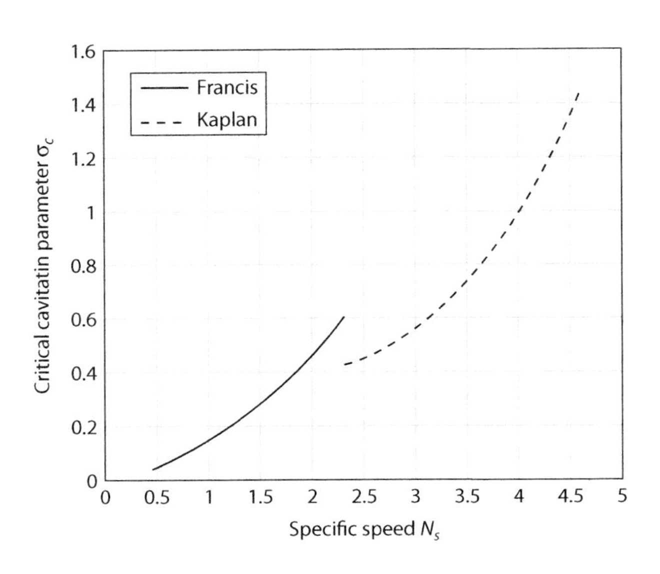

FIGURE 4.17
Typical critical cavitation parameters for reaction turbines. (Modified from Dougherty et al., *Fluid Mechanics with Engineering Applications*, 8th edn, McGraw-Hill, New York, 1985.)

Also, H' is the so-called net positive suction head that is used widely in the study of cavitation of pumps (Chapter 5). The variables p_i and V_i refer to the pressure and velocity, respectively, at the inlet of the draft tube (or the exit of the turbine) where the lowest pressure is most likely to occur. The cavitation formula based on based on suction specific speed as given in Equation 4.22a is preferred by practicing engineers.

Example 4.3

A Francis turbine is placed 8 ft. above the tail water at a location where the elevation is 7500 ft. and the local temperature is 50°F. The specific speed is 1.6. What is the maximum allowable head to avoid cavitation? Compare the result if the turbine were set at sea level, where the water temperature is 80°F.

Solution: From Table A.1, the vapor pressure head at 50°F is 0.41 ft. At an elevation of 7500 ft. atmospheric pressure is 11.2 psia. Thus,

$$\frac{p_a}{\gamma} = \frac{(11.2)(144)}{62.4} = 25.8\,\text{ft.}; \frac{p_v}{\gamma} = 0.41\,\text{ft.} \tag{a}$$

From Figure 4.17, $\sigma_c = 0.32$ and

$$\sigma = \frac{(p_a - p_v)/\gamma - H_s}{H} = \frac{25.8 - 0.41 - 8}{H} = \frac{17.44}{H} \tag{b}$$

To avoid cavitation, $\sigma > \sigma_c$. Thus,

$$\frac{17.44}{H} > 0.32 \text{or } H < 54.5\,\text{ft.} \tag{c}$$

At sea level and at 80°F,

$$\frac{p_a}{\gamma} = \frac{(14.7)(144)}{62.4} = 33.9\,\text{ft.}; \frac{p_v}{\gamma} = 1.17\,\text{ft.} \tag{d}$$

$$\sigma = \frac{33.9 - 1.17 - 8}{H} = \frac{24.75}{H} > \sigma_c = 0.32 \tag{e}$$

Hence, $H < 77.4$ ft.

Comments: Thus, an increase in temperature significantly affects the allowable head since vapor pressure is strongly dependent on temperature.

Example 4.4

A reaction turbine produces 10,000 hp when the flow through the system is 900 cfs. The turbine speed is 200 rpm. What should the setting of the wicket gates be if their height is 2.5 ft. and water leaves the wheel with no whirl? The diameter of the rotor just inside the wicket gates is 14 ft.

Solution: The velocity triangles are shown in the following figure. Using the flow rate, radial velocity at the inlet can be calculated.

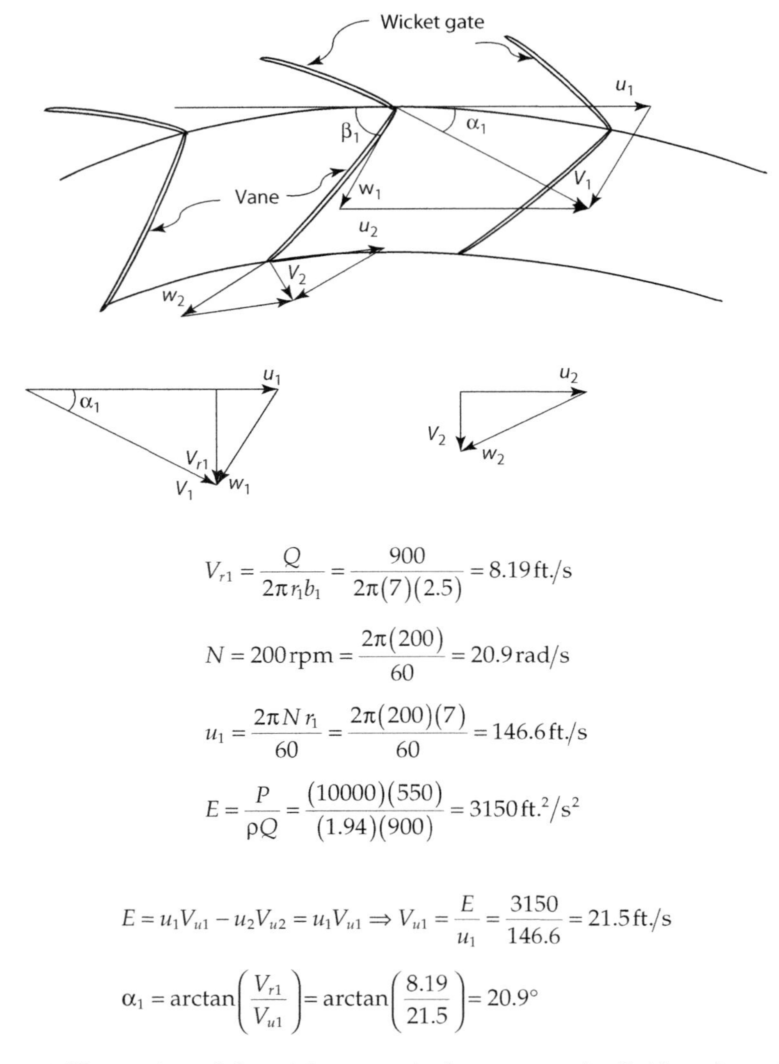

$$V_{r1} = \frac{Q}{2\pi r_1 b_1} = \frac{900}{2\pi(7)(2.5)} = 8.19\,\text{ft./s}$$

$$N = 200\,\text{rpm} = \frac{2\pi(200)}{60} = 20.9\,\text{rad/s}$$

$$u_1 = \frac{2\pi N r_1}{60} = \frac{2\pi(200)(7)}{60} = 146.6\,\text{ft./s}$$

$$E = \frac{P}{\rho Q} = \frac{(10000)(550)}{(1.94)(900)} = 3150\,\text{ft.}^2/\text{s}^2$$

$$E = u_1 V_{u1} - u_2 V_{u2} = u_1 V_{u1} \Rightarrow V_{u1} = \frac{E}{u_1} = \frac{3150}{146.6} = 21.5\,\text{ft./s}$$

$$\alpha_1 = \arctan\left(\frac{V_{r1}}{V_{u1}}\right) = \arctan\left(\frac{8.19}{21.5}\right) = 20.9°$$

Comment: The setting of the wicket gates is the same as the fluid angle at the inlet since the fluid enters the rotor through the wicket gates.

Example 4.5

Perform the preliminary design of the Nixon Rapids plant with the following specifications: $H = 162$ ft.; $P = 140{,}000$ hp; $N = 100$ rpm.

Solution: The specific speed is given by

$$N_s = \frac{N\sqrt{P}}{H^{5/4}} = \frac{100\sqrt{140000}}{(162)^{5/4}} = 65$$

The dimensionless value would be 1.49. From Figure 4.13, the following values are obtained:

$$\phi = 0.8; \frac{B}{D} = 0.42; \frac{D_t}{D} = 1.2; \alpha_1 = 24^\circ$$

$$\phi = 0.8 = \frac{u}{V_1} = \frac{u}{\sqrt{2gH}} \Rightarrow u_1 = 81.7 \text{ ft./s}$$

$$\omega = \frac{2\pi N}{60} = 10.47 \text{ rad/s}$$

$$u_1 = \frac{\omega D}{2} \Rightarrow D = \frac{2u_1}{\omega} = 15.6 \text{ ft.} \Rightarrow B = 6.55 \text{ ft.}$$

$$D_t = (1.2)(15.6) = 18.72 \text{ ft.}$$

The next step in the design process is to estimate the flow rate and determine whether the calculated dimensions can produce the required flow. Assuming the efficiency of the turbine to be 90%, the flow would be

$$Q = \frac{P}{\eta\gamma H} = \frac{(140000)(550)}{(0.9)(62.4)(162)} = 8463 \text{ cfs}$$

$$V_r = V_1 \sin\alpha_1 = \sqrt{2gH}\sin\alpha_1 = \sqrt{2(32.2)(162)}\sin 24 = 41.5 \text{ ft./s}$$

$$Q = \pi D B V_r = \pi(15.6)(6.55)(41.5) = 13321 \text{ cfs}$$

Since the flow being produced is more than the required value, the parameters need to be adjusted. The easiest dimension to adjust would be the blade height *B*. If it is reduced to 4.1 ft., the flow would be close to the required amount.

Comment: 1) The diameter of the runner (available from the manufacturer) is 19 ft. 2 in., which is quite close to the value calculated in the example, namely 18.72 ft. Thus, the rudimentary calculations give reasonably accurate dimensions. 2) Since the efficiency of hydraulic turbines is quite high, it could have been assumed to be 100% to estimate the flow rate.

Example 4.6

A Francis turbine has the following specifications: the inlet and exit diameters are 12 and 8 ft., respectively, and the flow rate is 700 cfs. The height of the blades is 1.5 ft. parallel to the axis of rotation. The inlet vanes are set such that the inlet velocity makes an angle of 15° with the blade tangent. The inlet and exit blade angles, β_1 and β_2, are 45° and −30°, respectively. Find the blade speed, pressure drop through the runner, degree of reaction, and the utilization factor.

Solution: Using the flow rate, radial velocities at the inlet and outlet can be calculated:

$$Q = 2\pi r_1 b_1 V_{r1} = 2\pi r_2 b_2 V_{r2}$$

$$V_{r1} = \frac{Q}{2\pi r_1 b_1} = \frac{700}{2\pi(6)(1.5)} = 12.37 \text{ ft./s} \tag{a}$$

$$V_{r2} = \frac{Q}{2\pi r_2 b_2} = \frac{700}{2\pi(4)(1.5)} = 18.56 \text{ ft./s}$$

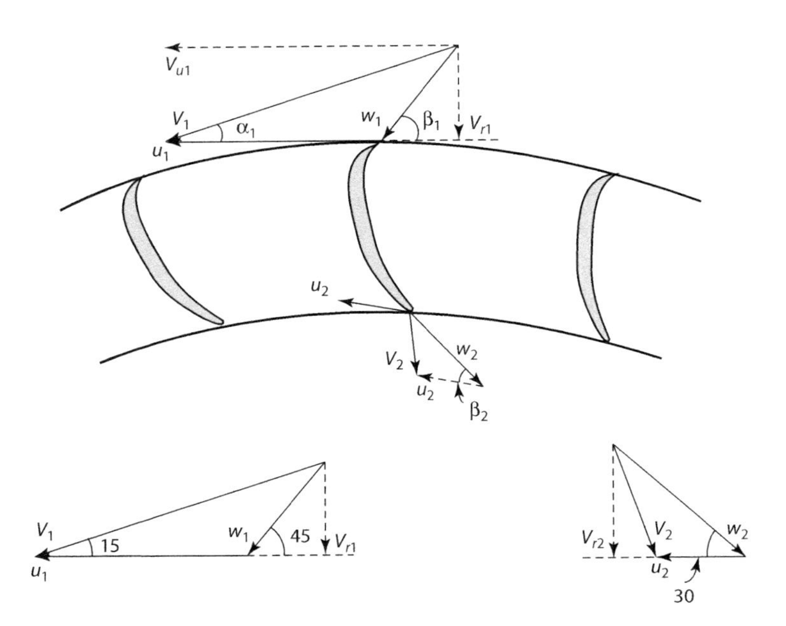

From the inlet velocity triangle, the following can be calculated:

$$V_1 = \frac{V_{r1}}{\sin(15)} = 47.8\,\text{ft./s};\ w_1 = \frac{V_{r1}}{\sin(45)} = 17.49\,\text{ft./s}$$
$$V_{u1} = V_1\cos(15) = (47.8)(\cos(15)) = 46.2\,\text{ft./s} \tag{b}$$
$$w_{u1} = w_1\cos(45) = (17.49)(\cos(45)) = 12.37\,\text{ft./s}$$

Thus,

$$u_1 = V_{u1} - w_{u1} = 46.2 - 7.15 = 33.83\,\text{ft./s}$$
$$\omega = \frac{u_1}{r_1} = \frac{33.83}{6} = 5.63\,\text{rad/s or } N = \frac{60\omega}{2\pi} = 53.84\,\text{rpm} \tag{c}$$
$$u_2 = \omega r_2 = (5.63)(4) = 22.52\,\text{ft./s}$$

Variables for the exit velocity triangle can be calculated as

$$w_2 = \frac{V_{r2}}{\sin(150)} = 37.12\,\text{ft./s}$$
$$w_{u2} = w_2\cos(150) = (37.12)(\cos(150)) = -32.16\,\text{ft./s} \tag{d}$$
$$V_{u2} = u_2 - |w_{u2}| = 22.52 - 32.16 = -9.64\,\text{ft./s}$$
$$V_2 = \sqrt{\left((V_{r2})^2 + (V_{u2})^2\right)} = 20.91\,\text{ft./s}$$

Alternatively, the exit velocity V_2 could be calculated from the law of cosines as

$$V_2 = \sqrt{\left((u_2)^2 + (w_2)^2 + 2u_2 w_2 \cos\beta_2\right)} = 20.91\,\text{ft./s}$$

Since all the velocities for the exit and inlet triangles are known, the required quantities can be calculated as follows:

$$E = u_1 V_{u1} - u_2 V_{u2} = (33.83)(46.2) - (22.5)(-9.61) = 1779.2\,\text{ft.} - \text{lbf/slug}$$

$$\text{or } E = \frac{V_1^2 - V_2^2}{2} + \frac{u_1^2 - u_2^2}{2} + \frac{w_2^2 - w_1^2}{2} = 1779.2\,\text{ft.} - \text{lbf/slug}$$

$$R = \frac{\dfrac{u_1^2 - u_2^2}{2} + \dfrac{w_2^2 - w_1^2}{2}}{E} = 0.48$$

$$\varepsilon = \frac{\left(V_1^2 - V_2^2\right) + \left(u_1^2 - u_2^2\right) + \left(w_2^2 - w_1^2\right)}{V_1^2 + \left(u_1^2 - u_2^2\right) + \left(w_2^2 - w_1^2\right)} = 0.89$$

$$\text{or } \varepsilon = \frac{E}{E + \dfrac{V_2^2}{2}} = 0.89$$

$$\text{or } \varepsilon = \frac{V_1^2 - V_2^2}{V_1^2 - RV_2^2} = \frac{(47.8)^2 - (20.9)^2}{(47.8)^2 - (0.48)(20.9)^2} = 0.89$$

The pressure drop through the rotor can be calculated from the energy equation for incompressible fluids as

$$h_1 + \frac{V_1^2}{2} = h_2 + \frac{V_2^2}{2} + E \Rightarrow \hat{u}_1 + \frac{p_1}{\rho} + \frac{V_1^2}{2} = \hat{u}_2 + \frac{p_2}{\rho} + \frac{V_2^2}{2} + E$$

Since the flows are isothermal, internal energy can be cancelled and the resulting equation can be written in terms of pressures only:

$$\frac{p_1}{\rho} - \frac{p_2}{\rho} = \frac{V_2^2}{2} - \frac{V_1^2}{2} + E = \frac{u_1^2 - u_2^2}{2} + \frac{w_2^2 - w_1^2}{2} = 853\,\text{ft.}^2/\text{s}^2$$

$$\Delta p = (853)(1.94) = 1655\,\text{psf} = 11.5\,\text{psi}$$

Alternatively, the definition of the degree of reaction can be used to calculate Δp:

$$R = \frac{\hat{h}_1 - \hat{h}_2}{\hat{h}_{01} - \hat{h}_{02}} = \frac{\left(\dfrac{p_1 - p_2}{\rho}\right)}{E}$$

$$\text{or } \frac{\Delta p}{\rho} = (R)(E) = (0.48)(1779) = 853\,\text{ft.}^2/\text{s}^2$$

$$\text{or } \Delta p = \frac{(853)(1.94)}{144} = 11.5\,\text{psi}$$

Comment: Degree of reaction is defined as the ratio of the static enthalpy to total (stagnation) enthalpy change across the turbine. Since the fluid is water, enthalpy change would be reflected in the pressure change across the rotor. Also, the various definitions of E, R, and ε are equivalent.

PROBLEMS

4.1 A cross section of a cylindrical draft tube proposed to be added to the exit of a reaction turbine is shown in the following image. The inlet velocity is 10 m/s and inlet diameter is 1 m. If the local temperature is 25°C, what is the value of H_s for incipient cavitation?

Ans: $H_s = 4.98$ m

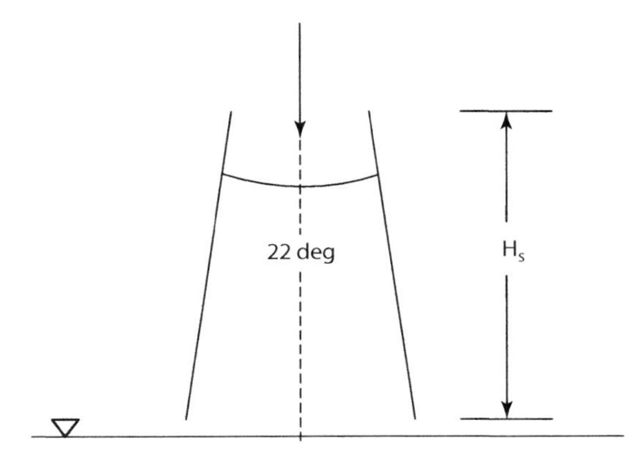

4.2 The following data refer to a mixed flow turbine:

If water leaves in a purely axial direction, what is the wicket gate setting? The turbine rotates at 200 rpm and produces 13,000 hp.

Ans: $\beta_1 = 84.3°$

Inlet diameter	11 ft.
Flow rate	750 cfs
Height of wicket gates	3 ft.

4.3 The mean radius of an axial flow reaction turbine is 2.6 ft. and the axial velocity is 25 ft./s. The wicket gates are turned at an angle such that relative velocity at the inlet flow makes an angle of −45° with the blade velocity. The wheel and hub radii are 3.6 ft. and 1.6 ft., respectively. The turbine rotates at 240 rpm. If water leaves the blade in a purely axial direction, calculate H and all relevant angles.

Ans: $H = 81.87$ ft.; $\alpha_1 = 31.8°$; $\beta_2 = 20.93°$

4.4 The inner and outer radii of a radial turbine are 0.75 and 1.25 m, respectively. The flow rate is 10 m^3/s and blade width is 0.25 m at both the inlet and outlet. If $\alpha_1 = 30°$, calculate (a) head, (b) theoretical power, (c) Δp through the runner, (d) degree of reaction, (e) utilization factor, and (f) blade angles. Water leaves the rotor radially and the turbine rotates at 75 rpm.

Ans: $H = 8.82$ m; $P = 866$ kW; $\Delta p = 70.75$ kPa; $R = 0.816$; $\varepsilon = 0.706$

4.5 A Francis turbine operates at 625 rpm when the flow rate through it is 40 cfs. The radius at the inlet is 2.5 ft. The flow at the inlet is set such that $\beta_1 = 80°$. If the blade height is 4 in., calculate the angle of the guide vanes α_1.

Ans: $\alpha_1 = 2.65°$

4.6 A Pelton wheel rotates at 550 rpm when the volumetric flow rate is 0.4 m^3/s and jet velocity is 100 m/s. The peripheral velocity factor (ϕ) is 0.47 and the jet turns through an angle of 170°. Calculate (a) wheel diameter, (b) power developed, and (c) kinetic energy of water leaving the buckets. Verify that the degree of reaction and pressure rise through the runner are zero. Also calculate the utilization factor.

Ans: $D = 1.63$ m; $P = 1977$ kW

4.7 Show that an ideal Pelton wheel ($\beta_2 = 180°$, that is, water turns in the bucket completely) develops the maximum power when wheel velocity u is half the jet velocity V_1.

4.8 A Pelton turbine rotates at 250 rpm and has a rotor diameter of 12 ft. The jet diameter is 8 in. and the exit blade angle is 165° (–15°). If the jet velocity is 180 ft./s, calculate the following: (a) torque, (b) power, (c) utilization factor, (d) pressure change through the rotor, (e) degree of reaction, (f) specific speed (make appropriate assumptions).

Ans: $P = 1566$ hp; $\mathcal{E} = 0.436$; $N_s = 0.268$

4.9 Obtain the principal dimensions of a turbine located in the Theodore Roosevelt Powerhouse in Arizona with the following specifications: power = 49,500 hp, head = 150 ft., and $N = 150$ rpm.

4.10 Based purely on theoretical considerations, show that the power from any hydraulic turbine is maximized when the frictional head losses are one-third of the net head. For frictional losses, use the Darcy–Weisbach equation:

$$h_f = f\frac{L}{D}\frac{V^2}{2g} = \frac{8fLQ^2}{\pi^2 g D^5}$$

4.11 Verify the design of the San Francisquito power plant, which has the following specifications: $P = 32{,}200$ hp, $H = 800$ ft., $N = 171.4$ rpm. Estimate the number of jets, if each jet is 14 in. in diameter.

Ans: The wheel diameter is 176 in. and the number of buckets is 22. These values are available in the public domain from the plant specifications.

4.12 A private home owner built a Pelton turbine on a creek in the backyard. The head and flow rate are 113 ft. and 350 gpm. It rotates at 514.3 rpm. It has two nozzles. Verify its design features.

Ans: Wheel diameter is given on the internet as 18 in. and the jet diameters are 1 in. each. The number of buckets is 18. Check the web site http://www.oldpelton.net/oldpp.html

4.13 Verify the design features of the Big Creek 2A plant, which houses a double-hung Pelton wheel. For this unit, $N = 300$ rpm, $H = 2200$ ft., and power for each wheel is 32,550 hp.

Ans: Wheel diameter is 137 in., single jet with diameter 8.5 in., 22 buckets; obtained from the internet.

4.14 An ideal Pelton wheel develops maximum power when the wheel velocity $\phi = u/V_1 = 0.5$. By calculating the efficiency and letting the jet velocity *be* $C_v(2gH)^{1/2}$, show that for maximum efficiency, $\phi = 0.5C_v$.

4.15 Draw the velocity triangles for the ideal inlet ($\alpha_1 = 0°$) of a Pelton wheel. Compare these triangles with a real situation when ($\alpha_1 = 10°$). Evaluate the difference in power output for these two scenarios.

4.16 Consider a reaction turbine with inlet and outlet diameters of 2 ft. and 8 in., respectively. The blade widths at the inlet and outlet are 3 in. and 2 in., respectively. It delivers 5000 gpm. The inlet blade angle $\beta_1 = 40°$ and the fluid exits radially. Calculate the head, pressure change through the rotor, degree of reaction, and utilization factor. The speed of the turbine is 400 rpm.

Ans: $R = 0.629$; $H = 65.49$ ft.; $\mathcal{E} = 0.81$; $\Delta p = 17.84$ psi

4.17 A Francis turbine, operating at a head of 350 ft., produces 13,500 hp while rotating at 200 rpm. How high can the turbine be placed to operate without cavitation when the ambient temperature is 80°F?

Ans: 24 ft.

4.18 Rework Problem 17 when the outside temperature varies from 35°F to 110°F.

Ans: 24.94 ft., 22.24 ft.

4.19 Obtain the design specifications for the Conowingo power plant on the Sesquehanna River with the following specifications: head = 89 ft., power = 54,000 hp, and $N = 81.8$ rpm.

4.20 The Chief Joseph hydroelectric power plant in Washington has the following specifications: power = 100,000 hp; $N = 100$ rpm, and head = 165 ft. What kind of turbine is it? Perform the preliminary design of the unit.

Ans: The discharge diameter for this unit is 197 in.

4.21 The specifications of the Bonneville 1 power plant in Oregon are: head = 60 ft.; power = 74,000 hp; $N = 75$ rpm. Indicate the type of turbine and its specifications.

Ans: The discharge diameter for this unit is 280 in.

4.22 The Grand Coulee hydroelectric power project on the Columbia River in Washington State has one of the largest reaction turbine runners in the world. It operates under a head of 285 ft. when rotating at 72 rpm. The plant has six turbines, with each producing 800,000 hp. Estimate the dimensions of the units and perform their preliminary design.

Ans: The units are Francis units, each with runner diameter of 32 ft.

4.23 As a project engineer, you are to decide what type of reaction turbine to pick at a location where the available head is 100 ft. River run-off data from several years indicate that a flow rate of 2300 cfs can be expected. What diameter and rpm would you recommend for the turbine if it is a (a) Francis and (b) Kaplan? Also, for each of these two cases, what would the safe settings of the runners above tail water be to avoid cavitation? The outside temperature varies from 40°F to 100°F.

4.24 A turbine is located at an elevation of 5000 ft. It operates under a head of 50 ft., and has a specific speed of 1.8. The turbine setting is 9 ft. above the tails. If the local temperature is 50°F, answer the following: (a) Is the setting safe from cavitation? (b) If it is not safe, what would the recommended setting be to prevent cavitation?

(c) It is proposed to increase the head on the turbine to 60 ft. What should the new setting be to make it safe from cavitation?

Ans: No; 7.8 ft.; 3.8 ft.

4.25 A Kaplan turbine has the following specifications: $H = 25$ ft., $N = 400$ rpm; power = 650 hp. If the efficiency of the unit is 90%, estimate (a) the flow rate, (b) approximate diameter of the runner, (c) turbine setting to avoid cavitation if the location is at sea level, and (d) turbine setting to avoid cavitation if the location of the unit is at an elevation of 5000 ft. The local temperature is 60°F.

Ans: (a) $Q = 254$ cfs; (c) $H_s = 5.8$ ft.; (d) 0.12 ft.

4.26 This problem considers the effects of adding a draft tube to the overall performance of reaction turbines. The exit diameter of a turbine is 100 cm and the exit is set at a height 4 m above the tails. (a) If water is directly discharged into the atmosphere, what is the pressure at the turbine exit? (b) If a draft tube of constant diameter equal to 100 cm is added to the exit, what would the pressure be at the turbine exit and what would be the net head added to the turbine? Ignore losses. (c) Repeat (b), but include friction losses assuming a flow of 1.3 m^3/s and $f = 0.02$. (d) The straight draft tube is now replaced with a diverging tube with exit diameter of 160 cm: what would the pressure be at the exit of the turbine and what would the net added head be? Comment on the effect of including friction losses for the diverging tube.

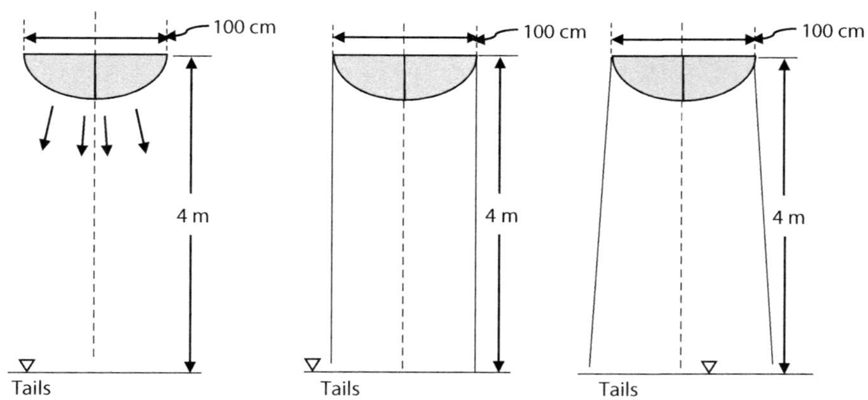

Ans: a) 101 kPa; b) 61.7 kPa; c) 61.8 kPa; d) 60. 5 kPa

4.27 A reaction turbine produces 5600 hp while operating at 260 rpm and a head of 60 ft. It is proposed to locate it in a place where the annual temperature variation is between 50°F and 108°F. What should the setting of the turbine be to avoid cavitation?

4.28 Perform the preliminary design of the Pelton wheel unit in Dixence, Switzerland. It operates under a net head of 5330 ft. and runs at 500 rpm. It is a double-hung unit with each wheel producing 25,000 hp.

Ans: Pitch diameter = 10.89 ft.; single jet of 3.71 in. diameter and jet velocity of 580 ft./s

5

Pumps

Apart from hydraulic turbines, the other class of machines that broadly falls under the incompressible category is pumps. There are remarkable similarities between centrifugal pumps and Francis turbines, both being broadly classified as radial/mixed flow types. These are useful for high heads and low flows. Similarly, axial flow pumps and Kaplan/propeller turbines have several common features, the most important being that they are ideally suited for high flows and low heads.

Upon completion of this chapter, the student will be able to

- Select various types of pumps based on specific speeds
- Understand the features of different types of pumps
- Analyze and perform preliminary design of various types of pumps
- Understand cavitation in the context of pumps

5.1 Introduction

Pumps are among the oldest devices known to humankind. The Archimedes screw pump that is used even today for handling solid/liquid mixtures has been in operation since 250 BC. The broadest definition of a pump would be any machine that delivers material from one location to another, with a possible pressure increase. Such devices can handle either solids, liquids, gases, or any combination of these. The devices that handle solids and liquids are called *pumps* whereas those that handle gases are called *fans*, *blowers*, or *compressors*. Typical gas pressure rise in fans and blowers would be of the order of a few inches to a few feet of water. Compressors could have higher gas pressure rises since they usually have several stages. Even though the pressure rise for each stage may be small, the overall pressure rise could be substantial.

Pumps can be classified as positive displacement pumps or dynamic pumps. Examples of positive displacement pumps include gear pumps, piston pumps, and screw pumps. The most common example of a positive displacement pump would be a human heart. All these devices usually have intermittent fluid flow and they pressurize the fluid by physically squeezing it. They also produce relatively small flow rates and high pressure rises. These, however, cannot be classified as turbomachines as they do not have rotating shafts and continuously moving fluid, which are requirements for turbomachines. Dynamic pumps that have rotating shafts with continuous fluid flow belong to the class of power-consuming turbomachines. They can be broadly classified as axial flow and centrifugal pumps. In centrifugal pumps, the rotation of the blades on the impeller imparts a whirling motion that causes higher pressure toward the circumference and lower pressure in the center. This forces the flowing fluid to enter near the center and flow outwards through

TABLE 5.1

Some Pump Installations and Their Details

	Description	H (ft.)	Flow (cfs)	N (rpm)	Ns
1	Rocky River	239	280	327	0.697373
2	Colorado River Aqueduct	444	200	450	0.509715
3	Gene Plant	310	200	400	0.593185
4	Iron Mountain Plant	146	200	300	0.782543
5	Cartersville Plant in Georgia	95	211	394	1.457074
6	Grand Coulee	344	1250	200	0.685809
7	Hiwasee Plant, TVA	205	3900	106	0.946585
8	Edmonston Pumping Plant	1970	315	600	0.278993

the volute, before being collected in the receiver. Axial pumps, on the other hand, do not have a volute and are generally placed in a pipe, with the inlet fluid having lower pressure than the exit pressure. There are millions of pumps in the world with varying flow rates and heads. Some pump installations, along with their specific speeds are given in Table 5.1.

5.2 Selection of Pumps

Similar to hydraulic turbines, pumps can be selected based on the specific speed. The range of specific speeds and the corresponding pump types are given in Table 2.2. Since pumps have been around for hundreds of years, and much experience has been gained in their design and selection, a chart based only on pumping head H and capacity Q has been developed. This is shown in Figure 5.1. The similarities between this figure and the turbine selection diagram shown in Figure 4.2 are striking. For high flow rates, axial flow pumps are most suitable (analogous to propeller or Kaplan turbines); however, for medium flow rates, centrifugal pumps are better suited (analogous to Francis turbines). From a most elementary perspective, an axial flow pump can be regarded as a Kaplan turbine, and a centrifugal pump can be considered to be a Francis turbine, with the flow direction reversed in both cases (albeit with very low efficiencies). Among pumps, there is nothing analogous to a Pelton wheel (suitable for very high heads). When large pressure rises are desired, positive displacement pumps are used.

From the definition of the specific speed $N_s = \dfrac{N\sqrt{Q}}{(gh)^{3/4}}$, it can be seen that higher specific speeds warrant higher flow rates with accompanying lower heads. Thus, as the specific speed increases, a marked change in the shape of the impellers can be seen, as shown in Figure 5.2. Centrifugal pumps produce relatively higher heads than axial flow pumps. This is because of the centrifugal effect, which tends to increase the pressure considerably. Since flow rate is the product of a velocity and an area, larger flows should imply larger areas. For axial flow pumps, the area is πD^2 (the area of the hub should be subtracted; however, the hub area would be a very small percentage of the axial area), whereas, for centrifugal pumps, the area would be πDt, where t is the thickness or height of the blades. Thus, $Q \sim D^2$ for axial pumps and $Q \sim D$ for centrifugal pumps; hence, axial flow pumps

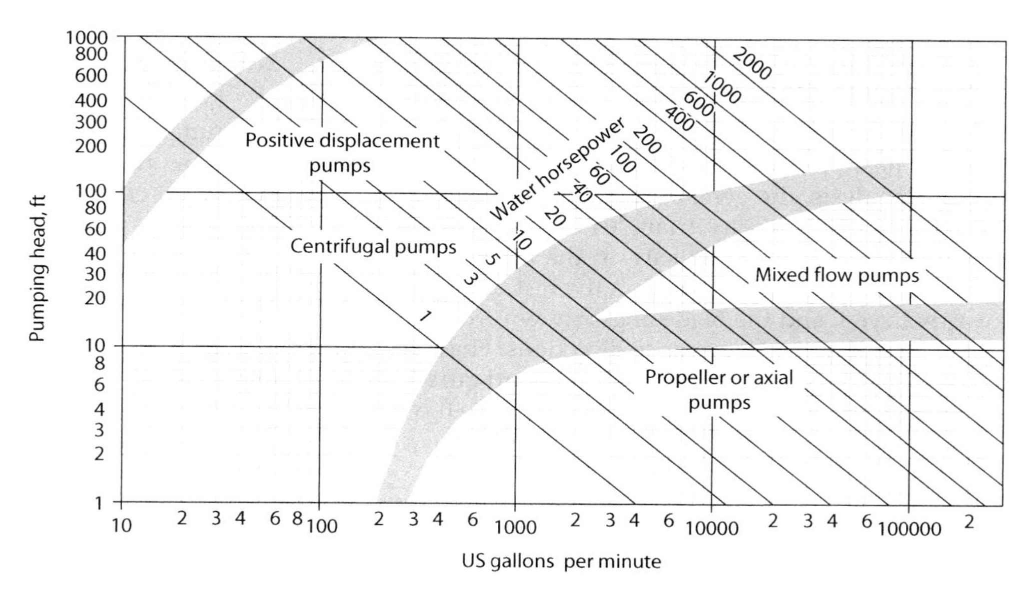

FIGURE 5.1
Pump selection diagram. (Modified from Streeter, V. L. and Wylie, E. B., *Fluid Mechanics*, 8th edn, McGraw-Hill, New York, 1985.)

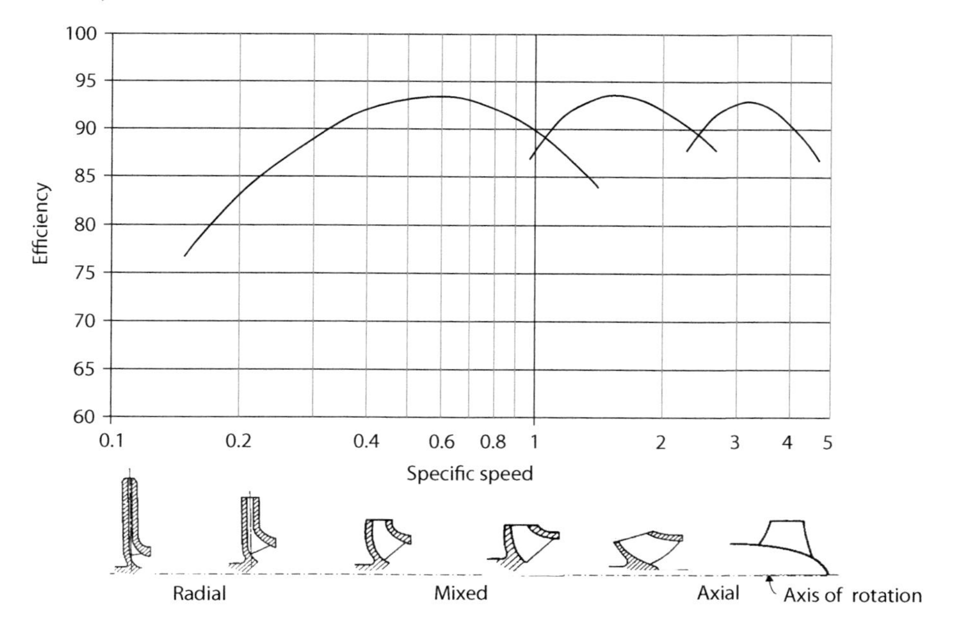

FIGURE 5.2
Effect of specific speed on impeller shape and efficiency for pumps.

produce higher flows. It also happens that centrifugal pumps exhibit high efficiencies at low flows and higher heads (low N_s), and vice versa for axial pumps. Thus, the impeller design changes from centrifugal to mixed to axial flow as the specific speed increases. This is shown in Figure 5.2.

5.3 Pump Characteristics

In the analysis of pumps, the three characteristics that play a dominant role are the variation of head, power, and net positive suction head (NPSH) with flow rate. From an elementary analysis, the general shape of the head versus flow rate (*H–Q*) curve can be determined. Typical velocity triangles for a centrifugal pump, along with a picture, are shown in Figure 5.3. For simplicity, in the following discussion, a slightly different sign convention has been adopted than in the rest of the text. The blade shown in Figure 5.3 is backward-curved, and the blade angle shown would be negative since the relative velocity and blade speed are in opposite directions. However, in the following discussion, this would be treated as positive. Thus, all backward-curved blades would have positive values of $\beta_2 < 90°$ and forward-curved blades would have $\beta_2 > 90°$.

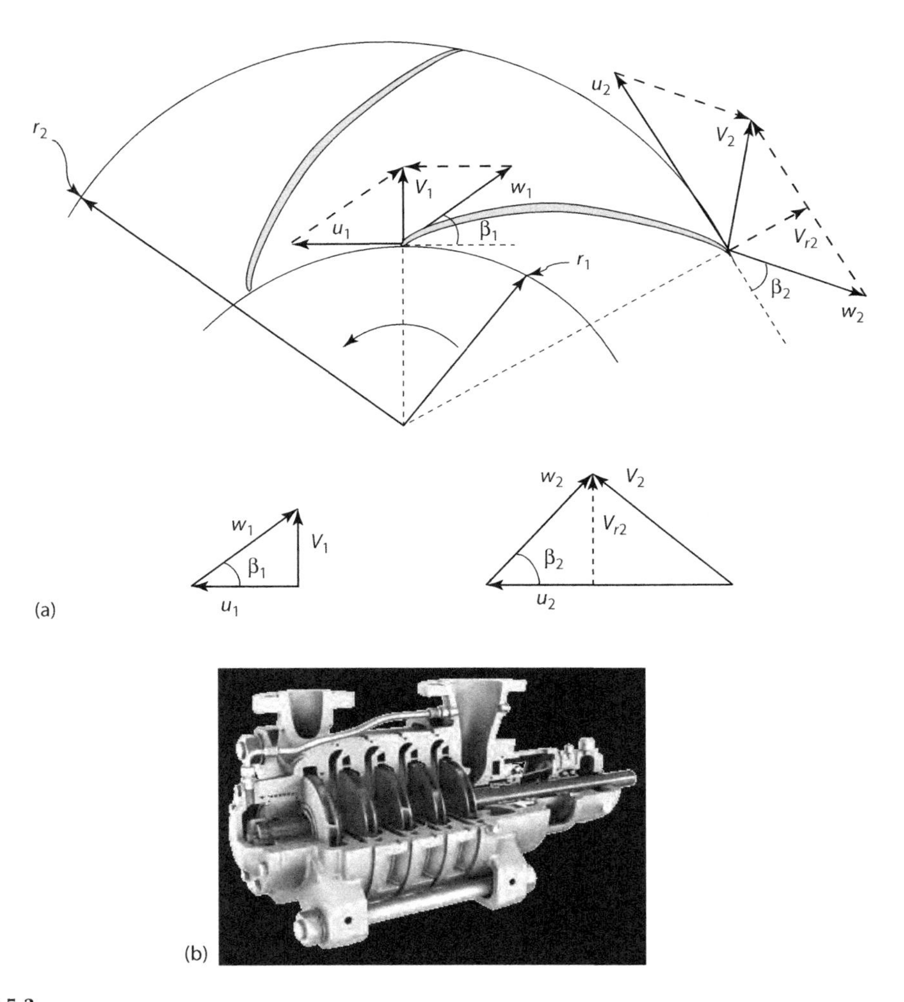

FIGURE 5.3
(a) Idealized inlet and exit velocity triangles for centrifugal pumps. (b) A typical multistage centrifugal pump. (Courtesy of Sulzer Pumps.)

The inlet flow is assumed to be purely radial. Hence, $V_{u1}=0$. Thus, from Euler's equation,

$$E = gH = u_1V_{u1} - u_2V_{u2} = -\ u_2V_{u2} \tag{5.1}$$

or

$$|H| = \frac{u_2V_{u2}}{g} = \frac{u_2}{g}\left[u_2 - V_{r2}\cot\beta_2\right] \tag{5.2}$$

However, from the equation for flow rate,

$$Q = 2\pi r_1 b_1 V_{r1} = 2\pi r_2 b_2 V_{r2} \tag{5.3}$$

or

$$V_{r2} = \frac{Q}{2\pi r_2 b_2} \tag{5.4}$$

Substituting Equation 5.4 into 5.2,

$$|H| = \frac{u_2}{g}\left[u_2 - \frac{Q}{2\pi r_2 b_2}\cot\beta_2\right] = \frac{u_2^2}{g} - \frac{u_2}{g}\frac{Q}{2\pi r_2 b_2}\cot\beta_2 \tag{5.5}$$

For simplicity, Equation 5.5 can be rewritten as

$$|H| = K_1 - K_2Q\cot\beta_2 \tag{5.6}$$

Since the sign of cotβ_2 depends on whether the blade angle is in the first or second quadrant, the shape of the *H–Q* curve changes quite significantly. Plots of Equation 5.6 are shown in Figure 5.4 for various values of β_2. If the angle $\beta_2 > 90°$, the blades are called forward-curved; if $\beta_2 = 90°$, the blades are radial; if $\beta_2 < 90°$, the blades are backward-curved. The sign convention adopted refers to the blade angles as the acute angles between *u* and *w*. Of these three, forward-curved vanes are never used, since this leads to instabilities and surging in pumps.

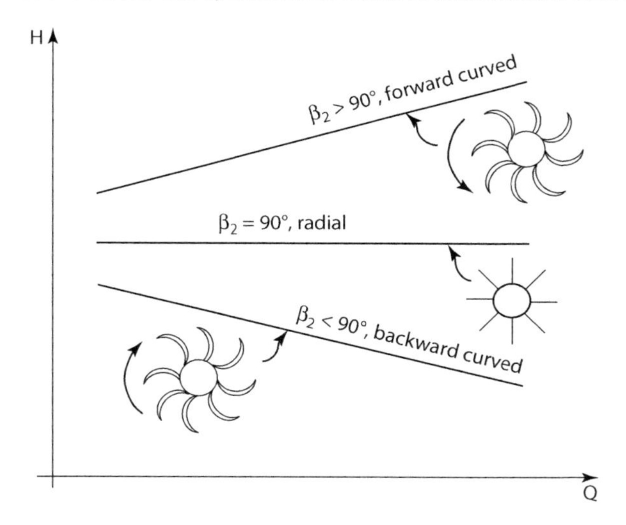

FIGURE 5.4
Variation of H–Q curve with shape of blades.

The analysis used so far did not account for the possible losses. Thus, there is a deviation in the actual shape of the pump curve from that predicted by Equation 5.6. There are several reasons for this. The first is *friction* loss, which varies as Q^2. The second loss is due to *circulatory flow*, resulting from leakage of the flow from the pressure side to the suction side of the blade, a process that occurs near the tip of the blade. This is similar to wingtip vortices that are formed on aircraft wings due to the leakage of flow from the high pressure below the wing to the low pressure above the wing. The relative velocity thus leaves with a value $w_2' > w_2$ at an angle $\beta_2' < \beta_2$. The result of this is that $w_{u2}' > w_{u2}$, which makes the value of $V_{u2}' < V_{u2}$, thereby reducing the head. This can be seen in Figure 5.5. The ratio of V_{u2}' to V_{u2} is called the *slip coefficient*, μ_s. It depends on the amount of circulation, which in turn depends on the number of blades and the geometry of the flow passage. An approximate relationship for the slip coefficient has been given by Shepherd (1956) as follows:

$$\frac{V_{u2}'}{V_{u2}} \equiv \mu_s = 1 - \frac{\pi u_2 \sin\beta_2}{V_{u2} n_B} \tag{5.6a}$$

where n_B is the number of blades.

There are several other correlations in the literature relating the slip coefficient to the number and geometry of blades, speed of the impeller, and pressure rise. They are particularly important in the study of compressors. As with pumps, they are semi-empirical in

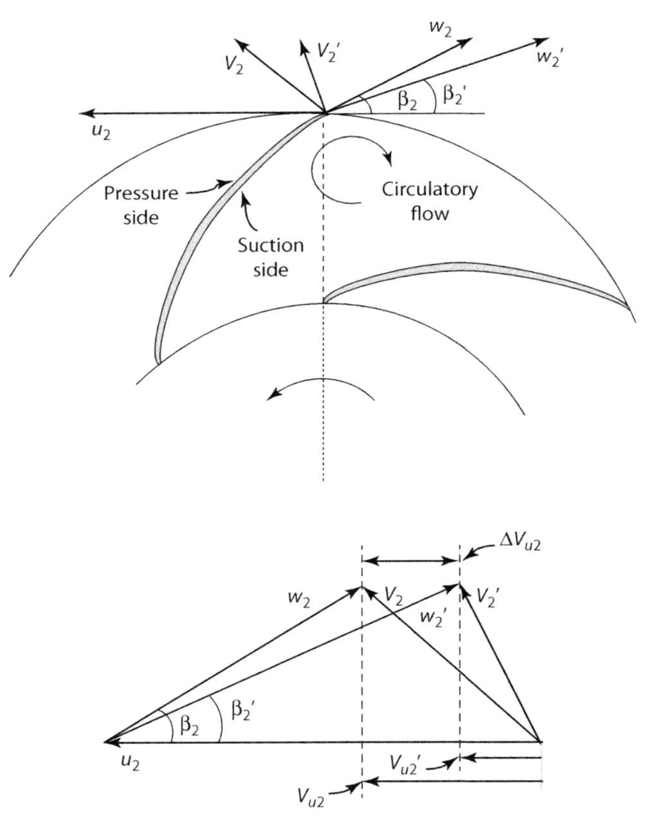

FIGURE 5.5
Circulatory losses at exit of pumps.

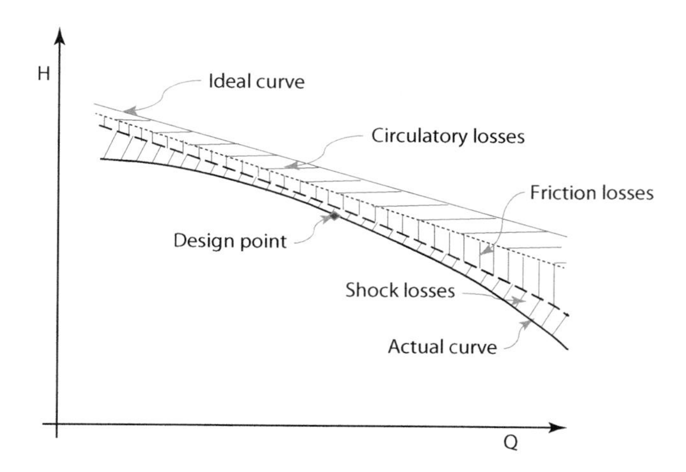

FIGURE 5.6
Ideal and actual H–Q curves with various types of losses.

nature, and their discussion is deferred to the later chapter on compressors. It is sufficient to mention that notable investigators in this area are Staniz, Stodola, and Balje, whose equations are used widely in compressor studies.

The third type of loss is called *shock* or *turbulence loss*. These losses arise as the fluid relative velocity is tangential to the blade only under design conditions of optimum efficiency. For other values of discharge, the flow is not tangential; hence the name *shock* losses. The final shape of the curve is shown in Figure 5.6.

5.4 Pumps and Piping Systems

For any application, the selection of a pump depends on the combination of the required flow rate and the head. Hence, the selection or installation of a pump in a piping system depends upon the head losses and elevation head that need to be provided by the pump. The piping system for delivery of a fluid between two points is based on the design by the piping engineer. In the simplest sense, any pump installation would connect two reservoirs. For simplicity, consider the flow system in Figure 5.7a. The piping layout, dimensions, and associated flow control devices would be designed by the engineer. By applying the steady-state energy equation for a single inlet and outlet (see Elger et al. (2012)) between the free surfaces, the following equation is obtained:

$$\frac{p_1}{\gamma} + z_1 + \frac{V_1^2}{2g} + h_p = \frac{p_2}{\gamma} + z_2 + \frac{V_2^2}{2g} + h_f + h_m \tag{5.7}$$

Writing the losses in terms of the velocities (or flow rates), the energy equation becomes

$$h_p = \Delta z + \frac{fl}{D}\frac{V^2}{2g} + \sum K \frac{V^2}{2g} \tag{5.7a}$$

In Equation 5.7a, h_p is the head required from the pump. In most pump applications, flow rate is a more relevant variable than velocity. Hence, using the relation $Q = V\frac{\pi D^2}{4}$, the energy equation can be rewritten in terms of flow rate as

$$h_p = \Delta z + \frac{8flQ^2}{\pi^2 gD^5} + \sum K\frac{8Q^2}{\pi^2 gD^4} = \Delta z + \left[\frac{8fl}{\pi^2 gD^5} + \sum K\frac{8}{\pi^2 gD^4}\right]Q^2 \tag{5.8}$$

In this equation, the first term is the static (hydrostatic) elevation head, the second is the loss due to friction given by the Darcy-Weisbach equation, and the third term represents minor losses. The right-hand side of the equation is called the *system head*, H_s, and a plot of H_s versus Q is called the *system curve*. It is customary to plot it on the same graph containing the pump curves, as shown in Figure 5.7b. The intersection of the pump curve and the system curve is the operating point, which gives the flow rate and head that would be available from the pump. The pump operating point should occur at the highest efficiency since pumps are power-consuming devices. If the operating point is not at the highest efficiency, the pump curve can be slightly shifted by changing the pump speed (illustrated in Example 5.3). The pump curve (h_p vs. Q) and efficiency curve (η vs. Q) are supplied by the pump manufacturer. A typical set of manufacturer's curves for a centrifugal pump

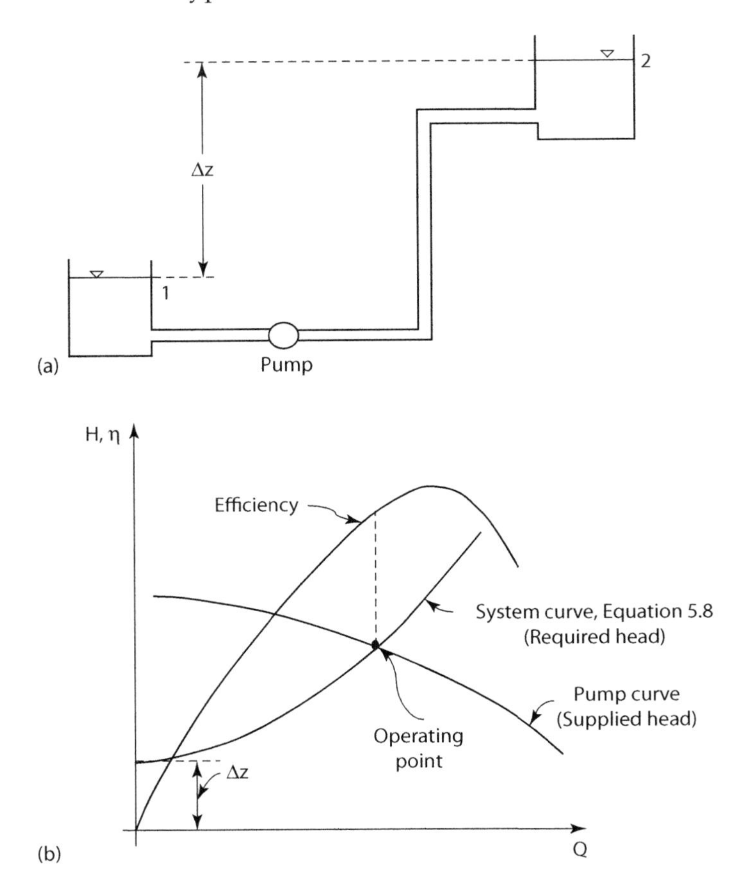

FIGURE 5.7
Typical pump in a piping system along with system and pump curves.

are shown in Figure 5.11. Usually, this set includes power and NPSH (related to cavitation, which is discussed later) curves as well.

Two points of interest are to be noted. First, given the head and efficiency curves (vs. flow) it is possible to obtain the power versus flow curve, since $P = \gamma QH/\eta$. Second, for geometrically similar pumps of different sizes and speeds, a single curve can be developed by plotting gH/N^2D^2 versus Q/ND^3. This would be a more convenient practice for pump companies to provide a single curve, especially if the pumps are similar. However, it has been the practice in the pump industry to provide a separate curve for each pump, rather than one for the homologous class.

In some situations, it may be advantageous to use two pumps in series or parallel. If two pumps are used in series, the performance curve would be the sum of the heads of the two machines at the same flow rate, as shown in Figure 5.8a. This configuration would be desirable for steep system curves that require high heads at relatively low flows. If two pumps are in parallel, then the performance curve would be the sum of the flow rates for the same head, as shown in Figure 5.8b. This situation would be desirable for shallow system curves that require high flows but at relatively low heads, as in cases of floods, where large quantities of water need to be moved. When pumps are installed in series or parallel,

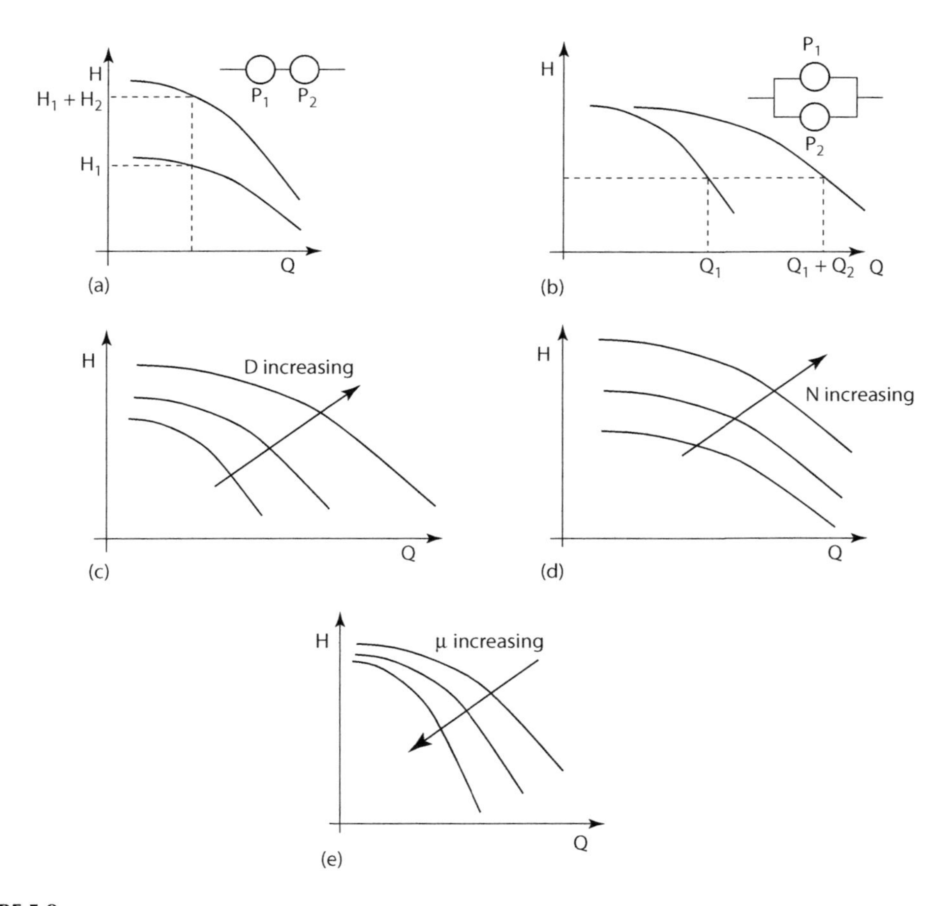

FIGURE 5.8
Head versus flow for various situations: (a) pumps in series, (b) pumps in parallel, (c) increasing impeller diameter, (d) increasing speed, (e) increasing viscosity.

care must be taken to see that the characteristics of both pumps are close. Also, the engineer must make sure that the selected pump will not be subject to cavitation.

In general, the selection of pumps depends on several factors, such as the head and flow requirements, space constraints, and speed. Thus, the operating point can be altered by changing the size (Figure 5.8c) or the speed (Figure 5.8d). If the performance curve for one pump is available, then the curves showing size or speed dependence can be obtained from the pump similarity laws. According these laws, for geometrically similar pumps,

$$\left(\frac{Q}{ND^3}\right)_1 = \left(\frac{Q}{ND^3}\right)_2; \quad \left(\frac{gH}{N^2D^2}\right)_1 = \left(\frac{gH}{N^2D^2}\right)_2 \tag{5.8a}$$

Although the majority of pumps are used for handling water, an important application of centrifugal pumps is in pumping fluids with significantly different densities and viscosities than water. Examples of such applications may be found in the petrochemical, food, and manufacturing industries. Variation of density can be accounted for easily since the specific weight of the fluid directly affects the head and, hence, the power requirements. However, effects of viscosity are more difficult to account for, since accurate estimates of viscous losses for flow fields inside the pump casing are impossible to obtain owing to the complexity of the flow. However, the effect of viscosity on pump performance can be described qualitatively, as shown in Figure 5.8e. It is seen that the head rapidly drops off with increasing viscosity. This is also accompanied by a rapid decrease in efficiency.

5.4.1 Pump Surge

It has already been pointed out that the rotation of pumps is such that the blades are backward facing. Forward-facing blades lead to a phenomenon in pumps called *surge*, which is explained in the following section. Consider the situation shown in Figure 5.9a, which corresponds to backward-facing vanes for which $dH/dQ < 0$. Consider a scenario where the operating point shifts from A to B. This could happen when a valve in the piping system is partially closed, resulting in decreased flow. Note that valve closure results in decrease in flow and hence decreased head, since head is proportional to Q^2. However, the decrease in the K factor for the valve is significantly higher. For instance, the K factor for a 50% closed gate valve is 28 times that of a fully open gate valve (see Elger et al. (2012)). Thus, a partly closed gate valve will give reduced flow but increased head loss. For the pump curve shown as in Figure 5.9a, a closure of the valve corresponds to the pump operating point shifting from A to B. Thus, the system needs a higher head to account for the higher losses, and the pump can accommodate this. This represents a stable situation.

Now consider the situation shown in Figure 5.9b, which corresponds to forward-facing vanes for which $dH/dQ > 0$. As in the previous scenario, let a valve in the piping system be partially closed. The operating point changes from C to D on the pump curve. This would mean reduced flow but higher system head requirements, and the pump needs to produce a higher head to maintain the flow. However, due to the positive slope of the H–Q curve, the pump produces decreased head. This reduces the flow further, and the operating point shifts further to the left along the pump curve, eventually reaching point E when the pump ceases to produce any flow. However, since it is still running and the valve is partially open, fluid flow begins again, and the operating point tends toward point F. Thus, the flow oscillates between points F and E, and the pump finally stops and produces no flow. This phenomenon is called *surge*, and it causes vibrations that result in pump damage. The increasing portion of the H–Q curve is thus unstable and needs to be avoided. Hence, the rotation of pumps is such that the blades are backward-facing.

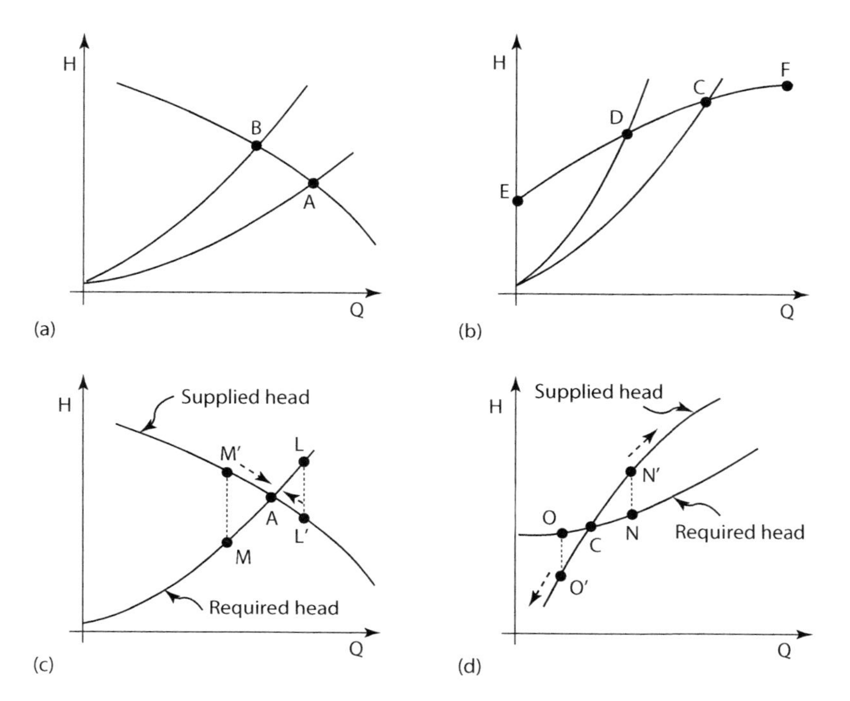

FIGURE 5.9
Pump surge; (a,c) stable *H–Q* curve and (b,d) unstable *H–Q* curve.

A similar argument can be used to explain this behavior by using stability of flow in the system and the pump. Consider the situation in Figure 5.9c. If, for some reason, the operating point moves from A to L, the required head H_L will be higher than the supplied head $H_{L'}$, that is, $H_L > H_{L'}$. Thus, the pump cannot supply the required head, the flow decreases, and the point L′ moves back to A. A similar situation exists with respect to point M. Thus, point A represents a *stable* point.

The exact opposite situation exists in Figure 5.9d. If the operating point moves from C to N, the supplied head $H_{N'}$ is greater than the require head, that is, $H_{N'} > H_N$. The pump is thus supplying more head than is required, and hence, the flow increases and the point N′ moves away from C. The situation is similar at point O. The point C represents an *unstable* point.

5.5 Cavitation in Pumps

As with hydraulic turbines such as the Francis and Kaplan types (Pelton wheels do not cavitate since they are open to the atmosphere), the major cause for repair of pumps is also cavitation. The definition and causes are the same as before; that is, the local pressure reaching vapor pressure owing to high fluid speeds. The consequences, namely, vibration, noise, loss of efficiency, and pitting, are also the same. The main difference between turbines and pumps regarding cavitation is that while the low-pressure point is the exit of the runner in the case of turbines, the corresponding point for pumps is the inlet to the impeller.

Cavitation is one of the major causes of pump failures. While loss of efficiency, noise, and the accompanying vibration can be tolerated, the damage done to impellers is extensive in the sense that the metal is "eaten away," leaving holes or pits; hence the term *pitting*. Under such circumstances, the only recourse available would be to replace the impeller, which would be very expensive. A picture of a pump impeller damaged by cavitation is shown in Figure 5.10a.

As was seen in Chapter 4, cavitation in hydraulic turbines can be quantified using the cavitation (Thoma) coefficient. Although the same method can be used here also, the more common parameter used to characterize cavitation in pumps is NPSH, which is defined as the difference between the stagnation pressure head at the inlet to the pump and vapor pressure. Consider the situation shown in Figure 5.10b, which shows the suction pipe of a pump.

From the energy equation relating the free surface and the inlet, the following equation-can be derived:

$$\frac{p_a}{\gamma} = \frac{p_s}{\gamma} + \frac{V_s^2}{2g} + H_s + h_l \tag{5.9}$$

By considering the difference between the inlet total pressure and vapor pressure, the available NPSH can be defined:

$$\text{NPSH} = \left(\frac{p_s}{\gamma} + \frac{V_s^2}{2g}\right) - \frac{p_v}{\gamma} \tag{5.10}$$

In the literature, the kinetic energy head is sometimes omitted, and NPSH is then defined as the difference between the pressure head at the inlet and vapor pressure head. However, in this book, velocity head is included, and Equation 5.10 is used to define NPSH.

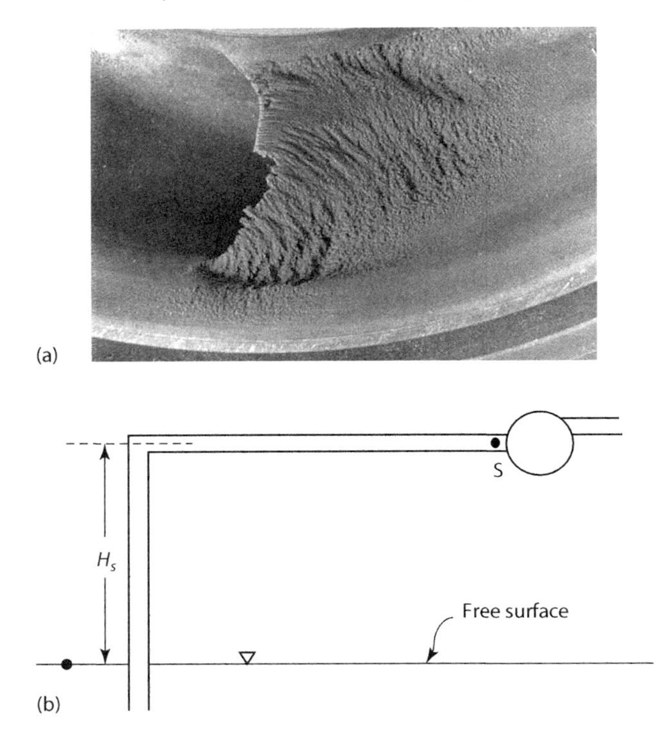

FIGURE 5.10
(a) Typical cavitation damage on an impeller blade. (b) Typical setting of a pump on the inlet or suction side. (Courtesy of Sulzer Pumps.)

Combining Equations 5.9 and 5.10,

$$\left(\frac{p_s}{\gamma}+\frac{V_s^2}{2g}\right)-\frac{p_v}{\gamma}=\frac{p_a}{\gamma}-\frac{p_v}{\gamma}-H_s-h_l=NPSH_A \tag{5.11}$$

This form of NPSH depends on the elevation of the pump inlet above the free surface. Although there is little control over the values of p_a and p_v (the former depends on the location and the latter on local temperature and the fluid; e.g., vapor pressure increases with temperature; similarly, highly volatile liquids such as gasoline have higher vapor pressures than water), H_s and h_l can be controlled. The setting of the pump determines the value of H_s. Pipe losses can be minimized by proper design. Since NPSH, as given in Equation 5.11, is a function of the available value of H_s, it is called *net positive suction head available*, or $NPSH_A$. It should be mentioned that flow control valves are placed in pipe lines to control volumetric flow rates. If these are placed on the inlet side of pumps, not only do they reduce the flow, they also reduce the suction pressure, making the pump susceptible to cavitation. This is one of the reasons that flow control valves are not placed on the inlet side of pumps.

In the preceding derivation, the simplistic assumption that the lowest pressure occurs in the suction pipe was made. In fact, the pressure in the eye of the impeller is likely to be lower. Hence, the required value of the NPSH is determined experimentally. In fact, three curves are provided for every pump by the manufacturer; namely, head versus flow rate (H vs. Q), efficiency versus flow rate (η vs. Q), and required NPSH versus flow rate ($NPSH_R$ vs. Q). Typical shapes of these curves are shown in Figure 5.11.

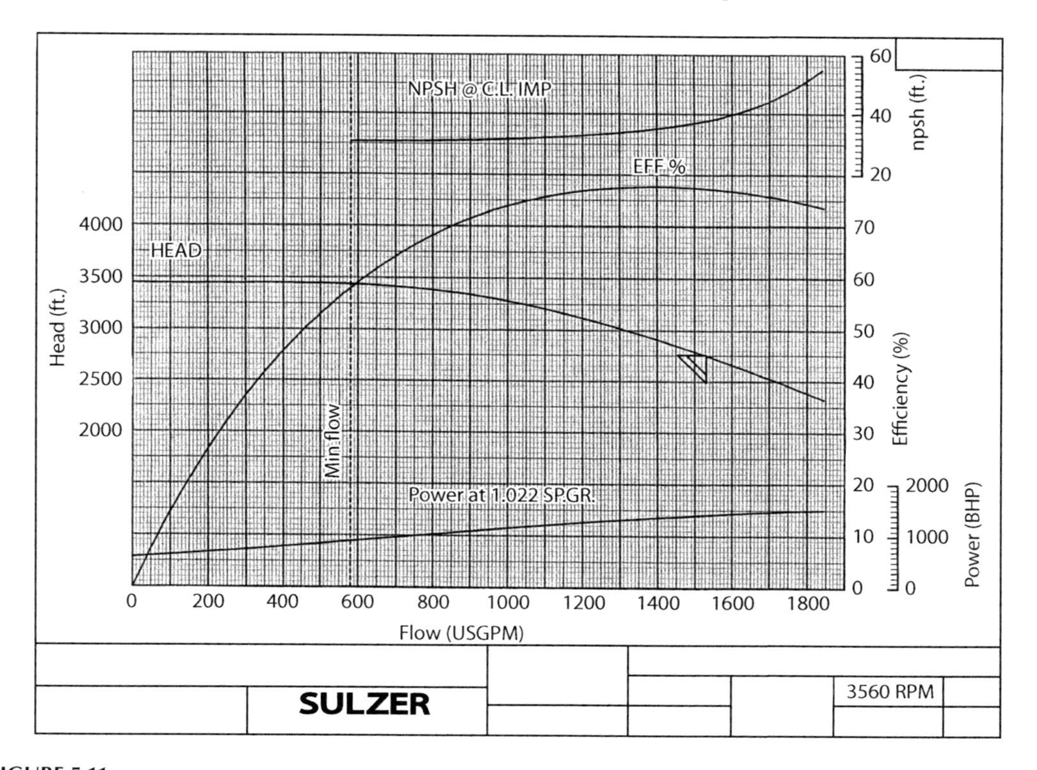

FIGURE 5.11
Typical pump characteristics. (Courtesy of Sulzer Pumps.)

For proper pump operation, it is necessary that

$$\text{NPSH}_A \geq \text{NPSH}_R \tag{5.12}$$

An alternative method of accounting for cavitation in pumps is using the cavitation coefficient σ. From Equation 5.10, NPSH is the difference between the total inlet pressure and vapor pressure. For incipient cavitation, local inlet pressure would be equal to vapor pressure. Thus,

$$\text{NPSH} = \left(\frac{p_s}{\gamma} + \frac{V_s^2}{2g}\right) - \frac{p_v}{\gamma} = \frac{V_s^2}{2g}$$

$$\text{and } \sigma \equiv \frac{V_s^2/2g}{H} = \frac{\left(\frac{p_s}{\gamma} + \frac{V_s^2}{2g}\right) - \frac{p_v}{\gamma}}{H} = \frac{\text{NPSH}}{H} \tag{5.13}$$

When the value of σ falls below a critical value σ_c, cavitation is likely to occur. The variable σ_c is called the *critical cavitation coefficient*. It depends on the specific speed and is determined experimentally. The variation of σ_c with specific speed N_s is shown in Figure 5.12.

Yet another approach for preventing cavitation is through the introduction of the suction specific speed, which is defined as

$$N_{ss} \equiv \frac{N\sqrt{Q}}{\left[g(\text{NPSH})\right]^{3/4}} \tag{5.14}$$

Based on the recommendation by the Hydraulic Institute (1994), for safe operation of pumps, N_{ss} should be less than 3.1.

5.6 Pump Design

As with turbines, only rudimentary details of the hydraulic design of pumps are presented. In turbines, the relevant variables are the required flow rate Q and power P. In pumps, the

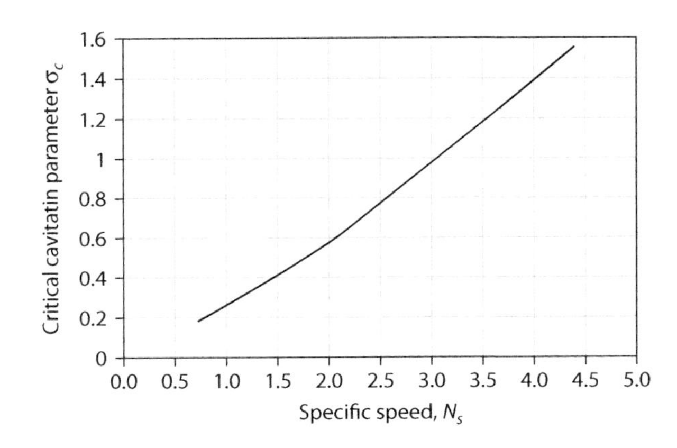

FIGURE 5.12
Variation of critical cavitation parameter with specific speed for pumps. (Modified from Dougherty, R. L. et al., *Fluid Mechanics with Engineering Applications*, 8th edn, McGraw-Hill, New York, 1985.)

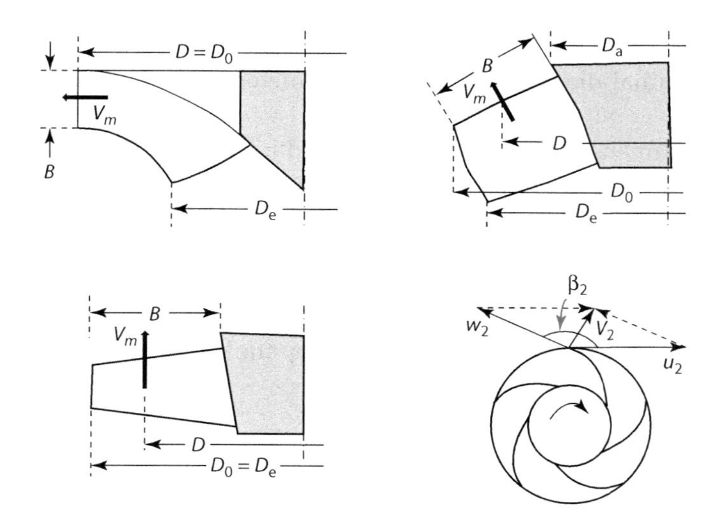

FIGURE 5.13
Dimensions of pumps.

relevant variables are the flow rate Q and required head H. For any application, since the pipe layout would be known, the required H to be supplied by the pump would also be known; as would the required flow rate, depending on demand. The procedure would be as follows (the geometry of various pumps and the relevant nomenclature are presented in Figure 5.13):

1. If the speed of rotation N is known, the specific speed N_s can be calculated.
2. From Figure 5.14, values for peripheral velocity factor ϕ, V_m/u_2, and B/D_0 can be chosen.
3. From the peripheral velocity factor ϕ, the blade speed u_2 can be calculated.

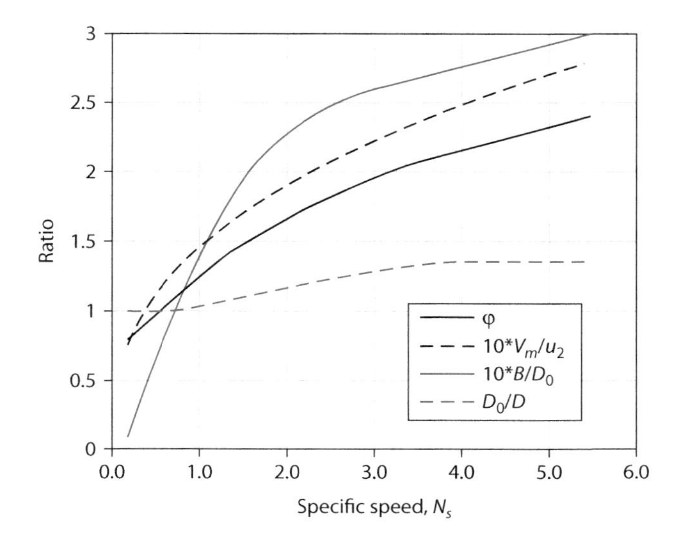

FIGURE 5.14
Pump proportions as function of specific speed. (Modified from Dougherty, R. L. et al., *Fluid Mechanics with Engineering Applications*, 8th edn, McGraw-Hill, New York, 1985.)

4. With a known blade speed at the exit, the nominal diameter D can be calculated.
5. From the nominal diameter D, impeller diameter D_0 and blade height B can be calculated.
6. Once all the dimensions are known, it should be verified whether the calculated dimensions will provide the required amount of flow. The blade height B can be adjusted depending on whether more or less flow is required.
7. For a centrifugal pump, the typical number of blades is between 5 and 12 (Jagdish Lal [1986]). A higher number of blades gives a flatter H–Q curve. The ideal head is obtained if the actual number of blades is infinite, which is an impossibility.

   An alternative procedure for calculating the number of blades is to use Pfliederer's equation, given as

$$n_B \equiv 6.5 \frac{D_2 + D_1}{D_2 - D_1} \sin \beta_m$$

$$\text{where } \beta_m = \frac{\beta_1 + \beta_2}{2}$$

   The value of β_1 is typically between 10° and 25°, with the assumption that the fluid enters the rotor with no momentum, that is, α_1 is 90°.
8. As a final check, the calculated diameter should be checked against that predicted by the Cordier diagram (Figure 2.2).

If pump speed is not known, a slight variation to this procedure is adopted. Using the pump selection diagram given in Figure 5.1, the type of pump can be determined. With this information, a suitable value of specific speed is picked that fixes the pump speed N. Once the speed is known, the procedure outlined can be used.

Example 5.1

A centrifugal pump is used to pump water from an open reservoir to an overhead tank. The suction pipe diameter is 6 in. and the flow rate is expected to be 1 ft.^{3}/s. The manufacturer's specification of $NPSH_R$ for the pump is 20 ft. and the local temperature is 80°F. For simplicity, friction losses can be neglected, and the only other losses through the system are due to a valve with a K factor of 18. What should the pump setting be to avoid cavitation?

Solution: At 80°F, specific weight and vapor pressure are 62.22 lbf/ft.^{3} and 0.507 psia. From Equation 5.11, $NPSH_A$ is given by

$$NPSH_A = \frac{p_a}{\gamma} - \frac{p_v}{\gamma} - z_s - h_l \tag{a}$$

Since head loss is given by

$$h_l = \Sigma K \frac{V^2}{2g} \tag{b}$$

with

$$V = \frac{Q}{A} = \frac{1}{(\pi/4)(6/12)^2} = 5.1 \text{ ft./s}$$

we get

$$h_l = 18\frac{(5.1)^2}{2(32.2)} = 7.27 \text{ ft.}$$

Equation (a) can now be written as

$$NPSH_A = \frac{(14.7)(144)}{62.22} - \frac{(0.507)(144)}{62.22} - z_s - 7.27 \text{ ft.}$$
$$= 25.58 - z_s$$

Since $\text{NPSH}_A \geq \text{NPSH}_R$,

$$25.58 - z_s \geq 20$$
$$\text{or} \quad z_s \leq 5.58 \text{ ft.}$$

Thus, the pump should not be set higher than 5.58 ft. to avoid cavitation.

Comments: Atmospheric pressure and vapor pressure are dependent on elevation and temperature, respectively. Hence, the setting would change if these variations are considered.

Example 5.2

The pump in Figure 5.11 is being considered for the following application. The reservoir surface is 16 ft. above the pump. The total pipe length is 20 ft., and the total value of the *K* factor for all minor losses is 2.0. The pipe diameter is 6 in. and the friction factor can be assumed to be a constant value of 0.02. Estimate the maximum amount of water that can be pumped without cavitation if the water is at 60°F. What would the allowable flow rate be if the temperature is increased to 140°F (Figure 5.15)?

Solution: The energy equation relating points 1 and *s* (the suction side of the pump) gives

$$\frac{p_1}{\gamma} + z_1 + \frac{V_1^2}{2g} + h_p = \frac{p_s}{\gamma} + z_2 + \frac{V_s^2}{2g} + h_t + h_f + h_m \quad \text{(a)}$$

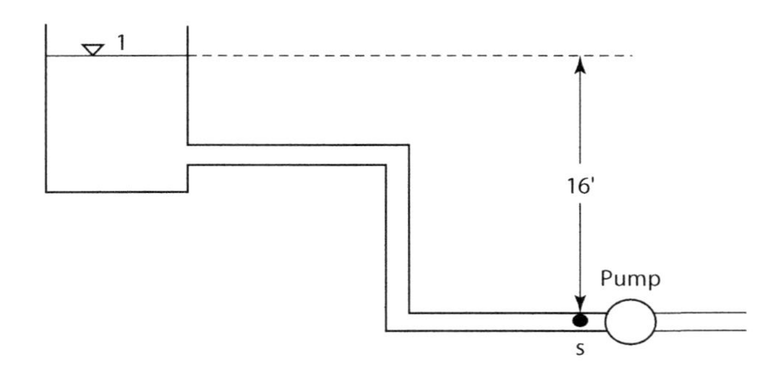

FIGURE 5.15
Schematic diagram of flow to pump inlet.

Substituting values for h_f and h_m,

$$\frac{p_a}{\gamma}+16=\frac{p_s}{\gamma}+\frac{V_s^2}{2g}+f\frac{L}{D}\frac{V^2}{2g}+\sum K\frac{V^2}{2g} \tag{b}$$

$$\frac{p_a}{\gamma}+16=\frac{p_s}{\gamma}+\frac{V_s^2}{2g}+\frac{8fLQ^2}{\pi^2 gD^5}+\sum K\frac{8Q^2}{\pi^2 gD^4}$$

At 60°F, $p_v=0.256$ psia and $\gamma=62.37$ lbf/ft.[3] (Table A.1). Subtracting p_v/γ from Equation (b),

$$\frac{p_a}{\gamma}+16-\frac{p_v}{\gamma}=\left(\frac{p_s}{\gamma}+\frac{V_s^2}{2g}-\frac{p_v}{\gamma}\right)+\frac{8fLQ^2}{\pi^2 gD^5}+\sum K\frac{8Q^2}{\pi^2 gD^4} \tag{c}$$

Since the quantity in parenthesis is NPSH, Equation (c) can be rewritten as

$$\frac{p_a}{\gamma}+16-\frac{p_v}{\gamma}=NPSH_A+\frac{8fLQ^2}{\pi^2 gD^5}+\sum K\frac{8Q^2}{\pi^2 gD^4} \tag{d}$$

or after substituting the values,

$$\frac{(14.7)(144)}{62.37}+16-0.591=NPSH_A+0.322Q^2+0.806Q^2 \tag{e}$$

Thus,

$$\mathrm{NPSH}_A=49.34-1.127Q^2 \tag{f}$$

The required NPSH (from the pump manufacturer's data) is plotted in the figure. On the same graph, the available NPSH is also shown. For flow rates below 1500 gpm, $NPSH_A \geq NPSH_R$, the requirement to avoid cavitation. Hence, the maximum allowable flow rate at 60°F would be 1500 gpm.

When the temperature is 140°F, $p_v=2.89$ psia and $\gamma=61.38$ lbf/ft.[3] and Equation (e) becomes

$$\frac{(14.7)(144)}{61.38}+16-6.78=NPSH_A+0.322Q^2+0.806Q^2$$

or

$$\mathrm{NPSH}_A=43.7-1.127Q^2 \tag{g}$$

From the figure, maximum flow rate can be seen to be approximately 1260 gpm (Figure 5.16).

Comments: As expected, the allowable flow rate to avoid cavitation would decrease with temperature. At 140°F, the fluid is closer to boiling than at 60°F.

Example 5.3

A pump operating at 1750 rpm has the following characteristics. It is used for pumping between two reservoirs. The total head loss through the piping system can be approximated as $h_l=15Q^2$, where h_l is in ft. and Q is in gpm. The elevation difference between the reservoirs is 33 ft. (water is being pumped from the lower tank to the upper tank).

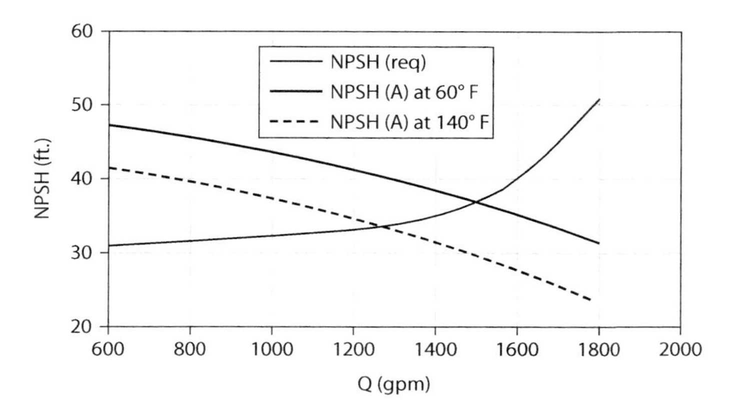

FIGURE 5.16
Variation of $NPSH_A$ and $NPSH_R$ with flow rate.

Q (gpm)	0	60	115	180	250	300	340
h_p (ft.)	100	97	92	81	61	40	19
Eff (%)	0	33	55	75	83	81	77

a. Find the head, flow rate, and efficiency at the operating point.
b. At the best efficiency, what would the flow and head be for a single pump? Under these conditions, what would the speed of the pump be?
c. Find the head and flow rate when two such pumps are connected in (i) parallel and (ii) series.
d. If the pump is now operating at 1500 rpm, what would the head and flow rate be for a single pump, two pumps in parallel, and two pumps in series?
e. Two pumps as in part (a) are connected in parallel. One of them is slowed down enough so that it does not produce any flow. At what speed would this occur?

Solution: From the given data, the pump curve and efficiency curves can be drawn. These are shown in the figure. The system curve can be obtained by applying the energy equation between the free surfaces of the reservoir.

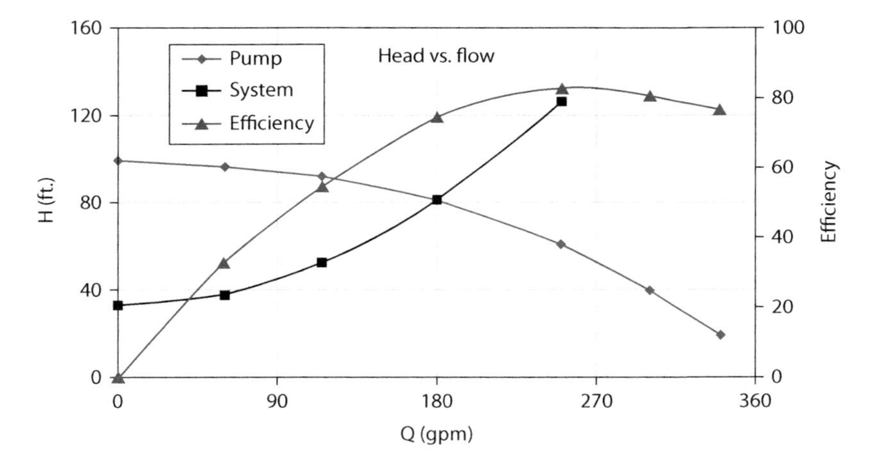

FIGURE 5.17
System and pump curves at 1750 rpm.

$$\frac{p_1}{\gamma} + z_1 + \frac{V_1^2}{2g} + h_p = \frac{p_2}{\gamma} + z_2 + \frac{V_2^2}{2g} + h_t + h_l \tag{a}$$

or

$$h_p = z_2 + h_l = 33 + 15Q^2 = h_s \tag{b}$$

The system curve given by $h_s = 33 + 15Q^2$ is plotted on the same figure.

a. At the operating point, the flow and head are 180 gpm and 80 ft., respectively. The efficiency at this point is approximately 73%. Since the peak efficiency is approximately 84%, the operating point needs to be changed so that peak efficiency is obtained (Figure 5.17).
b. From the figure, the operating point at 1500 rpm is approximately $H = 60$ ft. and $Q = 140$ gpm. This corresponds to an efficiency of 65% (point A on the efficiency graph). To obtain the curves at 1500 rpm, scaling laws have been used. Accordingly, the head is scaled as

$$\left(\frac{gH}{N^2D^2}\right)_{1500} = \left(\frac{gH}{N^2D^2}\right)_{1750}$$

or (d)

$$H_{1500} = H_{1750}\left(\frac{1500}{1750}\right)^2 = H_{1750}\frac{36}{49} = 0.7347H_{1750}$$

Similarly, the flow rate is scaled as

$$\left(\frac{Q}{ND^3}\right)_{1500} = \left(\frac{Q}{ND^3}\right)_{1750}$$

or (d)

$$Q_{1500} = Q_{1750}\left(\frac{1500}{1750}\right) = Q_{1750}\frac{6}{7} = 0.8571Q_{1750}$$

At the peak efficiency (which, from the graph, is approximately 83%), the flow and head are 250 gpm and 126 ft., respectively. To have the new operating point correspond to maximum efficiency, the pump curve needs to be shifted up. Since the scaling of head and scaling of flow rate lead to different speeds ($H \sim N^2$, whereas $Q \sim N$), the speed corresponding to maximum efficiency has to be determined approximately. Accordingly, the pump curve corresponding to 2200 rpm is shown in the figure. The operating point is almost at the maximum efficiency but is slightly lower. If the pump were operating at 2250 rpm, the operating point would be at the highest efficiency (Figure 5.18).

c. For two pumps connected in series, the head is doubled for the same flow. When two pumps are connected in parallel, the flow is doubled while the head is the same. These are shown for 1750 rpm in the figure. When the pumps are connected in parallel, the head and flow rate are 95 ft. and 205 gpm, respectively. When in series, the corresponding values are 125 ft. and 250 gpm.
d. Similar results can be obtained when the speed is changed to 1500 rpm. When the pumps are in series, H and Q are 95 ft. and 210 gpm; when in parallel, H and Q are 65 ft. and 160 gpm. The plot corresponding to 1500 rpm has not been shown in order not to clutter the figure (Figure 5.19).

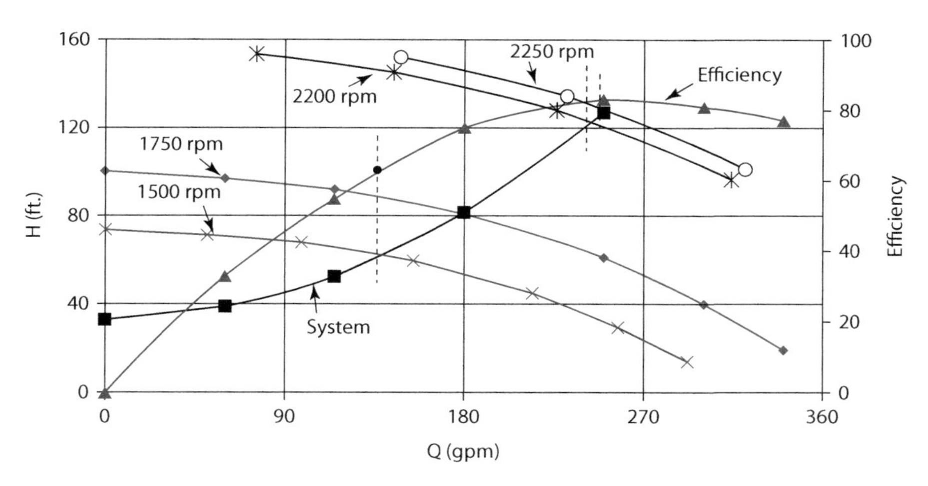

FIGURE 5.18
Operating points at various pump speeds.

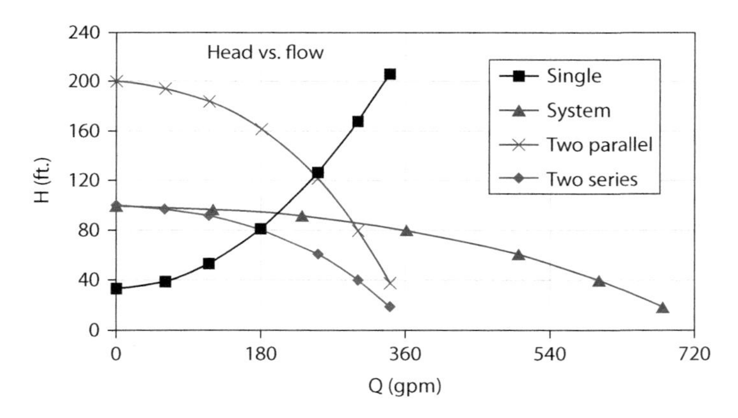

FIGURE 5.19
Two pumps at 1750 rpm in series and parallel.

e. At the operating point for two pumps in parallel at 1750 rpm, the head is 94 ft. For the pump that has the reduced speed, this needs to be the shut-off head. For a single pump, the shut-off head is 100 ft. The speed corresponding to the new shut-off head can be calculated from the pump similarity laws.

$$\frac{H_1}{H_2} = \frac{N_1^2}{N_2^2} \Rightarrow N_2 = N_1\sqrt{\frac{H_2}{H_1}} \Rightarrow N_2 = 1750\sqrt{\frac{94}{100}} = 1697\,\text{rpm}$$

Comments: The example shows the effect of variation of speed, pumps connected in series, and pumps connected in parallel on the head and flow rates.

Example 5.4

For a particular pump, the flow rate required is 70,000 gpm at a head of 225 ft. If the rotation is 600 rpm, calculate the principal dimensions of such a pump.

Solution: The first step is to calculate the specific speed:

$$Q = 70{,}000\,\text{gpm} = \frac{70000}{448.83} = 156\,\text{cfs};$$

$$N = 600\,\text{rpm} = \frac{2\pi(600)}{60} = 62.8\,\text{rad/s} \tag{a}$$

$$N_s = \frac{N\sqrt{Q}}{(gh)^{0.75}} = \frac{(62.8)\sqrt{156}}{((32.2)(225))^{0.75}} = 1.0$$

This corresponds to a centrifugal pump. From Figure 5.14, the following values are obtained for the various design variables:

$$\phi = 1.25; \quad \frac{D_0}{D} = 1.05; \quad \frac{B}{D_0} = 0.14; \quad \frac{V_{m2}}{u_2} = 0.145; \tag{b}$$

The remaining variables can now be computed. Thus,

$$\phi = \frac{u_2}{\sqrt{2gH}} = 1.25 \Rightarrow u_2 = (1.25)\sqrt{2(32.2)(225)} = 150\,\text{ft./s}$$

$$D = \frac{2u_2}{N} = \frac{2(150)}{62.8} = 4.77\,\text{ft.}; \qquad V_{m2} = (0.145)(150) = 21.75\,\text{ft./s}$$

$$D_0 = (1.05)(4.77) = 5.02\,\text{ft.}; \qquad B = (0.14)(5.02) = 0.7\,\text{ft.}$$

The final step in the design process is to verify that the dimensions given produce an adequate flow rate.

$$Q = \pi D_0 B V_{m2} = \pi(5.02)(0.7)(21.75) = 239\,\text{cfs}$$

This value is significantly higher than the required value of 156 cfs. Several choices are now available. The easiest would be to reduce the blade height B to reduce the available flow. If the blade height is reduced to, say, 0.5 ft., the flow rate would be

$$Q = \pi(5.02)(0.5)(21.75) = 170.8\,\text{cfs}$$

This is much closer to the required value. Hence, an appropriate solution would be to set the blade height to 0.5 ft. and to introduce a flow control valve on the discharge side of the pump.

Comment: The calculated dimensions should be adjusted to the nearest whole numbers. Thus, a centrifugal pump with an impeller diameter of 60 in. and blade height of 6 in. should be adequate.

PROBLEMS

5.1 Design a pump that handles 2000 gpm and produces a head of 200 ft. when running at 1800 rpm.

5.2 Design a pump that can handle 10,000 gpm at a head of 30 ft. when running at 1000 rpm.

5.3 A multistage pump is to be considered for pumping 1000 gpm of water with a head of 450 ft. The shaft rotation is restricted to 1500 rpm. Due to space limitations, the diameter is not to exceed 12 in. How many stages would be required for this application?

Ans: Five stages

5.4 The following data refers to a centrifugal pump:

Inlet diameter = 3 in.

Outlet diameter = 7 in.

Blade angles at inlet and outlet = 30°

Blade width at inlet = 2 in.

Blade width at outlet = 1 in.

If the flow rate is 900 gpm, find the (a) rotational speed, (b) head, (c) power, and (d) pressure rise in the impeller. Assume $\alpha_1 = 90°$.

Ans: N = 2026 rpm; H = 75.3 ft; Δ_p = 22.73 psi

5.5 The impeller of a centrifugal pump is 5 cm and 2 cm wide at inlet and outlet, respectively. The inlet and outlet radii are 10 cm and 25 cm, respectively. The blade angles at the inlet (β_1) and outlet (β_2) are 20° and 10°, respectively. For a rotation of 1750 rpm, find the (a) discharge, (b) power, and (c) pressure rise through the impeller. Assume no whirl velocity at the inlet.

Ans: 0.209 m^3/s; 77.3 kW; 337 kPa

5.6 At the best efficiency point for a centrifugal pump, the head and flow rate are 100 ft. and 500 gpm, respectively. The blades are equiangular, that is, $\beta_1 = \beta_2$. The inlet and outlet blade widths are 1 in. and 0.5 in., while the inlet and outlet radii are 2 in. and 6 in., respectively. For such a pump, find the blade angles and pump speed. Assume $\alpha_1 = 90°$.

Ans: 30.8°; 1228 rpm

5.7 Although air is compressible, at low speeds it can be treated as an incompressible fluid. Hence, blowers that handle air work on the same principles as pumps. One such blower has inside and outside radii of 22 cm and 25 cm, respectively, and a width of 45 cm. It handles 2 m^3/s of air at 25°C. If $\alpha_1 = 90°$, $\alpha_2 = 15°$, and it rotates at 1500 rpm, find the (a) blade angles, (b) pressure rise, and (c) theoretical power required.

Ans: 5.62°; 0.42 kPa; 0.97 kW

5.8 An air blower is designed to operate at 3600 rpm at 70°F. The impeller width is 40 in. and the blades are equiangular, that is $\beta_1 = \beta_2 = 30°$. What flow rate and pressure rise can be expected if the inner and outer radii are 10.5 in. and 12 in., respectively? Assume no whirl velocity at the inlet.

Ans: $\Delta_p = 0.543$ psia

5.9 Diffuser vanes are sometimes used in centrifugal pumps to increase the static pressure at discharge. A few such vanes are shown in the following figure. If the flow rate is 0.1 m^3/s and the height of blades is 2 cm, find the increase in static pressure. The inside and outside radii are 15 cm and 21 cm, respectively.

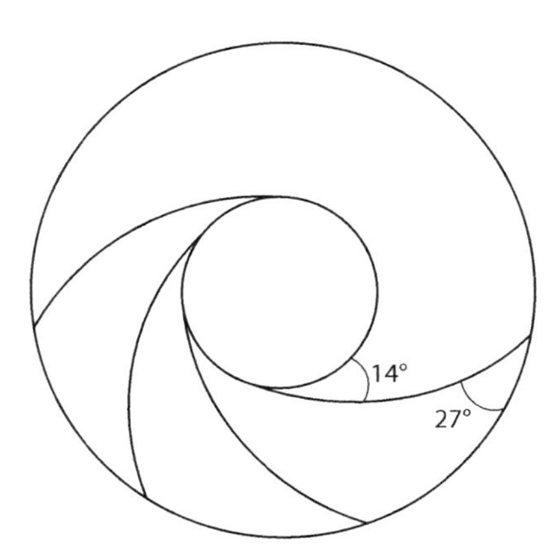

5.10 A pump operating with a head of 30 m has a critical cavitation parameter of 0.25. The local atmospheric pressure is 101 kPa and the local temperature is 35°C. If the inlet losses are 0.4 m, what should the pump setting be to avoid cavitation (where should the pump be placed in relation to the surface of the reservoir)?

Ans: 1.88 m above the water surface

5.11 A pump that is used to pump 70°F water from a reservoir has a specific speed of 1.4. If the head on the pump is 60 ft., what is the NPSH on this pump? Also, if the head loss in the inlet pipe is 2 ft., what should the safe setting of the pump be to avoid cavitation?

Ans: 8.34 ft. above the water

5.12 It is proposed to operate the pump in Problem 11 at a location where the elevation is 5000 ft. and the local temperature variation is from 40°F in winter to 100°F in summer. What should the pump settings be in winter and summer? If the pump is to operate year-round, what should the safe setting be?

Ans: 3.12 ft. in winter and 1.40 ft. in summer.

5.13 At its best efficiency point (design point), a pump has the following features: speed = 1800 rpm, head = 119 ft., and flow rate = 1600 gpm. It is operating at a location where the local temperature is 70°F. If friction losses are neglected and the suction pipe diameter is 10 in., what is the suction pressure at the pump inlet for incipient cavitation?

Ans: $p_s = 7.8$ psia; answer depends on σ_c from graph, could be 8–11 psia

5.14 Two reservoirs, A and B, are connected by a pipe that is 10,000 ft. in length and has a diameter of 2 ft. For simplicity, the friction factor can be taken to be 0.024 and other losses can be neglected. The elevation difference between the reservoirs is 40 ft and water at 60°F is flowing due to gravity from the upper to the lower reservoir. (a) What flow would be expected for these conditions? (b) A pump operating at 1750 rpm is available for use to double the flow rate. What would the specific speed be under these conditions? (c) What would the approximate impeller diameter be for this pump and (d) what type of pump would it be?

Ans: (a) $Q = 14.6$ cfs; (b) $N_s = 2.02$; (c) $D = 1.6$ ft.; $D_0 = 1.9$ ft.; mixed flow

5.15 The flow and head requirements for a particular application are 20 cfs and 100 ft., respectively. Two identical pumps with a rated speed of 1750 rpm are an option.

What would be the specific speed and the impeller diameters if the pumps are (a) connected in series and (b) if they are connected in parallel.

Ans: (a) $N_s = 3.22$ (b) $N_s = 3.22$, $D = 1.22$ ft.

5.16 What type of a pump would be most suited to delivering 30 cfs against a head of 300 ft. when rotating at 480 rpm? What would the impeller diameter be?

Ans: Centrifugal; 53 in.

5.17 What type of a pump would be most suited to delivering 300 cfs against a head of 30 ft. when rotating at 480 rpm? What would the impeller diameter be?

Ans: Axial; 48 in.

5.18 The following are the operating characteristics of a pump when operating at 1800 rpm:

Q (gpm)	H (ft.)	η (%)
0	78	0
400	75	30
800	70	53
1200	64	73
1600	55	85
2000	44	86
2400	30	70

This pump is installed in a piping loop that consists of a total length of 200 ft. The impeller diameter is 8 in. The minor losses to be considered are due to the entrance, exit, and an elbow (K = 1.5). For simplicity, assume that the friction factor is constant at 0.02. The system connects two reservoirs with a difference in elevation of 10 ft. and the fluid is being pumped up. (a) At the operating point, what are the head and flow rate? (b) What is the efficiency at the operating point? (c) In your opinion, would this be a good choice of pump? (d) To get the best efficiency, should be pump be running faster or slower?

Ans: At operating point, $\eta = 73\%$, $H = 35$ ft.; $Q = 2250$ gpm $= 5.01$ cfs

5.19 Consider the piping system shown in the following diagram. The pump operates at 1750 rpm.

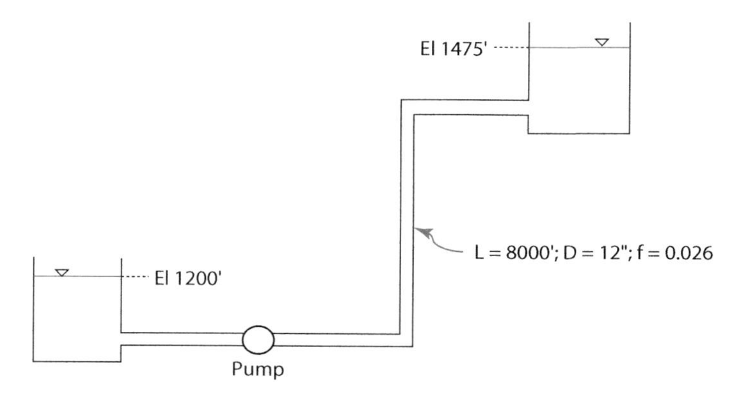

The pump curve is given by $h_p = 350 - 24Q^2$ where h_p is given in ft. and Q in cfs. What would the head and flow rate delivered be under these conditions?

Ans: $Q = 160$ cfs; $h_p = 288$ ft

5.20 Solve the previous problem for two identical pumps connected in (a) series and (b) parallel.

Ans: a) $Q = 2.83$ cfs, $h_p = 315$ ft; b) $Q = 2.58$ cfs; $h_p = 310$ ft

5.21 If the pump in problem 19 is slowed down to 1500 rpm, what would the head and flow rate be for (a) a single pump, (b) two pumps in series, and (c) two pumps in parallel? (d) If one of the pumps is slowed down until it produces no flow, at what speed would this occur?

Ans: a) $Q = 2.12$ cfs, $h_p = 300$ ft

5.22 Consider the problem of cavitation of highly volatile fluids. A certain pump rotating at 1500 rpm delivers 0.04 m^3/s of gasoline from an open tank. The specific gravity of gasoline is 0.82, and its vapor pressure is 32 kPa at this particular location. The suction pipe ($f = 0.022$) is 50 m long and has a diameter of 10 cm. If the gasoline is being delivered at a head of 30 m, what should the pump setting be to avoid cavitation? Include only the friction losses in the suction pipe.

Ans: Setting of −27.03 m

5.23 Obtain the approximate dimensions of the Rocky River pump installed by Worthington Corporation. It delivers 279.5 cfs of water at 238.84 ft. of head when running at 327 rpm.

Ans: Dimensions as they exist; $D = 90$ in., $B = 9$ in.

5.24 Obtain the principal dimensions of the pumps at the Grand Coulee plant with the following specifications: $Q = 1250$ cfs; $H = 344$ ft.; $N = 200$ rpm.

Ans: Dimensions as they exist; $D = 167.75$ in., $B = 19.5$ in.

5.25 The Colorado River Aqueduct, which delivers water to the Los Angeles area, has pumps in the Hayfield plant to deliver 200 cfs at 444 ft. when running at 450 rpm. Estimate the principal dimensions of these pumps.

5.26 Consider two pumps connected in parallel when the combined flow rate between them is 3000 gpm. Their respective characteristics are $H = 40 - 0.8Q^2$ and $H = 35 - 0.6Q^2$ (where Q is in 1000 gpm and H is in ft.). Calculate the discharge and head in each pump.

5.27 The following test results were obtained for a centrifugal pump with a 16 in. diameter impeller running at 1200 rpm:

H (ft.)	40	37	31	20	5
Q (gpm)	0	1	2	3	4
efficiency, η (%)	0	50	75	81	71

a. Using curve fitting (exact polynomial fit), obtain a fourth-degree curve for a H–Q curve and an η–Q curve. Plot them on a graph.

b. What is the peak efficiency of the pump? What are the values of Q and H at the peak efficiency?

c. Working under the assumption that the same pump were operated at 1800 rpm, obtain the H–Q data and graph them.

d. If a geometrically similar pump of 20 in. diameter is to be tested, what H–Q data would be obtained? Plot the results.

5.28 In a certain application, a pump is required to deliver 1000 gpm against a head of 300 ft. when running at 1800 rpm. Due to space limitations, the pump impeller is to be less than 10 in. Estimate the number of stages required under these circumstances.

Ans: Four stages

5.29 Qualitatively explain the shifting of performance curves in Figure 5.8 c, d, and e. These curves represent the effect of increase in diameter, speed, and viscosity of the fluids.

6

Fans and Blowers

Although fans and blowers handle gases, the pressure rise is quite small, and hence they can be treated using the same principles as pumps. They are primarily used for the movement of large volumes of gases with minimal pressure rise.

Upon completion of this chapter, the student will be able to:

- Understand the similarities between pumps and blowers
- Understand the features of different types fans and blowers
- Analyze and perform the preliminary design of blowers and fans

6.1 Introduction

The primary use of a fan is to move large volumes of air or gases with a very small pressure rise, usually a few inches of water. On the other hand, compressors, while also handling gases, encounter large enough pressure rises that there is a measurable increase in density. Another name used for a fan in the turbomachinery literature is a *blower.* It delivers gases with a slightly higher pressure rise than regular fans. In contrast to fans and blowers, a compressor develops a higher pressure rise that is expressed in terms of pressure ratio.

Fans can be classified as either axial, centrifugal, or mixed flow types. Most of the low-pressure fans are of the axial type whereas most of the higher-pressure type are centrifugal. As in the case of pumps, centrifugal fans produce relatively higher pressures per stage at lower flow rates, while axial fans and blowers deal with higher flow rates and lower pressures per stage. Typical examples of axial flow fans are ceiling, table, or ventilation fans. Induced draft or high draft fans used in mines and industrial furnaces are also called blowers. They are used in several industrial applications such as cooling towers, the cooling of motors, generators, and internal combustion engines, air conditioning systems, and mine ventilation. The major need in all these applications is to move large volumes of gas, a requirement to which axial fans are most suited.

The operational principles of fans and pumps are similar although the pressure rise in fans is orders of magnitude smaller than in pumps. Another difference relates to the measurement of flow rates in fans. Measurement of flow rates is more difficult for gases and vapors than for liquids. Since the pressure rise produced is small, it is difficult to use obstruction devices such as orifice meters, venturimeters, or flow nozzles. Hence, it is common practice to use the so-called pitot traverse in the delivery duct to calculate flow rates. Since the pressure rises are small across fans, the exit velocity, and hence the dynamic pressure at the exit $\left(=\frac{1}{2}\rho V_e^2\right)$ can be a significant fraction of the pressure rise. Hence, it is necessary to clarify the basis on which pressure measurements are made and plotted against flow rates. Another feature of interest is that the efficiency of fans can be expressed using either static pressure rise or stagnation pressure rises. These differences are either minimal or non-existent in pumps due to high pressure rises.

6.2 Axial Fans

A simplified version of an axial fan consists of a rotor on which several blades are mounted and connected to a hub. It is run by an electric motor, which increases the stagnation pressure across the rotor, which is mounted in a cylindrical casing. Flow is received through a converging passage and discharged through a diffuser. A schematic of a single stage of such a fan is shown in Figure 6.1.

Unlike compressors, which use well-designed airfoil sections for the blades, it is not uncommon to used curved sheet metal blades in axial fans. The number of blades varies quite widely. It can be as low as two and as high as fifty. Some of the parameters that are used in the analysis of fans include work, pressure rise, pressure coefficient, and degree of reaction, as they pertain to a single stage. These are discussed in this chapter. It is quite

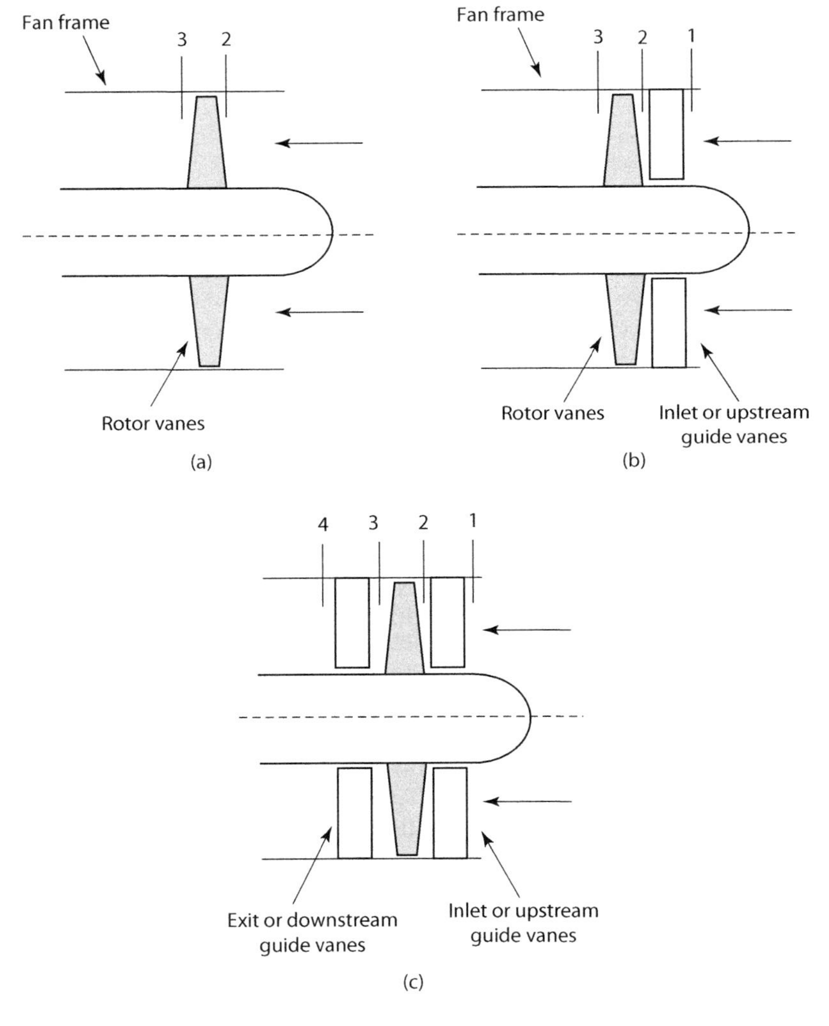

FIGURE 6.1
Schematic of an axial flow fan (a) without guide vanes, (b) with upstream guide vanes, (c) with upstream and downstream guide vanes.

usual to have some form of upstream and downstream guide vanes. These, along with the numbering scheme, are shown in Figure 6.1. From Euler's equation, the stage work for a fan without guide vanes (Figure 6.1a) is given by

$$h_2 + \frac{V_2^2}{2} = h_3 + \frac{V_3^2}{2} + E \Rightarrow h_{02} = h_{03} + E$$

$$E = h_{02} - h_{03} \Rightarrow E = \Delta h_0 = \frac{\Delta p_0}{\rho} \tag{6.1}$$

$$E = u\left(V_{u2} - V_{u3}\right)$$

Since the pressure rise is quite small, the flow will be isothermal and, hence, the static enthalpy rise would be equal to the stage energy change, as shown in Equation 6.1. The mass flow rate and power are given by

$$\dot{m} = \rho A V_a = \rho \frac{\pi}{4}\left(D_t^2 - D_h^2\right) V_a$$

$$P = \dot{m} E = \dot{m}(\Delta h_0) = \dot{m}(C_p \Delta T_0) = \dot{m} u\left(V_{u2} - V_{u3}\right) \tag{6.2}$$

where:
D_t is the tip diameter
D_h is the hub diameter

The stage reaction is defined as the ratio of the pressure rise in the rotor to the pressure rise in the entire stage. Thus,

$$R = \frac{(\Delta p)_{rotor}}{(\Delta p)_{stage}} \tag{6.3}$$

The degree of reaction for fans can vary from zero to values higher than one. The pressure coefficient is defined as

$$C_p = \frac{(\Delta p)_{stage}}{\frac{1}{2}\rho u^2} \tag{6.4}$$

The pressure coefficient can be defined either in terms of the static or stagnation pressure rise across the stage. Two other variables that are of relevance in the context of axial fans are the flow coefficient and blade loading coefficient. These have been introduced in Chapter 2, but are modified slightly in the context of axial fans. Thus,

$$\phi = \frac{V_a}{u}; \qquad \psi = \frac{E}{u^2} \tag{6.4a}$$

These are discussed in more detail in the chapters on axial flow gas turbines and compressors as they are used extensively in their discussion.

There are some situations, especially with inlet guide vanes, when the degree of reaction for fans can be greater than unity. This concept is illustrated in the following example.

Example 6.1

Consider an axial flow fan stage with upstream guide vanes, as shown in Figure 6.1b. The blade velocity is 200 m/s and the inlet blade angle is −60°. The exit blade angle is −45° and the fluid leaves the rotor axially. Assume that the axial velocity is a constant and the stage is a repeating stage. Calculate the degree of reaction.

Solution: The velocity triangles are shown.

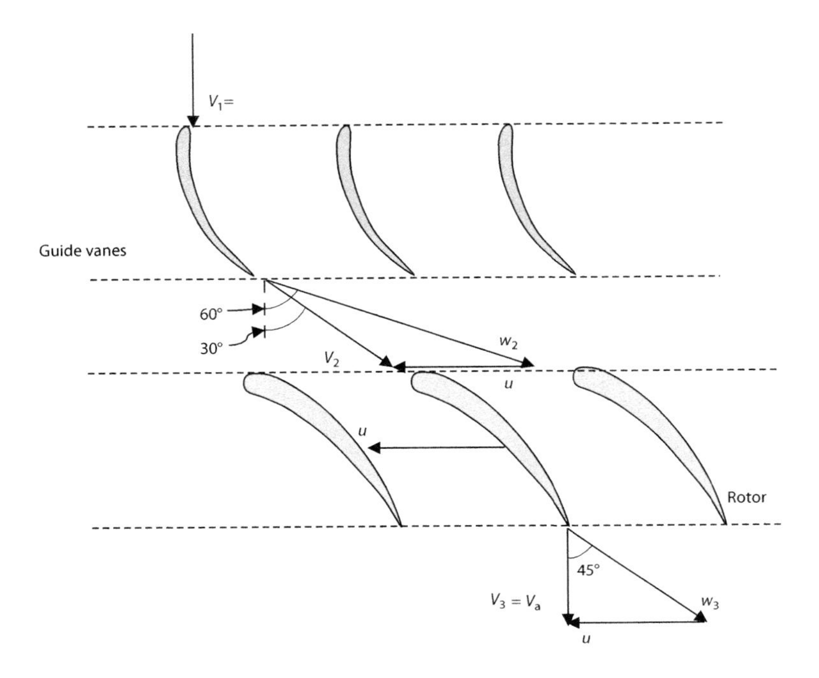

From the exit velocity triangle, the following can be written:

$$V_3 = V_a = \frac{u}{\tan\beta_3} = \frac{200}{\tan 45} = 200\,\text{m/s}$$

$$w_3 = \left(V_3^2 + u^2\right)^{0.5} = 282.8\,\text{m/s}$$

$$w_2 = \frac{V_a}{\cos\beta_2} = \frac{200}{\cos 60} = 400\,\text{m/s}$$

$$w_{u2} = w_2 \sin\beta_2 = 400 \sin 60 = 346.4\,\text{m/s}$$

$$V_{u2} = w_{u2} - u = 346.4 - 200 = 146.4\,\text{m/s}$$

$$V_2 = \left(V_{u2}^2 + V_a^2\right)^{0.5} = 247.9\,\text{m/s}$$

With the details of the velocity triangles, the degree of reaction can be used to calculate the numerical value. For this, the energy equation is applied across the rotor and the stage. Thus, across the rotor,

$$h_2 + \frac{V_2^2}{2} = h_3 + \frac{V_3^2}{2} + E = h_3 + \frac{V_3^2}{2} + \frac{V_2^2 - V_3^2}{2} + \frac{w_3^2 - w_2^2}{2}$$

$$\text{or } h_2 = h_3 + \frac{w_3^2 - w_2^2}{2} \quad \Rightarrow \quad h_2 - h_3 = \frac{w_3^2 - w_2^2}{2} = -40000\ \text{J/kg}$$

$$\text{or } |\Delta h|_{stat} = 40000\ \text{J/kg}$$

Across the entire stage,

$$h_1 + \frac{V_1^2}{2} = h_3 + \frac{V_3^2}{2} + E \quad \Rightarrow \quad h_{01} = h_{03} + E \quad \Rightarrow \quad h_{02} = h_{03} + E$$

$$h_{01} - h_{03} = |\Delta h_0|_{stage} = |E| = u(V_{u2} - V_{u3}) = uV_{u2} = (200)(146.4) = 29281\ \text{J/kg}$$

The degree of reaction is given by

$$R = \frac{|\Delta h|_{stat}}{|\Delta h_0|_{stage}} = \frac{40000}{29281} = 1.37$$

Comments: The degree of reaction is greater than one. Instead of using enthalpies, the degree of reaction can also be calculated in terms of the pressure rises or kinematical quantities such as velocities.

Example 6.2

An axial flow fan stage with both upstream and downstream guide vanes is shown in Figure 6.1c. For such a fan, if $V_1 = V_4$, and $\alpha_2 = \alpha_3$, show that the degree of reaction is unity. Also, show that the stage pressure rise is given by $(\Delta p)_{st} = 2\rho u^2(\varphi \tan\beta_2 - 1)$ and $\psi = 2(\varphi \tan\beta_2 - 1)$ where ϕ and ψ are the flow coefficient and blade loading coefficient, respectively.

Solution: The velocity triangles are shown.

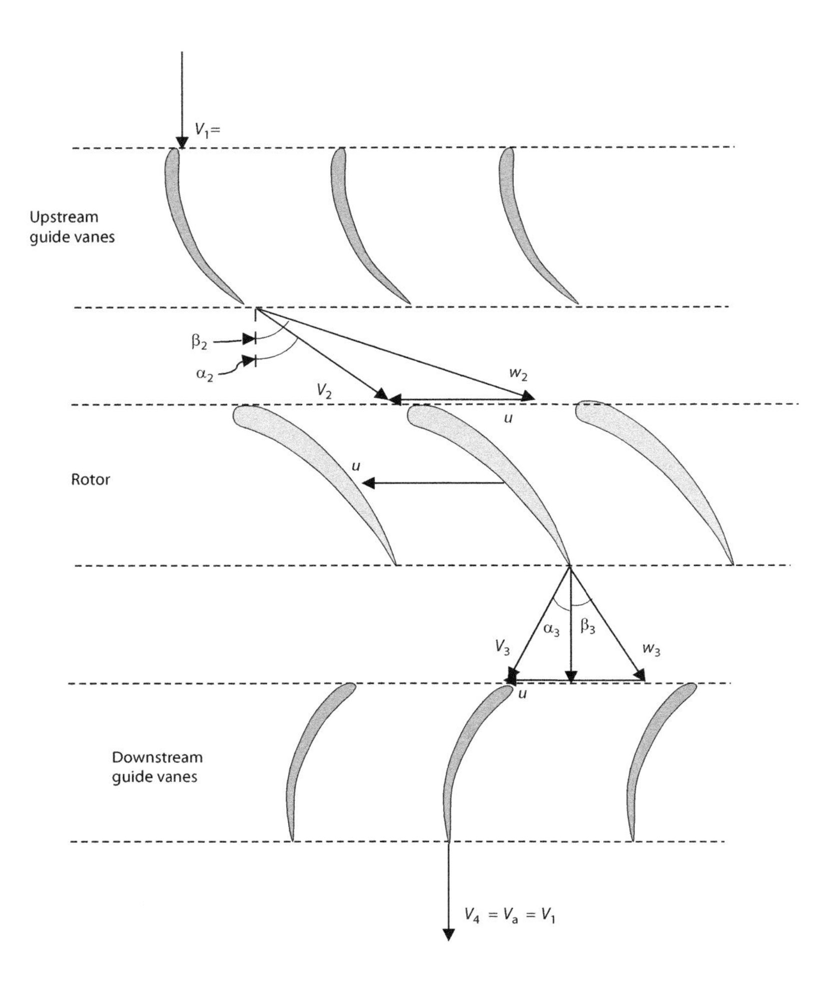

Since $\alpha_2 = \alpha_3$ ($V_{u2} = V_{u3}$) and axial velocity is constant, the following relationships can be written from the velocity triangles:

$$V_{u2} = w_{u2} - u = V_a \tan\beta_2 - u = u\left(\frac{V_a}{u}\tan\beta_2 - 1\right) = u(\phi\tan\beta_2 - 1) = V_{u3}$$

Euler's equation, applied across the rotor, gives

$$E = u(V_{u2} + V_{u3}) = 2uV_{u2} = (2u)\left[u(\phi\tan\beta_2 - 1)\right] = 2u^2(\phi\tan\beta_2 - 1)$$

$$\psi = \frac{E}{u^2} = 2(\phi\tan\beta_2 - 1)$$

Energy equations, applied across the upstream guide vanes, rotor, and downstream guide vanes, will respectively look like

$$h_1 + \frac{V_1^2}{2} = h_2 + \frac{V_2^2}{2} \Rightarrow h_{01} = h_{02}$$

$$h_2 + \frac{V_2^2}{2} = h_3 + \frac{V_3^2}{2} + E \Rightarrow h_{02} = h_{03} + E$$

$$h_3 + \frac{V_3^2}{2} = h_4 + \frac{V_4^2}{2} \Rightarrow h_{03} = h_{04}$$

After simplification (since $V_1 = V_4$),

$$h_1 + \frac{V_1^2}{2} = h_4 + \frac{V_4^2}{2} + E \Rightarrow h_1 = h_4 + E \Rightarrow h_1 - h_4 = E$$

$$\text{or } \Delta h = E \Rightarrow \frac{\Delta p}{\rho} = E \Rightarrow (\Delta p)_{stage} = 2\rho u^2(\phi \tan\beta_2 - 1)$$

Also,

$$h_1 + \frac{V_1^2}{2} = h_4 + \frac{V_4^2}{2} + E \Rightarrow h_{01} = h_{04} + E \Rightarrow (\Delta h)_0 = E$$

Thus, the static and stagnation enthalpies are the same; hence, the degree of reaction is unity.

Comments: The same result would be obtained by writing Euler's equation in terms of velocities.

Typical performance curves for an axial flow fan are shown in Figure 6.2. These differ markedly from a typical centrifugal fan and will be discussed later in the chapter. The power curve exhibits significant differences as it decreases continuously with increasing flow. Hence, it is usually not possible to overload the drive for an axial flow fan. The primary deficiency of the axial fan is the changing slope of the pressure characteristic in certain ranges of the flow rates. This could result in surging and accompanying vibration

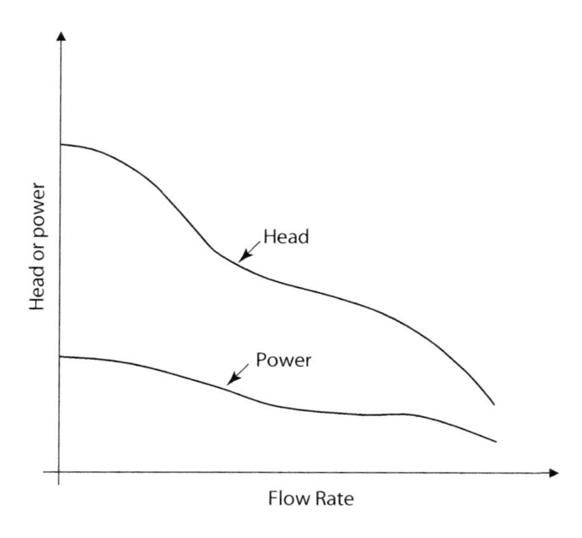

FIGURE 6.2
Typical characteristics of an axial flow fan.

under certain flow conditions. Additionally, since such fans operate at higher speeds, they can be noisy.

In contrast to fans in which the increase in stagnation pressure is limited to a few inches of water, axial or centrifugal flow blowers have a relatively higher pressure rise and spin at higher speeds. But the pressure is still not large enough to consider effects of compressibility. Hence, the analysis methods that are used for pumps are applicable here also. Axial blowers can either be single stage, consisting of a set of guide vanes (which may not always be present) and rotor vanes, or could consist of multiple stages of stator and rotor vanes. These are used mostly in ventilation applications since the flow rates tend to be quite large. An axial flow blower is considered in the following example.

Example 6.3

An axial flow blower spins at 4000 rpm when handling air flow at 1.2 lbm/s. The tip and hub diameters are 8 in. and 5 in., respectively. The blower has no inlet guide vanes and the static pressure and temperature at the inlet are 14.696 psia and 75°F, respectively. The rotor blades turn the flow inward (towards the axial direction) by 25°. Calculate the blade angles, degree of reaction, and static and stagnation pressure rises through the rotor. Also, calculate the power input into the blower. The absolute velocity at the inlet is axial.

Solution: The inlet and exit states of the rotor are designated by 2 and 3, respectively. The velocity triangles are shown.

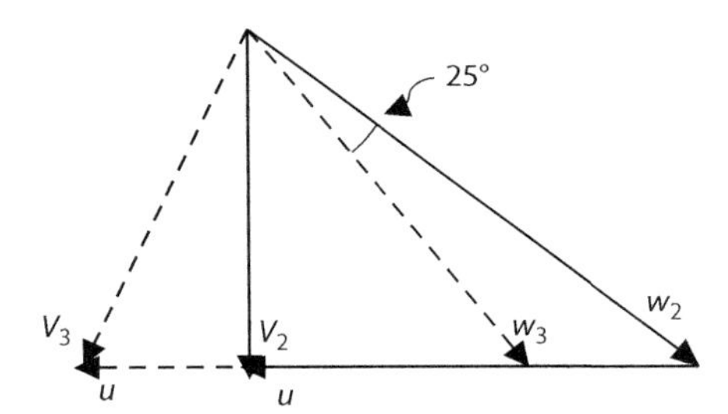

The annular area and the blade speed can be calculated as

$$A = \frac{\pi\left(d_t^2 - d_h^2\right)}{4} = 0.213\,\text{ft.}^2;\;\; d_m = \frac{\left(d_t + d_h\right)}{2} = 0.542\text{ ft.}$$

$$N = 4000\text{ rpm} = 418.8\text{ rad/s} \;\Rightarrow\; u = \left(418.8\right)\frac{\left(0.542\right)}{2} = 113.4\text{ ft./s}$$

$$\dot{m} = 1.2\text{ lbm/s} = \frac{1.2}{32.2} = 0.0373\text{ slugs/s}$$

Since the inlet pressure and temperature are known, the density can be calculated. Also, from the mass flow rate, the axial velocity can be obtained. Thus,

$$\rho_2 = \frac{p_2}{RT_2} = \frac{(14.696)(144)}{1716(75+460)} = 0.00231\,\text{slugs/ft.}^3$$

$$\dot{m} = \rho_2 V_a A \;\Rightarrow\; V_a = \frac{\dot{m}}{\rho_2 A} = \frac{0.0373}{(0.00231)(0.213)} = 76\,\text{ft./s} = V_2$$

From the velocity triangles,

$$w_2 = \sqrt{u^2 + V_2^2} = 136.6\,\text{ft./s}; \quad \beta_2 = \arctan\left(\frac{u}{V_a}\right) = \arctan\left(\frac{113.4}{76}\right) = 56.2°$$

Since the rotor turns the relative velocity by 25° inward, $\beta_3 = 56.2° - 25° = 31.2°$. From the exit velocity triangle,

$$w_{u3} = V_a \tan(56.2°) = 46\,\text{ft./s} \quad \Rightarrow \quad w_3 = \sqrt{w_{u3}^2 + V_a^2} = 88.8\,\text{ft./s}$$

$$V_{u3} = u - w_{u3} = 67.45\,\text{ft./s}; \quad V_3 = \sqrt{V_{u3}^2 + V_a^2} = 101.6\,\text{ft./s}$$

The energy transfer and degree of reaction can be calculated as

$$E = u_2 V_{u2} - u_3 V_{u3} = u\left(V_{u2} - V_{u3}\right) = -u V_{u3} = -7652\,\text{ft.}^2/\text{s}^2$$

$$R = \frac{(w_3^2 - w_2^2)/2}{E} = \frac{-5377}{-7652} = 0.70$$

For incompressible fluids, enthalpy depends on pressure and not on temperature. Thus,

$$\Delta h_0 = \frac{\Delta p_0}{\rho} \quad \Rightarrow \Delta p_0 = \rho\,\Delta h_0 = (0.00231)(7652.) = 17.63\,\text{psf}$$

$$\Delta h = \frac{\Delta p}{\rho} \quad \Rightarrow \Delta p = \rho\,\Delta h = (0.00231)(5377..) = 12.40\,\text{psf}$$

$$\text{P} = \dot{m}\text{E} = (0.0372)(7652.) = 0.657\,\text{ft}\cdot\text{lb/s}$$

Comments: The pressure rise is very small (it will be in inches of water when converted), as expected. The degree of reaction can be calculated from the velocities as explained in Chapter 3 or as the ratio of static and stagnation pressure rises. Both would yield the same answer.

6.3 Centrifugal Fans

Applications requiring higher pressures than possible with axial fans use centrifugal fans. Their shapes and principles of operation are quite similar to those for centrifugal pumps. A typical centrifugal fan is shown in Figure 6.3. It consists of an impeller that is connected to an electric motor. The impeller on which blades (radial, backward-, or forward-curved) are mounted is surrounded by a volute casing in which the gas collects circumferentially and is delivered to the outlet. The blades are fabricated by welding curved or straight sheet metal to the side walls of the rotor, which makes fan manufacture relatively simple and inexpensive. The inlet to the fan is near the central portion, which is called the *eye*.

Analogous to centrifugal pumps, centrifugal fans can also have backward-curved, radial, or forward-curved blades. However, unlike pumps, wherein backward-curved blades are used almost exclusively, all three types are used in centrifugal fans. Typical shapes of such blades are shown in Figure 6.4. Backward-swept blades are used for lower-pressure applications and lower flow rates. They are most suitable for continuous operation. The width to diameter ratio

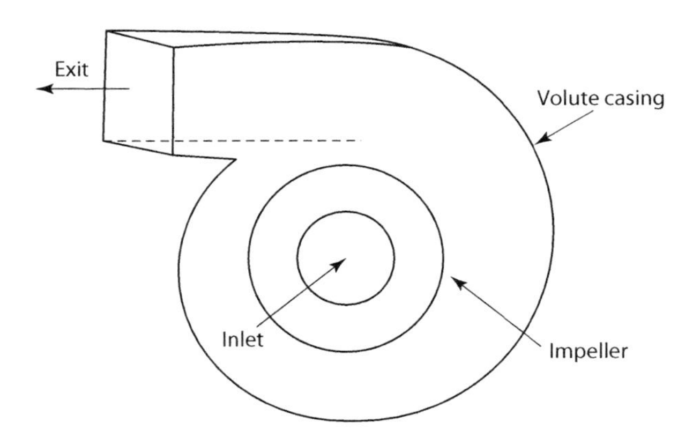

FIGURE 6.3
Schematic diagram of a centrifugal fan or blower.

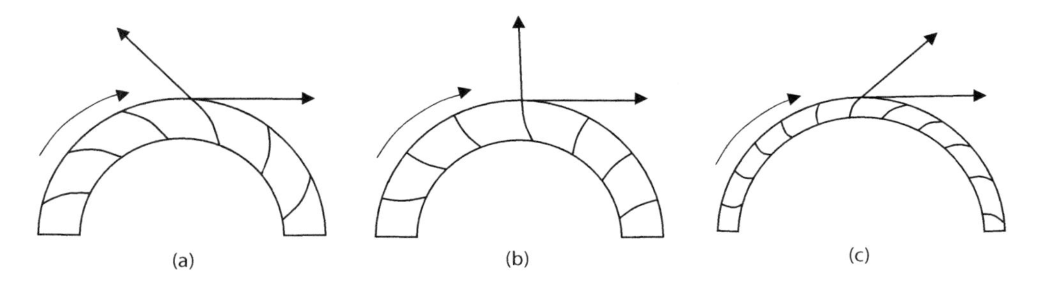

FIGURE 6.4
Blade shapes for centrifugal fans: (a) backward-curved, (b) radial, and (c) forward-curved.

of such fans is between 0.02 and 5, and the number of blades varies between 6 and 17 (see Yahya (2011)). Radial blades are less susceptible to blockage and dust erosion and hence most suitable in industrial ventilation, where air or gases are usually mixed with dust. Forward-curved blades have the highest number of blades of the three and are preferred when small size and initial cost of installation are important. Such fans have lower maintenance requirements. Also, forward-curved blades require lower tip speeds and hence produce less noise than the other two. These are used for heating and air conditioning applications.

As is the case in pumps, the pressure rise is approximately proportional to the absolute velocity at the exit of the rotor. Hence, the characteristic curves produced by these shapes differ from each other. Typical shapes for pressure rise and power requirements are shown in Figure 6.5. Backward-curved blades typically have power curves that reach a maximum and decrease at higher flow rates. Thus, if the fan drive is sized properly to correspond to peak power, overloading will not be a problem for such fans. For radial and forward-curved blades, the power increases continuously with flow and overloading can be an issue.

6.4 Fan Laws and Design Parameters

As has been noted already, fans and blowers produce low pressure rises and the fluid can be considered incompressible. For such flows, simple relationships exist between quantities such as speed, flow rate, pressure rise, and power. For geometrically similar

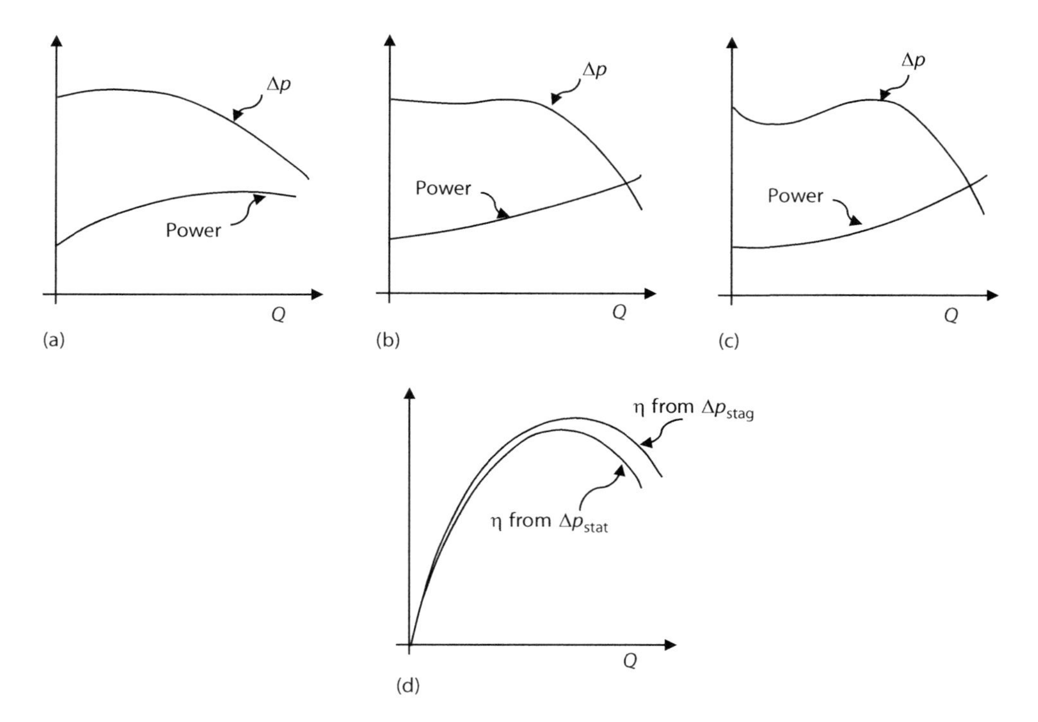

FIGURE 6.5
Performance curves for (a) backward-curved, (b) radial, and (c) forward-curved blades, and (d) efficiency of fans.

machines, these equations were derived in Chapter 2. The situation becomes somewhat simplified when the performance of the same machine (fan or blower) is considered at different speeds. The equations are then called *fan laws, affinity laws,* or *similarity laws*. They can be summarized as follows:

1. The volumetric flow rate Q is directly proportional to speed N, that is, $Q/N = \text{constant}$.
2. The stagnation pressure rise Δp_0 is proportional to the square of the speed N, that is, $\Delta p_0/N^2 = \text{constant}$.
3. The power input P is proportional to the cube of speed N, that is, $P/\rho N^3 = \text{constant}$.

   Since the pressure changes are small, density is constant, and hence the law becomes $P/N^3 = \text{constant}$.

It was mentioned in the chapters on hydraulic turbines and pumps that their designs have become quite routine because of their long existence and extremely high efficiencies. The same reasoning also applies to fans. They have been around for quite a while and are very efficient. Hence, their design has also become an art, and, as such, many experience-based empirical correlations are used. Some of the design features include the following:

1. The typical speed of rotation varies between 300 and 3600 rpm.
2. The diameter ratio influences the length of the blades. According to Eck (1973), the diameter ratio $d_1/d_2 = 1.2\phi^{1/3}$
3. The recommended value of blade width to diameter ratio, b_1/d_1, is 0.2 for forward-curved and radial blades; for backward tipped blades, $b_1/d_1 < 0.2$.

4. The blades, whether straight or curved, are usually made from sheet metal and are either welded or riveted to the impeller. The optimum blade angle at entry is about 35° (Yahya 2011).
5. There is a wide variation in the number of blades in centrifugal fans. Too few blades imply wasted power due to reduced flow delivery and too many blades imply higher friction losses. An empirical relation recommended by Eck (1973) is

$$n_B = \frac{8.5 \sin \beta_2}{1 - \dfrac{d_1}{d_2}} \tag{6.5}$$

There are other correlations available in the literature, the most famous of them being credited to Pfleirder, according to whom

$$n_B = 6.5 \left(\frac{d_2 + d_1}{d_2 - d_1} \right) \sin \frac{\beta_1 + \beta_2}{2} \tag{6.6}$$

Stepanoff (1965) suggested the following equation for number of blades:

$$n_B = \frac{\beta_2}{3} \tag{6.7}$$

Example 6.4

The following data refer to a centrifugal fan:

Speed	= 1200 rpm
Outer diameter	= 18 in.
Inner diameter	= 16 in.
Width of vanes	= 3 in.
Flow rate	= 110 ft.^{3}/s
Exit blade angle	= −65°
Power input	= 2.5 hp

The inlet flow can be assumed to be purely radial and the blade width is constant. Air can be assumed to be at atmospheric pressure and 72°F. Calculate the efficiency based on stagnation conditions and static conditions. Also estimate the fluid angle at the exit.

Solution: The inlet and exit velocity triangles are show in the following figure:

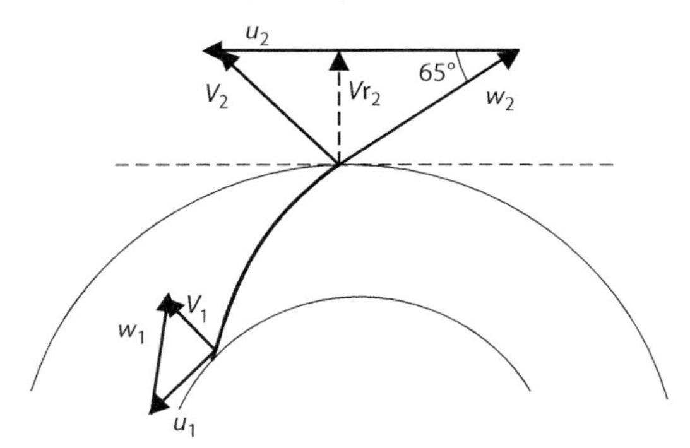

From the flow rate and the area, the radial velocity at the exit can be calculated. Thus,

$$A_2 = 2\pi r_2 b = 2\pi(1.5/2)(0.25) = 1.178\,\text{ft.}^2; \quad V_{r2} = \frac{Q}{A_2} = \frac{110}{1.178} = 93.4\,\text{ft./s}$$

$$N = 1200\,\text{rpm} = 125.7\,\text{rad/s} \Rightarrow u_2 = (125.7)(1.5/2) = 94.3\,\text{ft./s}$$

Since the pressure and temperature are known, the density can be calculated. Thus,

$$\rho_2 = \frac{p_2}{RT_2} = \frac{(14.7)(144)}{1716(72+460)} = 0.00232\,\text{slugs/ft.}^3$$

$$\dot{m} = \rho_2 Q = (0.00232)(93.4) = 0.255\,\text{slugs/ft.}^3$$

From the exit velocity triangle,

$$w_{u2} = \frac{V_{r2}}{\tan\beta_2} = \frac{93.4}{\tan 65} = 43.5\,\text{ft./s}; \quad V_{u2} = u_2 - w_{u2} = 94.2 - 43.5 = 50.7\,\text{ft./s}$$

$$w_2 = \sqrt{V_{r2}^2 + w_{u2}^2} = \sqrt{93.4^2 + 43.5^2} = 103.0\,\text{ft./s}$$

$$V_2 = \sqrt{V_{r2}^2 + V_{u2}^2} = \sqrt{93.4^2 + 50.7^2} = 106.3\,\text{ft./s}$$

$$\alpha_2 = \arctan\left(\frac{V_{r2}}{V_{u2}}\right) = \arctan\left(\frac{93.4}{50.7}\right) = 61.5°$$

The energy transfer and degree of reaction can be calculated as (incoming flow is radial)

$$E = u_1 V_{u1} - u_2 V_{u2} = -u_2 V_{u2} = -(94.2)(50.7) = -4779\,\text{ft.}^2/\text{s}^2 = \Delta h_0$$

The negative sign indicates that the device is power consuming. For incompressible fluids, enthalpy depends on pressure and not on temperature. Thus,

$$\Delta h_0 = \frac{\Delta p_0}{\rho} \Rightarrow \Delta p_0 = \rho\,\Delta h_0 = (0.00232)(-4779) = -11.1\,\text{psf}$$

Also, from the energy equation,

$$h_1 + \frac{V_1^2}{2} = h_2 + \frac{V_2^2}{2} + E \Rightarrow \Delta h = h_1 - h_2 = E + \left(\frac{V_2^2}{2} - \frac{V_1^2}{2}\right)$$

$$V_1 = V_{r1} = \frac{Q}{2\pi r_1 b_1} = \frac{110}{2\pi(1.33/2)(0.25)} = 10.5\,\text{ft./s}$$

$$\Delta h = -4779 + \left(\frac{106.3^2}{2} - \frac{105^2}{2}\right) = -4651\,\text{ft.}^2/\text{s}^2$$

$$\Delta h = \frac{\Delta p}{\rho} \Rightarrow \Delta p = \rho\,\Delta h = (0.00232)(-4651) = -10.79\,\text{psf}$$

Efficiency based on stagnation enthalpy can be calculated as

$$\eta\Big|_{stag} = \frac{\dot{m}\left|\Delta h_0\right|}{P} = \frac{(0.255)(4779)}{(2.5)(550)} = 88.8\%$$

Similarly, efficiency based on static enthalpy can be calculated as

$$\eta\big|_{stat} = \frac{\dot{m}|\Delta h|}{P} = \frac{(0.255)(4651)}{(2.5)(550)} = 86.3\%$$

Comments: As can be seen, the efficiency based on static enthalpy rise is lower than the efficiency based on stagnation enthalpy rise.

PROBLEMS

6.1 An axial flow fan rotates at 1000 rpm and the mean radius is 1.2 ft. The blade height is 8 in. and air is being drawn from the atmosphere. The axial velocity is a constant, and fluid exits the rotor axially. The fan is provided with upstream or inlet guide vanes and the flow at the inlet of these is purely axial. If the blade angles at the inlet and exit of the rotor are −54.5° and −45°, respectively, calculate the following: (a) degree of reaction, (b) pressure rise in the rotor, and (c) power input into the fan.

Ans: $R = 1.2$; $\Delta p_{\text{stat}} = 0.123$ psi

6.2 Consider an axial flow fan with a speed of 1000 rpm, with tip and hub diameters of 2 ft. and 1 ft., respectively. The flow coefficient is 0.3. The fan consists only of an enclosed rotor, without any guide vanes at the inlet or exit. The blade angle at the exit of the rotor is −12°, and the inlet static conditions are 14.7 psi and 600 R. The flow enters the rotor with no whirl and the axial velocity can be considered to be a constant. Calculate the following: (a) static and stagnation pressure rise across the rotor, (b) degree of reaction, (c) specific speed, (d) blade and nozzle angles at the inlet and exit, (e) theoretical power transferred to the fluid.

Ans: $R = 0.554$; $P = 0.7$ hp; $N_s = 1.135$; $\Delta p_{\text{stat}} = 0.048$ psi; $\Delta p_{\text{stag}} = 0.087$ psi

6.3 Show that for an axial fan without guide vanes, the stagnation pressure rise is given by the following expression:

$$\Delta h_0 = \frac{\Delta p_0}{\rho} = u^2(1 - \phi \tan\beta_3)$$

where:

ϕ is the flow coefficient
β_3 is the exit blade angle

Subscripts 1 and 2 indicate the inlet and exit for the upstream guide vane (non-existent here) while subscripts 2 and 3 indicate the inlet and exit for the rotor (see Figure 6.1). Assume that the inlet to the rotor is purely axial.

6.4 Consider an axial flow blower spinning at 4200 rpm with tip and hub diameters of 25 cm and 12.5 cm, respectively. The blower has no inlet guide vanes and the inlet and exit blade angles to the rotor are −75° and −55°, respectively. The inlet velocity is purely axial. The mass flow rate is 0.466 kg/s and the static conditions at the inlet are 101.3 kPa and 35°C. Calculate the degree of reaction and stagnation and static pressure rises through the rotor.

Degree of reaction	= 0.6
Exit diameter	= 60 cm
Blade width at exit	= 6.9 cm
Blade width at inlet	= 7.5 cm
Power input into the fan	= 4.5 kW
Flow rate	= 3 m^3/s
Exit blade angle	= −72°

Ans: $R = 0.691$; $\Delta p_{stat} = 891$ Pa; $\Delta p_{stag} = 1202$ Pa

6.5 An axial flow fan has a hub to tip diameter of 0.30, with the hub diameter being 16 cm. When rotating at 3600 rpm, the power input is 115 watts. The inlet blade angle at the mean radius is 68° measured with respect to the axial direction. If the inlet air is at standard conditions, and inlet velocity is axial, calculate the static and stagnation pressure rises across the fan. Assume that the efficiency of the fan based on static conditions is 85%. Also calculate the efficiency based on stagnation conditions.

6.6 Obtain the relationship between pressure coefficient, defined in terms of the stagnation conditions, and blade loading coefficient, as given by Equations 6.4 and 6.4a.

6.7 Assuming the same inlet conditions (with no whirl) and the same blade speed, show that the absolute velocity at the exit increases from backward-curved, to radial, to forward-curved blades. Draw the velocity triangles for each of these cases.

6.8 Air at atmospheric pressure and at 25°C enters a centrifugal blower when the rotor is spinning at 900 rpm. The inner and outer diameters of the rotor are 50 cm and 60 cm, respectively. Air enters the rotor radially and makes angle of −68° with the wheel tangent at the exit. The flow rate expected is 1.8 m^3/s. If the efficiency, based on stagnation enthalpy, is 88.1%, calculate the power input, efficiency based on static enthalpy, and the exit fluid angle.

Ans: $\eta = 55.7\%$; exit fluid angle = 270°

6.9 A centrifugal fan with inner and outer diameters of 8 in. and 9 in., respectively, spins at 1500 rpm. It draws air at atmospheric pressure and 70°F at a mass flow rate of 1 lbm/s. The relative velocities at the inlet and exit are 65 ft./s and 70 ft./s, respectively, while the absolute velocities at the inlet and exit are 55 ft./s and 82.64 ft./s, respectively. Calculate the degree of reaction, static and stagnation pressure rises, and power input into the fan.

Ans: $R = 0.5$; $\Delta p_{stat} = 0.016$ psi; $\Delta p_{stag} = 0.031$ psi

6.10 The following data refer to a centrifugal fan spinning at 1200 rpm:

Atmospheric air enters the fan radially at 25°C. Calculate the blade angle and inlet at the radius, and the efficiencies based on both static and stagnation conditions.

Ans: $\eta_{stag} = 89.6\%$; $\eta_{stat} = 53.7\%$

6.11 A centrifugal fan with radial vanes delivers 4 m^3/s of atmospheric air at 30°C when spinning at 2400 rpm. The total pressure rise across the fan is equivalent to 120 cm of water. If air enters the rotor radially, calculate the specific speed and the fluid angle at the exit of the fan.

Ans: $\alpha_2 = 8.97°$; $N_s = 0.497$

7

Radial Gas Turbines

Up to this point, the discussion has centered on machines that handle incompressible fluids. Even in those handling air, such as fans and blowers, the pressure changes were so low that they were studied using the same principles as for pumps and hydraulic turbines. From this chapter onward, the working fluid will undergo significant pressure changes. Thus, temperature will be an integral part of the analysis. These machines are broadly classified as gas turbines and compressors. Based on the direction of flow, these machines can be further classified as radial or axial. This chapter deals with radial-type gas turbines.

Upon completion of this chapter, the student will be able to

- Analyze velocity diagrams for radial flow gas turbines
- Predict the stage performance of gas turbines
- Perform analysis and preliminary design of such turbines

7.1 Brayton Cycle

Gas turbine engines are based on the Brayton cycle, which was proposed by George Brayton around 1870. Although the earliest gas turbines can be traced to the Englishman John Barber around the same decade, they did not produce any net power since the compressors (and the turbines) had low efficiencies. Also, the turbine inlet temperatures were limited due to lack of availability of materials that could withstand high temperatures. Serious attention was paid to the development of gas turbine engines after World War II, especially for use in aircraft propulsion. The components of an open cycle gas turbine engine used for propulsion are a compressor, a combustor, and a turbine. For a closed cycle, these components would be combined with a heat exchanger after the turbine. The open and closed cycle components are shown in Figure 7.1.

In an open cycle, the heat exchanger is missing and it is used primarily for propulsion. The turbine produces enough power to run the compressor and auxiliary equipment, with the remaining power being used for propulsion or the production of electric power. A closed cycle is used only for power generation. The thermodynamic processes corresponding to the cycle are shown in Figure 7.2. The ideal compression and expansion processes in the compressor and turbine are indicated by 1-2′ and 3-4′ while the actual processes are labeled 1-2 and 3-4. Flows through the combustor and heat exchanger can be approximated as isobaric or constant-pressure processes. The thermal efficiency is defined as the ratio of net work output from the cycle to thermal energy input. Thus,

$$\eta_{th} = \frac{w_{net}}{q_{in}} = \frac{W_t - |W_c|}{Q_H} \tag{7.1}$$

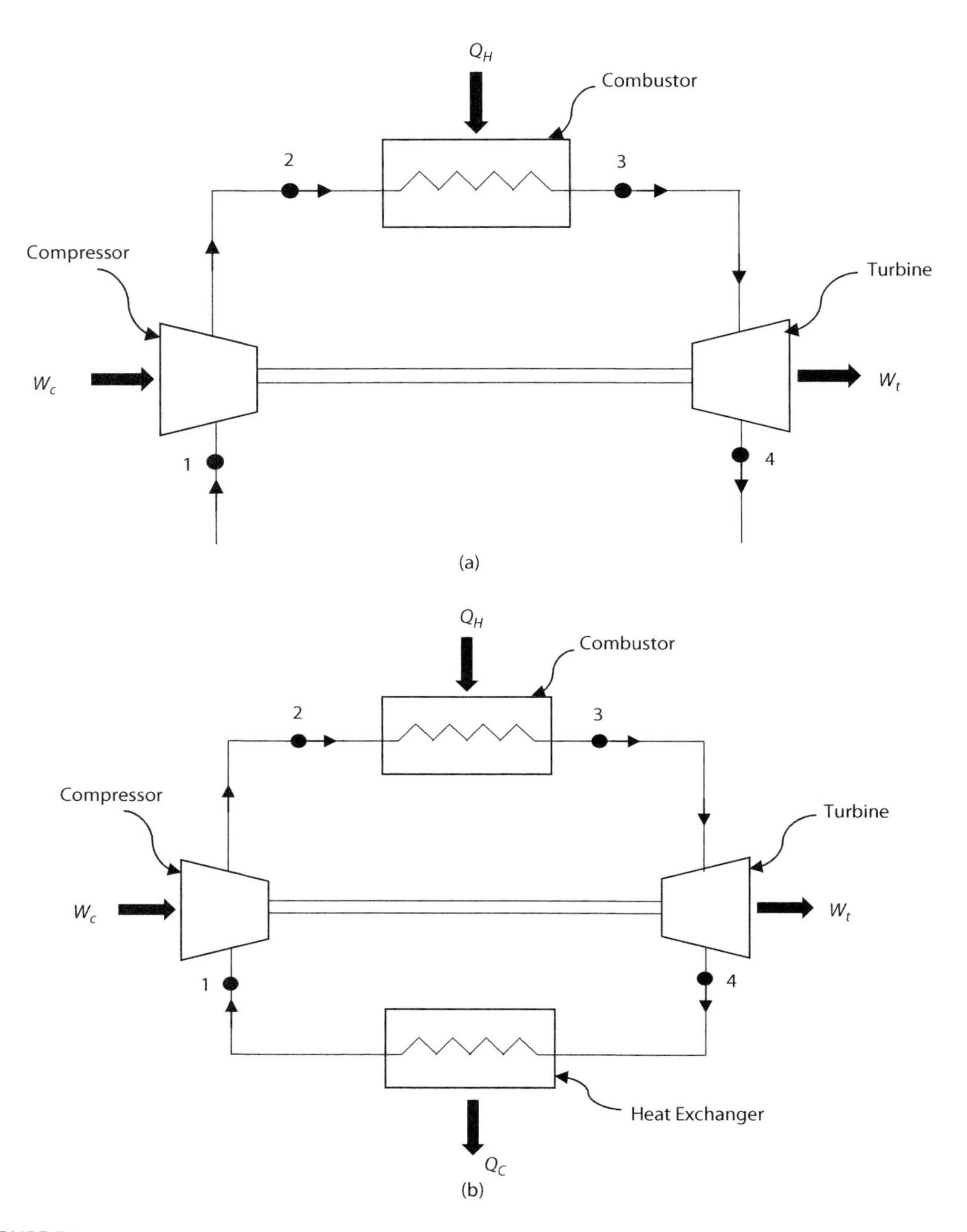

FIGURE 7.1
Brayton cycle components: (a) open cycle and (b) closed cycle.

By assuming ideal conditions, the cycle efficiency can be shown to be dependent on the cycle pressure ratio $\frac{p_2}{p_1}\left(=\frac{p_3}{p_4}\right)$ and the cycle temperature ratio T_3/T_1 (see Example 7.2). From the design perspective, the inlet temperature to the turbine T_3 is raised to as high a level as possible within the limits of the blade materials of the stage. This, and the cycle

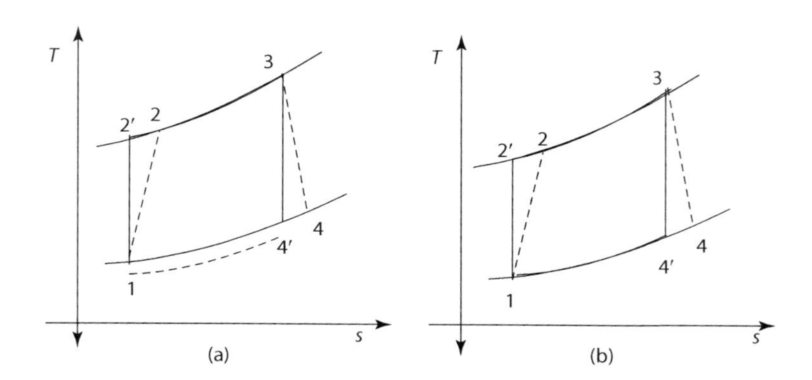

FIGURE 7.2
Thermodynamic processes for the Brayton cycle: (a) open cycle and (b) closed cycle.

pressure ratio, are the preliminary variables picked during the design of gas turbines, as these fix the inlet thermodynamic state to the turbine.

Another variable that is relevant in the Brayton cycle is the so-called *back work ratio,* defined as the ratio of compressor work to turbine work$\left(=|W_c|/W_t\right)$. A power plant with a high back work ratio requires a larger turbine to provide the additional power requirements of the compressor. This is partly because the compressor and turbine are usually mounted on the same shaft, thus having the same speed. However, this is not true with modern transport aircraft, since they use bypass air for a significant fraction of their propulsive power, creating the need for multiple spools for the compressor and the turbine (see Ahmed El-Sayed (2008) for details). Since they are part of the same cycle, the mass flows and pressure ratios are roughly equal.

The discussion of the thermodynamic cycle itself has been brief, since our interest is mainly in the turbine and compressor analysis and not in the details of the cycle; that is, the reheating, regenerating, intercooling, and so on. Interested readers can find these in Van Wylen and Sonntag (1985). A typical example of the Brayton cycle follows:

Example 7.1

The following data refer to a gas turbine power plant operating on the ideal Brayton cycle (1-2′-3-4′):

Pressure ratio = 10
Compressor inlet temperature = 25°C
Turbine inlet temperature = 1000°C
Find the thermal efficiency and the back work ratio.

Solution: The temperature versus specific entropy diagram for the cycle is shown in Figure 7.2b.

Making the standard assumptions of steady, constant specific heats, and an air standard Brayton cycle, the following computations can be made:

Process 1-2′: Isentropic compression in compressor

$$T_{2'} = T_1\left(\frac{P_2}{P_1}\right)^{(k-1)/k} = (25+273)(10)^{(1.4-1)/1.4} = 575.3\,\text{K} \tag{a}$$

Process 3-4′: Isentropic expansion in turbine

$$T_{4'} = T_3\left(\frac{P_4}{P_3}\right)^{(k-1)/k} = (273 + 1000)(1/10)^{(1.4-1)/1.4} = 660.3\,\text{K} \tag{b}$$

Compressor work input is given by

$$w_c = h_{2'} - h_1 = C_p\left(T_{2'} - T_1\right) = \left(1.004\frac{\text{kJ}}{\text{kg}\cdot\text{K}}\right)(575.3 - 298)\text{K} = 278.4\,\text{kJ/kg}$$

Turbine work output is given by

$$w_t = h_3 - h_{4'} = C_p\left(T_3 - T_{4'}\right) = \left(1.004\frac{\text{kJ}}{\text{kg}\cdot\text{K}}\right)(1273 - 660.3)\text{K} = 615.2\,\text{kJ/kg}$$

Heat input would be

$$q_{in} = h_3 - h_2 = C_p\left(T_3 - T_2\right) = \left(1.004\frac{\text{kJ}}{\text{kg}\cdot\text{K}}\right)(1273 - 575.3)\text{K} = 700.49\,\text{kJ/kg}$$

The net work for the cycle is the difference between turbine output and compressor input, given by

$$w_{net} = w_t - w_c = (615.2 - 278.4)\ \text{kJ/kg} = 336.8\ \text{kJ/kg}$$

Thus, thermal efficiency is

$$\eta_{th} = \frac{w_{net}}{q_{in}} = \frac{336.8\,\text{kJ/kg}}{700.49\,\text{kJ/kg}} = 0.48\% \text{ or } 48\% \tag{c}$$

Back work ratio is the ratio of compressor work to turbine work. Thus,

$$\text{Back work ratio} = \frac{w_c}{w_t} = \frac{278.4\ \text{kJ/kg}}{615.2\ \text{kJ/kg}} = 0.45\% \text{ or } 45\%$$

Comment: Thus, almost half of the work produced by the turbine is used up in driving the compressor. Hence, high efficiencies are critical for compressors and turbines so that there is positive net work.

In addition to high efficiencies, the inlet temperature and pressure are very important in the performance of gas turbines. The importance of the temperature and the pressure ratios T_3/T_1 and p_3/p_1 on the performance on the cycle are illustrated in the next example.

Example 7.2

Considering the Brayton cycle shown in Figures 7.1 and 7.2, obtain an expression for the temperature ratio T_3/T_1 in terms of the pressure ratio p_3/p_1.

Solution: Compressor efficiency can be written as

$$\eta_c = \frac{h_{2'} - h_1}{h_2 - h_1} = \frac{T_{2'} - T_1}{T_2 - T_1} \tag{a}$$

or

$$T_2 - T_1 = \frac{T_{2'} - T_1}{\eta_c} = \frac{T_1\left(\frac{T_{2'}}{T_1} - 1\right)}{\eta_c} \tag{b}$$

Since process 1-2′ is isentropic, Equation (b) becomes

$$T_2 - T_1 = \frac{T_1\left(\left(\frac{p_{2'}}{p_1}\right)^{\frac{k-1}{k}} - 1\right)}{\eta_c} = \frac{T_1\left(\left(\frac{p_2}{p_1}\right)^{\frac{k-1}{k}} - 1\right)}{\eta_c} \tag{c}$$

Similarly, the efficiency of the turbine can be written as

$$\eta_t = \frac{h_3 - h_4}{h_3 - h_{4'}} = \frac{T_3 - T_4}{T_3 - T_{4'}} \tag{d}$$

or

$$T_3 - T_4 = \eta_t\left(T_3 - T_{4'}\right) = \eta_t T_3\left(1 - \frac{T_{4'}}{T_3}\right) \tag{e}$$

Again, since 3-4′ is isentropic,

$$T_3 - T_4 = \eta_t T_3\left(1 - \left(\frac{p_{4'}}{p_3}\right)^{\frac{k-1}{k}}\right) = \eta_t T_3\left(1 - \left(\frac{p_4}{p_3}\right)^{\frac{k-1}{k}}\right) \tag{f}$$

Letting r_p be the pressure ratio, Equations (c) and (f) can be written as

$$T_2 - T_1 = \frac{T_1\left(\left(r_p\right)^{\frac{k-1}{k}} - 1\right)}{\eta_c} \tag{g}$$

and

$$T_3 - T_4 = \eta_t T_3\left(1 - \left(\frac{1}{r_p}\right)^{\frac{k-1}{k}}\right) \tag{h}$$

Since net work is the difference between compressor work and turbine work,

$$w_{net} = C_p\left(T_2 - T_1\right) - C_p\left(T_3 - T_4\right) \tag{i}$$

By setting the net work to zero, an expression for temperature ratio in terms of the pressure ratio can be obtained. Using Equations (g) and (h),

$$\frac{T_1\left(\left(r_p\right)^{\frac{k-1}{k}} - 1\right)}{\eta_c} = \eta_t T_3\left(1 - \left(\frac{1}{r_p}\right)^{\frac{k-1}{k}}\right) \tag{j}$$

or

$$\frac{T_3}{T_1} = \frac{1}{\eta_c\eta_t}\left(\left(r_p\right)^{\frac{k-1}{k}} - 1\right) \Bigg/ \left(1 - \left(\frac{1}{r_p}\right)^{\frac{k-1}{k}}\right)$$

After simplification,

$$\frac{T_3}{T_1} = \frac{(r_p)^{\frac{k-1}{k}}}{\eta_c \eta_t} \tag{k}$$

By making a simplifying assumption that the turbine and compressor efficiencies are equal, a plot of temperature ratio versus pressure ratio can be made, as shown in the next figure. It can be seen that for a pressure ratio of eight, the ratio of maximum to minimum temperatures (T_3/T_1) should be about three with compressor and turbine efficiencies around 80%. Thus, if the inlet temperature is around room temperature (300 K), the turbine inlet temperature must be about 900 K just to break even in terms of the turbine and compressor work. Usually, the temperatures need to be much higher to provide any net work output from the cycle. The absence of materials capable of withstanding such high temperatures has been the main drawback of the Brayton cycle in power production.

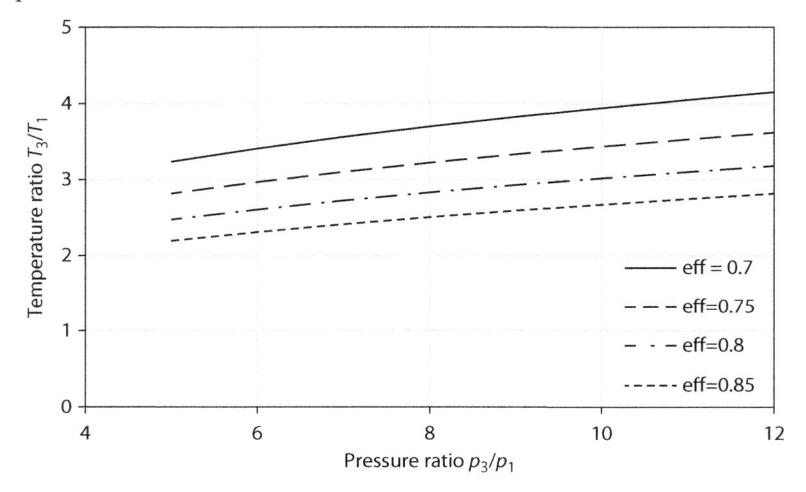

Pressure and temperature ratios for brayton cycle.

Comment: The inlet temperature to the turbine should be significantly higher than described to produce an appreciable amount of net work. It is for this reason that turbine blade cooling and turbine material selection are active areas of gas turbine research.

7.2 Radial Flow Gas Turbines

Gas turbines can be broadly classified as radial and axial types, with the latter being used predominantly for propulsion. Because of the nature of flow and the mechanisms of pressure rise, there are significant differences between them. In this section, the preliminary design and analysis of radial flow turbines are presented.

The earliest radial gas turbine was the so-called Ljungström turbine, which is a radially *outward flow* type. Although it has never been used in the United States, it has been used quite extensively in Europe. It consists of alternate rows of blades rotating in opposite directions, with each row driving a separate generator. Further details can be found in Shepherd (1956) and Dixon and Hall (2010). The Ljungström turbine uses steam (compressible) as the working fluid. For water, the analogous machine (although it is a radially *inward flow* type) is called the *Francis turbine*, which was studied previously in Chapter 4. A turbomachine with outward flow would be a centrifugal pump, but it is power consuming rather than power producing. Radial flow turbines using compressible fluids are similar in principle to the Francis turbine, but are smaller in size and run faster. These types of turbines are used widely in turbochargers, small turboprop aircraft engines, auxiliary power units, natural gas processing, air liquefaction, and waste heat and geothermal applications. The majority of radial flow turbines for both compressible and incompressible flows are of the inward flow type; hence the name *inward flow radial* (IFR) turbine. This is the primary device in the small power generation units used for space power generation.

The basic features of an IFR are shown in Figure 7.3. It is strikingly similar to a centrifugal compressor except for the direction of flow. In a compressor, the flow is radially outward, whereas, in the turbine, it is radially inward. The features of centrifugal compressors and axial flow compressors are discussed in detail in later chapters. Hence, the discussion here will be confined to radial flow turbines.

As seen in the longitudinal view, the flow enters radially through the volute, which contains a series of vanes placed circumferentially that act like nozzles. These vanes are called *stator vanes*, and they expand the gas to a velocity V_2 before it enters the rotor. The gas enters the rotor at an angle α_2 measured with respect to the radial direction. The relative

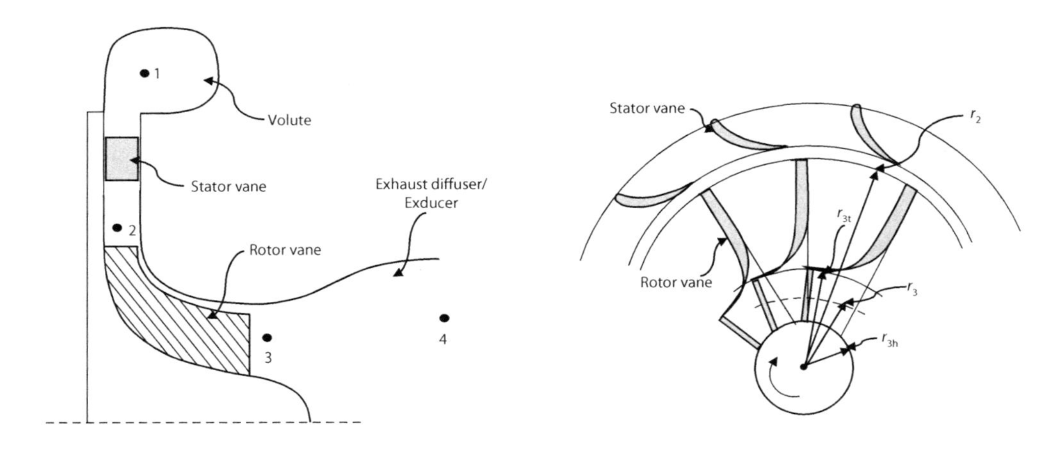

FIGURE 7.3
Longitudinal and transverse sections through an IFR turbine.

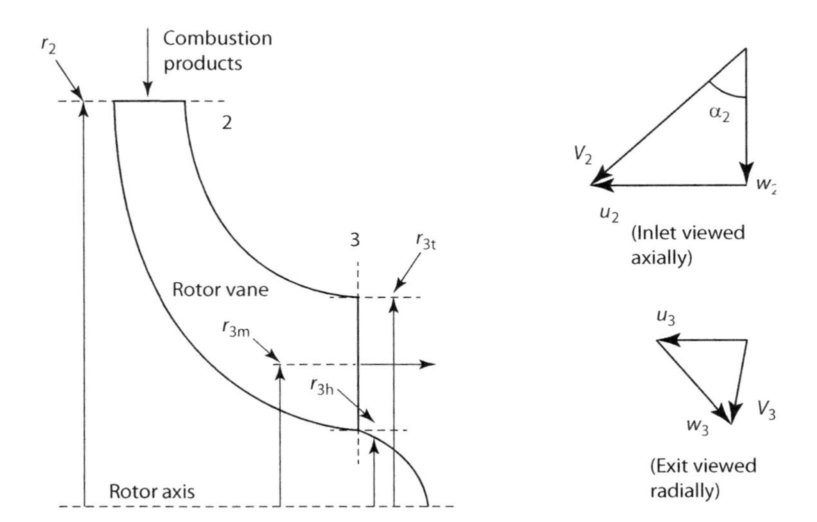

FIGURE 7.4
Rotor blade and velocity triangles at inlet and exit of rotor.

velocity w_2 enters the rotor radially at a radius r_2 and turns to exit the rotor axially at a smaller radius r_3. The trailing portion of the rotor vanes are curved such that the relative velocity at the exit w_3 will have both radial and tangential components. The gas expands through the rotor to a pressure p_3 and finally exits through a diffuser (sometimes called the *exducer*) to the ambient atmospheric pressure p_4.

The velocity triangles at the inlet and exit of the rotor are shown in Figure 7.4. It is assumed that the flow enters radially (hence, the relative velocity at inlet state 2 is purely radial). Since the flow enters radially and exits axially, the velocity triangle at the inlet (state 2) is drawn while viewing the rotor *along the axis*. Hence, the blade velocity is shown to the left with relative velocity entering radially. At the exit of the rotor (state 3) the velocity triangle is drawn while viewing *along the radius*. The energy equation can now be applied to each component as follows: For the stator, the energy equation is

$$h_1 + \frac{V_1^2}{2} = h_2 + \frac{V_2^2}{2}$$
$$\text{or} \quad h_{01} = h_{02} \tag{7.2}$$

Assuming constant specific heats (which is a reasonable assumption to simplify the calculations), the equation becomes $T_{01} = T_{02}$, that is, the stagnation temperature (enthalpy) is constant across the stator. This should not be surprising since there is no energy extraction or rejection in the stator, as it simply acts as a nozzle to direct the fluid toward the rotor.

Across the rotor, the energy equation becomes

$$h_2 + \frac{V_2^2}{2} = h_3 + \frac{V_3^2}{2} + E$$
$$\text{or} \quad h_{02} = h_{03} + E \quad \text{or} \quad E = C_p\left(T_{02} - T_{03}\right) \tag{7.3}$$
$$\text{and} \quad E = u_2 V_{u2} - u_3 V_{u3}$$

The last of these three equations follows from Euler's formula. Finally, across the diffuser, the energy equation gives

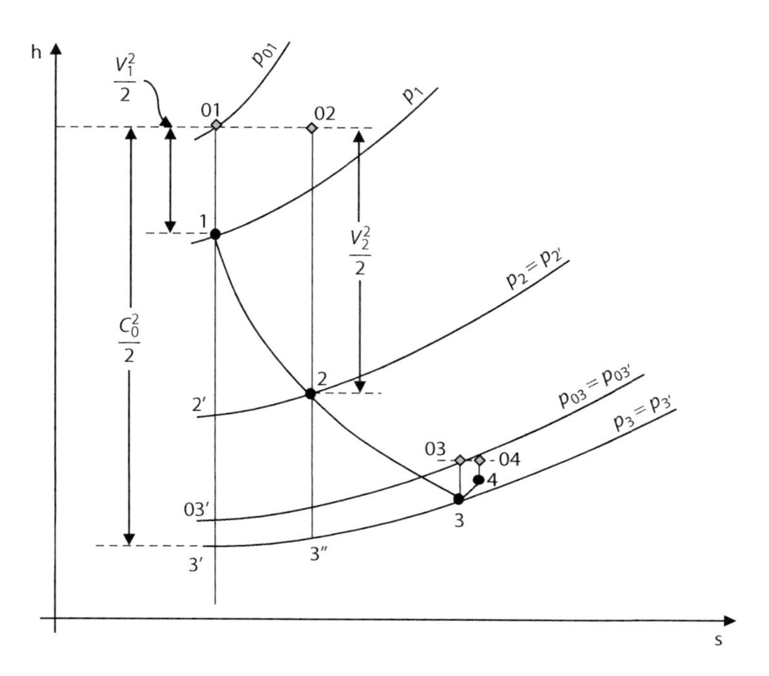

FIGURE 7.5
Thermodynamic states for the radial flow turbine.

$$h_3 + \frac{V_3^2}{2} = h_4 + \frac{V_4^2}{2}$$
$$\text{or} \quad h_{03} = h_{04} \quad \text{or} \quad T_{03} = T_{04} \tag{7.4}$$

again, assuming constant specific heats.

The corresponding thermodynamic states for the turbine are shown in Figure 7.5. The relationships between the static and stagnation states are shown only for some points in order not to clutter the figure.

The static and stagnation states are related by the absolute velocity, while the static and relative stagnation states are related by the relative velocity at any point in the turbine. Thus,

$$T_{01} = T_1 + \frac{V_1^2}{2C_p}$$
$$T_{02R} = T_2 + \frac{w_2^2}{2C_p} \tag{7.5}$$

where subscript R denotes properties measured with respect to relative velocity. The pressures and temperatures for the isentropic processes in Figure 7.5 can be related isentropically. For example, process 1-2′ is isentropic, hence

$$\frac{p_1}{p_{2'}} = \left[\frac{T_1}{T_{2'}}\right]^{\frac{k}{k-1}} = \frac{p_1}{p_2} \tag{7.6}$$

The last part of Equation 7.6 follows since p_2 and p_2' are on a constant-pressure line. The two processes 03-3 and 03′-3′ are isentropic. Hence,

$$\frac{p_{03}}{p_3} = \left[\frac{T_{03}}{T_3}\right]^{\frac{k}{k-1}} \text{ and } \frac{p_{03'}}{p_{3'}} = \left[\frac{T_{03'}}{T_{3'}}\right]^{\frac{k}{k-1}} \tag{7.6a}$$

However, the pressure ratios p_{03}/p_3 and $p_{03'}/p_{3'}$ are equal, since $p_3 = p_{3'}$ and $p_{03} = p_{03'}$. Hence,

$$\frac{T_{03}}{T_3} = \frac{T_{03'}}{T_{3'}} \tag{7.7}$$

The following efficiencies are defined in the context of gas turbines and compressors. The first is the total to static efficiency, defined as

$$\eta_{ts} = \frac{h_{01} - h_{03}}{h_{01} - h_{3'}} = \frac{E}{h_{01} - h_{3'}} = \frac{T_{01} - T_{03}}{T_{01} - T_{3'}}$$
$$\text{or } \eta_{ts} = \frac{h_{02} - h_{03}}{h_{02} - h_{3'}} = \frac{E}{h_{02} - h_{3'}} = \frac{T_{02} - T_{03}}{T_{02} - T_{3'}} \tag{7.8}$$

The total to total efficiency is defined as

$$\eta_{tt} = \frac{h_{01} - h_{03}}{h_{01} - h_{03'}} = \frac{E}{h_{01} - h_{03'}} = \frac{T_{01} - T_{03}}{T_{01} - T_{03'}}$$
$$\text{or } \eta_{tt} = \frac{h_{02} - h_{03}}{h_{02} - h_{03'}} = \frac{E}{h_{02} - h_{03'}} = \frac{T_{02} - T_{03}}{T_{02} - T_{03'}} \tag{7.9}$$

A related variable called the *spouting velocity* (sometimes called the *isentropic velocity*) is defined as

$$\frac{c_0^2}{2} = h_{01} - h_{3'} \tag{7.10}$$

An interesting physical interpretation based on the flow of an incompressible fluid can be given using flow from a reservoir, as shown in Figure 7.6. If losses are ignored, then the fluid issuing from the spout will exit with a velocity of $c_0 = \sqrt{2gH}$, an application of Bernoulli's theorem. Again, from the same theorem, the rise of the fluid will be to a height H, the head available in the tank. In the absence of losses, the maximum velocity in the spout will occur at the exit of the spout, analogous to isentropic expansion in a nozzle; hence the name *spouting* velocity. Thus, the spouting velocity is the maximum velocity that would be obtained in a nozzle during an ideal expansion from p_{01} to $p_{3'}$; that is, expansion from the highest possible pressure to the lowest possible pressure in the cycle.

In the case of ideal (frictionless) expansion and complete recovery of exit kinetic energy, $V_{u2} = u_2$, $V_{u3} = 0$, and the ideal energy transfer would be

$$E = u_2 V_{u2} = u_2^2 = \frac{c_0^2}{2}$$

Therefore,

$$\frac{u_2}{c_0} = \sqrt{\frac{1}{2}} = 0.707$$

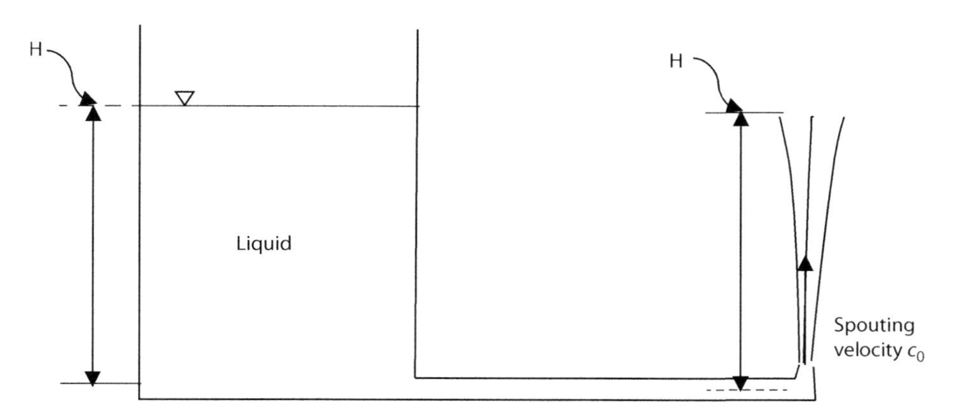

FIGURE 7.6
Spouting velocity of an incompressible fluid from a tank.

In practice, it is found that the best overall efficiency is obtained when this ratio is in the range $0.68 < u_2 < c_0 < 0.71$ (Sarvanamuttoo et al., 2009).

Using the definitions of total to total efficiency, η_{tt}, and total to static efficiency, η_{ts}, the following relationship can be obtained (see Problem 7.11).

$$\eta_{tt} = \frac{1}{\frac{1}{\eta_{ts}} - \frac{V_{3'}^2}{2E}} \quad (7.11)$$

where $V_{3'}$ is defined by the equation

$$\frac{V_{3'}^2}{2} = h_{03'} - h_{3'} \quad (7.12)$$

There are several places in the radial turbine where losses take place, the most prominent being the nozzle (stator) and rotor passages. These are conveniently expressed in terms of nozzle loss coefficient and rotor loss coefficient. Besides these, the flow entering the rotor is not at the design inlet angle and differs by a small amount, called the *angle of incidence*, giving rise to incidence losses (also called *shock losses*). This occurs during off-design conditions.

A simplified version of Figure 7.5 showing only the expansion process in the nozzle is shown in Figure 7.7a. Processes 1-2 and 1-2′ are the actual and ideal expansions. Using these states, the nozzle loss coefficient can be defined as the ratio of the loss in kinetic energy between the ideal expansion and actual expansion in a nozzle to the actual kinetic energy at the nozzle exit. In other words,

$$\lambda_n = \frac{(h_2 - h_{2'})}{\frac{1}{2}V_2^2} \quad (7.13)$$

The nozzle loss coefficient λ_n can be defined in terms of the nozzle velocity coefficient C_n, which is the ratio of the actual to ideal velocities at the nozzle exit; that is,

$$C_n = \frac{V_2}{V_{2'}} = \sqrt{\frac{(h_{02} - h_2)}{(h_{01} - h_{2'})}} = \sqrt{\frac{(h_{02} - h_2)}{(h_{02} - h_{2'})}} \quad (7.14)$$

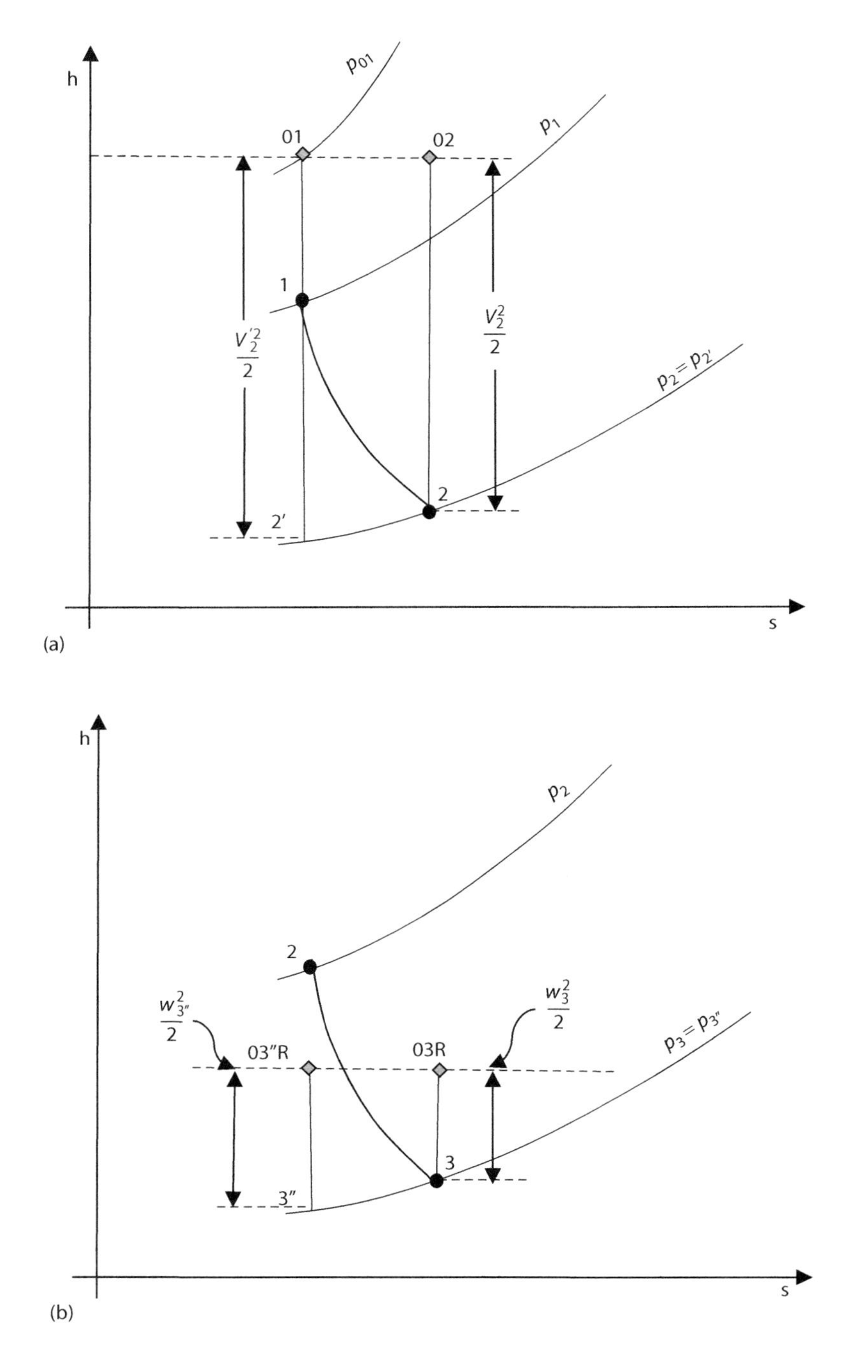

FIGURE 7.7
Rotor expansion process to define (a) nozzle loss coefficient and (b) rotor loss coefficient.

Also,

$$h_{02} = h_2 + \frac{1}{2}V_2^2$$
$$h_{02} = h_{01} = h_{2'} + \frac{1}{2}V_{2'}^2 \quad (7.15)$$

The three preceding equations can be combined to obtain the following relationship between the nozzle loss coefficient and nozzle velocity coefficient (see problem 7.19).

$$\lambda_n = \frac{1}{C_n^2} - 1 \quad (7.16)$$

Since nozzles are very efficient devices, the values of C_n are very high (well above 90%). Typical values of the nozzle loss coefficient are between 0.06 and 0.23 (Dixon and Hall, 2010).

The losses in the rotor are expressed using the rotor loss coefficient (for notation see Figure 7.7b):

$$\lambda_r = \frac{(h_3 - h_{3''})}{\frac{1}{2}w_3^2} \quad (7.17)$$

Similar to the nozzle, the rotor velocity coefficient can be defined as

$$C_r = \frac{w_3}{w_{3''}} \quad (7.18)$$

The two coefficients can be combined (details are left as an exercise, Problem 7.9) to get

$$\lambda_r = \frac{1}{C_r^2} - 1 \quad (7.19)$$

7.3 Design Features of Radial Turbines

In this section, the basic design features of radial flow turbines will be discussed. As with hydraulic turbines and pumps, the discussion of details will be limited, since a lot of the equations are based on empiricism and extensive experimentation. Typically, the inlet stagnation conditions and the exit pressure (p_{01}, T_{01}, and p_3) would be available to the designer. Based on these values, the dimensions of the turbine, number of blades, rotor and nozzle loss coefficients, and velocity ratios can be estimated.

The nominal design of an IFR turbine assumes that the inlet blades are purely radial and the flow exiting the rotor has no whirl. Stated alternatively, inlet relative velocity is purely radial at the entrance and exit absolute velocity is purely axial (i.e., $V_{u2} = u_2$ and $V_{u3} = 0$). The exit condition corresponding to $V_{u3} = 0$ can be obtained from Figure 7.4 by letting the exit absolute velocity V_3 be purely axial, that is, making the exit triangle a right triangle. The energy transfer for such a turbine would be

$$E = u_2^2 \quad (7.20)$$

The Cordier diagram that was introduced in Chapter 2 is a useful tool to predict the initial dimensions of hydraulic machines, especially pumps. In a similar manner, the Balje diagram gives the variation of specific speed N_s and specific diameter D_s for various values of total to static efficiency. The diagram is shown in Figure 7.8.

As introduced in Chapter 2, specific speed and specific diameter are defined as

$$N_s = \frac{N\sqrt{Q}}{(gH)^{3/4}}$$

$$D_s = \frac{D(gH)^{1/4}}{\sqrt{Q}}$$

A couple of questions regarding head and flow rate immediately arise in connection with these definitions. The first relates to the quantity gH, which is the total energy available for hydraulic turbines with no losses. Since it is not possible to define head for gas turbines, it is customary to use the kinetic energy associated with the spouting velocity, $\frac{1}{2}c_0^2$, in its place. Notice that the definition of spouting velocity assumes an isentropic expansion from the highest inlet pressure to the lowest exit pressure. The second question is regarding the volumetric flow rate through the gas turbine. Unlike the mass flow rate, which remains constant, (a consequence of the steady flow assumption) this quantity is obviously not a constant through the turbine. However, it is customary to use the flow rate at the turbine exit in the equations. The diameter is taken to be the inlet diameter of the rotor. Thus, the equations will be replaced as

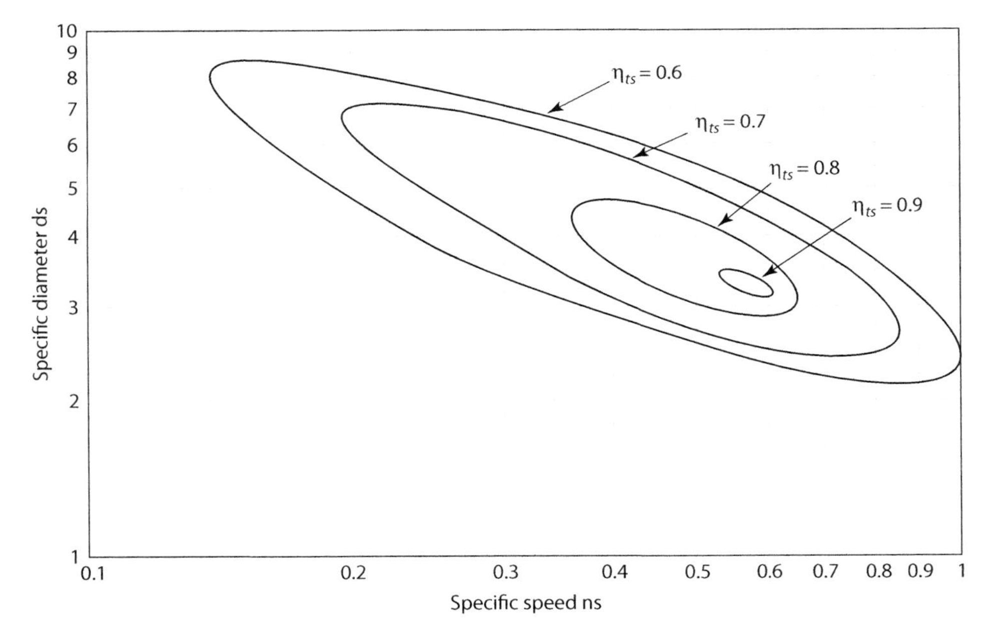

FIGURE 7.8
Variation of specific diameter versus specific speed for the design of radial turbines. (Modified from Balje, O.E., *Turbomachines: A Guide to Design, Selection, and Theory*, Wiley, New York, 1981.)

$$N_s = \frac{N\sqrt{Q_3}}{\left(\frac{c_0^2}{2}\right)^{3/4}}$$

$$D_s = \frac{D_2\left(\frac{c_0^2}{2}\right)^{1/4}}{\sqrt{Q_3}} \tag{7.21}$$

Taking the product of the specific speed and specific diameter, the resulting equation is

$$N_s D_s = \frac{ND_2}{\left(\frac{c_0^2}{2}\right)^{1/2}} = \frac{2^{3/2}u_2}{c_0} = 2.83\frac{u_2}{c_0} \tag{7.22}$$

Note that the product of angular velocity and diameter is twice the blade speed. Using the Balje diagram in conjunction with Equation 7.22, the preliminary selection of D_s, N_s, η_{ts}, and the ratio u_2/c_0 can be made. Assuming a suitable value of η_{ts}, values of D_s and N_s can be selected from the Balje diagram. From these, the ratio u_2/c_0 can be calculated using Equation 7.22. Since the inlet and exit pressures are known, c_0, and hence u_2, can be calculated. According to Logan (1993), typical values of u_2/c_0 lie between 0.55 and 0.80. However, the range recommended by Sarvanamuttoo (2009) is 0.68–0.71. The accepted practice is to pick D_s and N_s for the best values of η_{ts} from the Balje diagram (Figure 7.8) and to calculate the ratio u_2/c_0. The blade speed (u_2) is restricted to 1600–1700 ft./s (Logan (1993)).

The nozzle and rotor loss coefficients are related to the respective velocity coefficients through Equations 7.16 and 7.19. The nozzle velocity coefficient is quite high, since nozzles convert pressure to velocity, a very efficient process, unlike diffusers, which convert velocity to pressure (consequently, they are less efficient). According to Logan (1993), the range for nozzle loss coefficient is

$$0.06 \le \lambda_N \le 0.24$$

and the corresponding range for rotor loss coefficient is

$$0.4 \le \lambda_R \le 0.8$$

By performing a simplified analysis using the specific speed defined in Equation 7.21, Dixon obtained an upper bound for the ratio of exit shroud diameter D_{3s} and rotor disk diameter D_2. His analysis is based on the relationship between the specific speed N_s and the flow coefficient, which can be shown to be (see Problem 7.20)

$$N_s = 2.83\sqrt{\left(\frac{Q_3}{ND_2^3}\right)} \tag{7.23}$$

Using this relationship, he showed that $D_{3s}/D_2 \le 0.7$.

There are several correlations available to determine the number of blades n_B and its dependence on blade parameters. Glassman (1976) used an empirical relationship between the nozzle angle α_2 and the number of blades. It is given as

$$n_B = 0.105(110 - \alpha_2)\tan\alpha_2 \quad \text{for } \alpha_2 \text{ in degrees} \tag{7.24}$$

TABLE 7.1

Design Variables for 90° IFR Gas Turbine

Variable Description	Symbol	Range	Source
Rotor loss coefficient	λ_R	0.4–0.8	Dixon and Hall (2010)
Nozzle loss coefficient	λ_N	0.06–0.24	Dixon and Hall (2010)
Rotor inlet flow angle	α_2	68°–75°	Rohilk (1968)
Rotor exit blade angle	β_3	50°–70°	Whitfield and Baines (1990)
Blade speed to spouting velocity	u_2/c_0	—	Balje diagram Figure 7.8
Relative velocity ratio	w_3/w_2	2–2.5	Logan (1993)
Exit velocity to blade speed	V_3/u_2	0.15–0.5	Whitfield and Baines (1990)
Exit hub to tip diameter ratio	D_{3h}/D_{3s}	<0.4	Rohlik (1968)
Exit tip to inlet rotor diameter	D_{3s}/D_2	<0.7	Rohlik (1968)
Exit mean to inlet diameter ratio	D_3/D_2	0.53–0.66	Logan (1993)
Inlet blade width to inlet diameter	b_2/D_2	0.05–0.15	Logan (1993)
Inlet Mach number	M_1	<0.2	Walsh and Fletcher (1998)
Exit Mach number without exducer	M_3	~0.3	Walsh and Fletcher (1998)
Blade speed at inlet	u_2	1600–1700 ft./s	Logan (1993)

By considering radial equilibrium and assuming the relative speed varies linearly between blades, Jamieson (1955) extended Stanitz's (1952) inviscid analysis and obtained the relationship $nB = 2\pi \tan\alpha_2$. Unfortunately, this seriously over-predicts the number of blades. Rohlik (1968) made a detailed study of the effects of geometry on efficiency for radial inflow turbines and recommended several limits on several parameters. These, along with other findings given in this section, are summarized in Table 7.1.

7.4 Design Procedure for 90° IFR Turbines

A brief procedure for the preliminary design of radial flow turbines is presented in this section. It is based largely on the guidelines provided by Logan (1993), with modifications. The most common inputs are the power required, inlet conditions (either or both temperature and pressure), and often the rotational speed of the runner. If rotational speed is not given, a value needs to be assumed (in tens of thousands of rpm). The steps can be outlined as follows:

1. From the Balje diagram Figure 7.8, obtain suitable values for N_s and D_s. A good choice of values is 0.6 and 3.0, respectively (their product should be approximately equal to 2.0). This gives a value of η_{ts} above 80%. Also, pick suitable values for u_2, α_2, β_3, the ratio D_{3s}/D_2, D_{3h}, and the nozzle loss coefficient λ_n. Many of these values can be picked as the design process progresses.

2. Calculate the ratio u_2/c_0 using Equation 7.22; pick a suitable value of u_2 (between 1600 and 1700 ft./s); calculate the spouting velocity c_0; calculate E $(=u_2^2)$ and mass flow rate from power; calculate D_2 (from blade velocity u_2 and angular velocity); assume a value for the ratio D_{3s}/D_2 according to the guidelines in Table 7.1 and calculate D_{3s}; calculate T_{03} from Equation 7.3.
3. Calculate $T_{3'}$ using Equation 7.10; since the process 01 to 3′ is isentropic (Figure 7.5), calculate p_{01}.
4. Pick inlet fluid angle α_2 (within the guidelines of Table 7.1); complete inlet velocity triangle, that is, calculate w_2 and V_2.
5. From T_{02} $(=T_{01})$ and V_2, calculate T_2 and M_2.
6. Pick a suitable value of λ_n; from value of λ_n, calculate $T_{2'}$ from Equation 7.17; from isentropic process 01 to 2′, calculate $p_{2'}$ $(=p_2)$ and hence ρ_2.
7. Using ρ_2 and mass flow rate, calculate b_2 and check that b_2/D_2 is as according to Table 7.1.
8. Pick a suitable value of D_{3h} and check that D_{3h}/D_{3s} is within the guidelines of Table 7.1.
9. Calculate D_3 from D_{3h} and D_{3s} using the formula $D_3 = \sqrt{\left(\frac{D_{3s}^2 + D_{3h}^2}{2}\right)}$ and check that D_3/D_2 is in within the limits of Table 7.1. Calculate u_3.
10. Pick suitable value for exit blade angle β_3 and complete the exit velocity triangle; verify that the ratios w_3/w_2 and V_3/u_2 are within the limits of Table 7.1.
11. Calculate the number of blades using Equation 7.24.

These steps are illustrated in Example 7.4.

Example 7.3

The following data refers to an IFR turbine. For simplicity, take the atmospheric pressure to be 100 kPa. The blades are radial at the inlet, and at the exit, flow leaves the rotor axially. There is no exit diffuser and the flow exits into the atmosphere.

Rotor inlet diameter = 36 cm
Rotor speed = 30,000 rpm
Nozzle inlet stagnation pressure = 400 kPa
Nozzle inlet stagnation temperature = 1200 K
Fluid angle at inlet of rotor, $\alpha_2 = 70°$
Pressure at nozzle (stator) exit = 215 kPa
Mean diameter at exit of rotor = 23 cm
Blade angle at exit of rotor = −61°

If the required power is 850 kW, calculate the (1) required width of the blades at the inlet, (2) Mach number at the nozzle exit, (3) nozzle (stator) loss coefficient, (4) nozzle velocity coefficient, (5) total to total efficiency, (6) total to static efficiency, (7) degree of reaction, and (7) spouting velocity.

Solution:

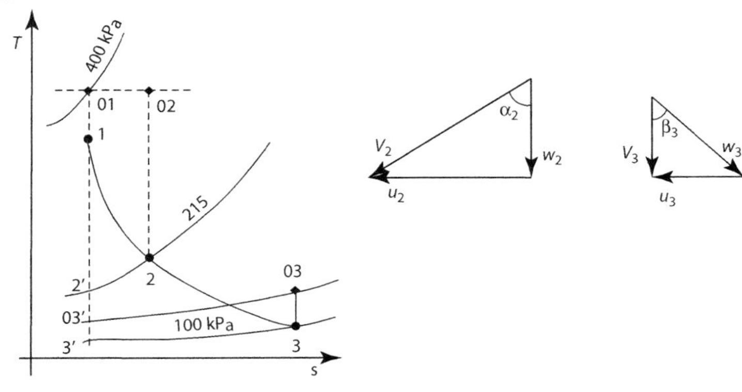

$$E = u_2 V_{u2} - u_3 V_{u3} = u_2^2$$

$$N = 30{,}000\,\text{rpm}$$

$$\omega = \frac{2\pi N}{60} = \frac{2\pi(30{,}000)}{60} = 3141.6\,\text{rad/s}$$

Hence,

$$u_2 = \omega r_2 = \frac{\omega D_2}{2} = \frac{(3141.6)(0.36)}{2} = 565.5\,\text{m/s}$$

$$E = u_2^2 = (565.5)^2 = 319{,}774\ \text{J/kg}$$

Since power is given, the mass flow rate can be calculated:

$$\dot{m} = \frac{P}{E} = \frac{(850)(1000)}{319744} = 2.66\,\text{kg/s}$$

Since $\alpha_2 = 70°$, from trigonometry,

$$V_2 = \frac{u_2}{\sin\alpha_2} = 601.8\,\text{m/s}$$

$$w_2 = \sqrt{\left(V_2^2 - u_2^2\right)} = 205.8\,\text{m/s}$$

Now the temperatures can be calculated:

$$T_{01} = T_{02} = 1200\,\text{K (given)}$$

$$T_2 = T_{02} - \frac{V_2^2}{2C_p} = 1200 - \frac{(601.8)^2}{2(1004)} = 1019.7\,\text{K}$$

The process 01-2′ is isentropic and $p_2 = p_{2'}$ = 215 kpa (given). Hence,

$$\frac{p_{2'}}{p_{01}} = \left(\frac{T_{2'}}{T_{01}}\right)^{\frac{k}{(k-1)}} \Rightarrow T_{2'} = 1005\,\text{K}$$

$$h_{01} = h_{02} = h_{2'} + \frac{(V_{2'})^2}{2} \quad \Rightarrow \quad V_{2'} = \sqrt{2C_p\left(T_{01} - T_2{}'\right)} = 626\,\text{m/s}$$

The nozzle loss coefficient and nozzle velocity coefficient can now be calculated.

$$C_n = C_v = \frac{V_2}{V_{2'}} = 0.96 \text{ and}$$

$$\lambda_n = \frac{1}{C_n^{\,2}} - 1 = 0.082$$

Since 2 to 02 is isentropic, from isentropic relations, the temperatures can be written as

$$\frac{T_{02}}{T_2} = 1 + \frac{k-1}{2} M_2^2 \Rightarrow M_2 = 0.94$$

From the mass flow rate and the velocities, the blade height can be calculated:

$$\rho_2 = \frac{p_2}{RT_2} = \frac{(215)(1000)}{(287)(1019)} = 0.735\,\text{kg/m}^3$$

$$2.66 = \dot{m} = \rho_2 A_2 (V_{rad})_2 = \rho_2 A_2 w_2 = \rho_2 2\pi r_2 b_2 w_2$$

$$b_2 = \frac{\dot{m}}{\rho_2 2\pi r_2 w_2} = \frac{2.66}{(0.735)2\pi(0.18)(205.8)} = 0.0155\,\text{m} = 1.55\,\text{cm}$$

To calculate the efficiencies and the degree of reaction, the parameters for the exit triangles need to be calculated. Thus,

$$u_3 = \omega r_3 = \frac{\omega D_3}{2} = \frac{(3141.6)(0.23)}{2} = 361.3\,\text{m/s}$$

$$V_3 = \frac{u_3}{\tan\beta_3} = \frac{361.3}{\tan 61} = 200.3\,\text{m/s and } w_3 = \frac{u_3}{\sin\beta_3} = \frac{361.3}{\sin 61} = 413.1\,\text{m/s}$$

Process 01-3′ is isentropic. Hence,

$$\frac{p_{3'}}{p_{01}} = \left(\frac{T_{3'}}{T_{01}}\right)^{\frac{k}{(k-1)}} \Rightarrow T_{3'} = 807.5\,\text{K}$$

$$E = C_p\left(T_{01} - T_{03}\right) \Rightarrow T_{03} = T_{01} - \frac{E}{C_p} \Rightarrow T_{03} = 881.7\,\text{K}$$

$$T_3 = T_{03} - \frac{V_3^2}{2C_p} = 881.7 - \frac{(200.3)^2}{2(1004)} = 861.7\,\text{K}$$

$$\frac{T_{3'}}{T_{03'}} = \frac{T_3}{T_{03}} \Rightarrow T_{03'} = T_{3'}\frac{T_{03}}{T_3} = (807.5)\frac{881.7}{861.7} = 826.2\,\text{K}$$

The efficiencies and spouting velocities can now be calculated:

$$\eta_{ts} = \frac{T_{01} - T_{03}}{T_{01} - T_{3'}} = 0.811$$

$$\eta_{tt} = \frac{T_{01} - T_{03}}{T_{01} - T_{03'}} = 0.851$$

$$c_0 = \sqrt{2C_p(T_{01} - T_{3'})} = \sqrt{2(1004)(1200 - 807.5)} = 887.9\,\text{m/s}$$

Degree of reaction can be calculated as

$$R = \frac{T_2 - T_3}{T_{02} - T_{03}} = \frac{T_2 - T_3}{T_{01} - T_{03}} = \frac{1019.7 - 861.7}{1200.0 - 881.7} = 0.496\% \text{ or } 49.6\%$$

Comment: Degree of reaction can also be calculated from the velocities using Equation 3.14 with subscripts 1 and 2 replaced by 2 and 3 to correspond to the inlet and exit of the rotor.

Example 7.4

An IFR turbine is to be designed to produce 500 kW when rotating at 40,000 rpm. The combustor gases enter the turbine at 900 K and are exhausted to the atmosphere. Assume that the local atmospheric pressure is 100 kPa and that the fluid at the exit has no swirl. Assume $k = 1.33$ and $R = 287$ J/kg·K.

Solution: The thermodynamic processes and the velocity triangles for a radial flow 90° IFR turbine are as follows:

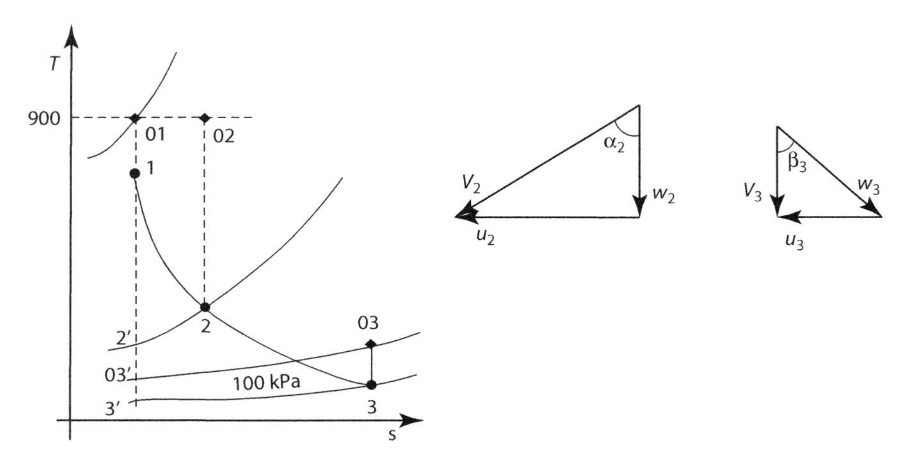

$$P = 500\,\text{kW}; \qquad k = 1.33; \; R = 287\,\text{J/kg}\cdot\text{K}$$

$$C_p = \frac{kR}{k-1} = 1157\,\text{J/kg}\cdot\text{K}$$

$$T_{01} = T_{02} = 900\,\text{K}; \quad p_3 = p_{3'} = p_{3''} = 100\,\text{kPa}$$

$$N = 40{,}000\,\text{rpm} = 4188\,\text{rad/s}$$

Pick $N_s = 0.6$ and $D_s = 2.9$. Hence,

$$N_s D_s = 1.74 = 2.83\frac{u_2}{c_0} \quad \Rightarrow \quad \frac{u_2}{c_0} = 0.61$$

$$\text{Pick } u_2 = 475\,\text{m/s} \Rightarrow c_0 = 772\,\text{m/s}$$

$$\text{But } C_p\left(T_{01} - T_{3'}\right) = \frac{c_0^{\;2}}{2} \Rightarrow T_{3'} = 642\,\text{K}$$

The process 01–3′ is isentropic and hence

$$\frac{p_{3'}}{p_{01}} = \left(\frac{T_{3'}}{T_{01}}\right)^{\frac{k}{(k-1)}} \Rightarrow p_{01} = 390\,\text{kPa}$$

$$E = u_2^2 = (475\,\text{m/s})^2 = 225{,}625\,\text{J/kg}$$

$$\dot{m} = \frac{P}{E} = \frac{(500)(1000)}{225{,}625} = 2.21\,\text{kg/s}$$

Since the blade and rotation speeds are known, the rotor dimensions can be calculated:

$$u_2 = \omega r_2 \Rightarrow r_2 = \frac{u_2}{\omega} = \frac{475}{4188} = 0.113\,\text{m or } D_2 = 0.226\,\text{m} = 22.6\,\text{cm}$$

Now, some of the parameters need to be picked:

$$\text{Pick } \frac{D_{3s}}{D_2} = 0.7 \Rightarrow D_{3s} = 15.8\,\text{cm}$$

Pick $\alpha_2 = 71.5^0$ and now complete the inlet velocity triangle

$$w_2 = \frac{u_2}{\tan\alpha_2} = \frac{475}{\tan 71.5} = 159\,\text{m/s};\; V_2 = \frac{u_2}{\sin\alpha_2} = \frac{475}{\sin 71.5} = 501\,\text{m/s}$$

The inlet Mach number can now be calculated:

$$T_{01} = T_{02} = T_2 + \frac{V_2^2}{2C_p} \Rightarrow T_2 = T_{01} - \frac{V_2^2}{2C_p} = 900 - \frac{(501)^2}{2(1157)} = 791.6\,\text{K}$$

$$a_2 = \text{Speed of sound} = \sqrt{kRT_2} = \sqrt{(1.33)(287)(791.6)} = 549.6\,\text{m/s}$$

$$M_2 = \frac{V_2}{a_2} = 0.91$$

To calculate $T_{2'}$, the value of λ_n needs to be picked:

$$\text{Pick } \lambda_n = 0.1 = \frac{T_2 - T_{2'}}{\left(\frac{V_2^2}{2C_p}\right)} \Rightarrow T_2 - T_{2'} = (0.1)\frac{V_2^2}{2C_p} = 10.8\,\text{K}$$

$$T_{2'} = T_2 - 10.8 = 791.6 - 10.8 = 780.8\,\text{K}$$

Since the process 01 to 2′ is isentropic,

$$\frac{p_{2'}}{p_{01}} = \left(\frac{T_{2'}}{T_{01}}\right)^{\frac{k}{(k-1)}} \Rightarrow p_{2'} = 219\,\text{kPa} = p_2$$

$$\text{Hence, } \rho_2 = \frac{p_2}{RT_2} = 0.968\,\text{kg/m}^3$$

$$\dot{m} = \rho_2 (V_{rad})_2 A_2 = \rho_2 w_2 A_2 \Rightarrow A_2 = \frac{\dot{m}}{\rho_2 w_2} = 0.014\,\text{m}^2$$

$$\text{But, } A_2 = \pi D_2 b_2 \Rightarrow b_2 = \frac{A_2}{\pi D_2} = \frac{A_2}{2\pi r_2} = 0.02\,\text{m} = 2\,\text{cm}$$

Now *check* to see if the ratio of b_2 to D_2 is within limits. Thus, $b_2/D_2 = 0.089$, which is acceptable.

Pick an acceptable value of D_{3h}, say $D_{3h} = 7$ cm.

Check the ratio $\frac{D_{3h}}{D_{3s}} = \frac{7}{15.8} = 0.44$, which is within limits.

Now, the exit velocity triangle needs to be constructed at the mean root mean square (rms) diameter at the exit.

$$D_3 = \text{mean rms diameter at rotor exit}$$

$$= \sqrt{\frac{D_{3s}^2 + D_{3h}^2}{2}} = 0.123\,\text{m} = 12.3\,\text{cm}$$

Check the ratio of the rotor exit to rotor inlet diameters; that is, $\frac{D_3}{D_2} = \frac{12.3}{22.6} = 0.54$, which is within limits.

Pick an acceptable value of rotor exit blade angle. Choose $\beta_3 = 52°$.

The remaining parameters and the exit velocity triangle can now be calculated.

$$E = C_p\left(T_{01} - T_{03}\right) \Rightarrow T_{03} = T_{01} - \frac{E}{C_p} = 900 - \frac{225625}{1157} = 704.9\,\text{K}$$

$$u_3 = \omega r_3 = \omega\frac{D_3}{2} = 4188\frac{(0.123)}{2} = 258\,\text{m/s}$$

$$w_3 = \frac{u_3}{\sin\beta_3} = \frac{258}{\sin 52} = 326\,\text{m/s};\ V_3 = \frac{u_3}{\tan\beta_3} = \frac{258}{\tan 52} = 201\,\text{m/s}$$

Finally *check* if the ratios w_3/w_2 and V_3/u_2 are within limits. These values are found to be 2.05 and 0.422, respectively, within acceptable limits.

Finally, the number of blades can be calculated to be $n_B = 0.1047\,(110-\alpha_2)\tan\alpha_2 = 12.04$. Pick 12 blades.

Comment: The design given is not unique, and several changes can be made to suit the requirements.

Example 7.5

The preliminary analysis of an IFR turbine is to be performed for later design modifications. It has flat radial blades at the inlet. The nozzle ring at the inlet is designed so that $\alpha_2 = 65°$. The gas properties are those of a mixture of combustion gases, with $k = 4/3$ and $R = 286.2$ J/kg·K. The rotor speed is 41,000 rpm and the inlet stagnation conditions to the stator are $p_{01} = 240$ kPa and $T_{01} = 1200$ K. The inlet diameter is 15 cm while at the exit the hub and tip diameters are 3 and 12 cm, respectively. The fluid exits the rotor axially with a speed of 150 m/s. The total to static efficiency is 81% and the nozzle velocity coefficient is 96.4%. Estimate the following: (1) nozzle loss coefficient, (2) spouting velocity, (3) blade angles at the exit on the hub, tip, and mean diameter, (4) degree of reaction, and (5) power output.

Solution: The thermodynamic processes along with the velocity triangles are shown in the following diagram:

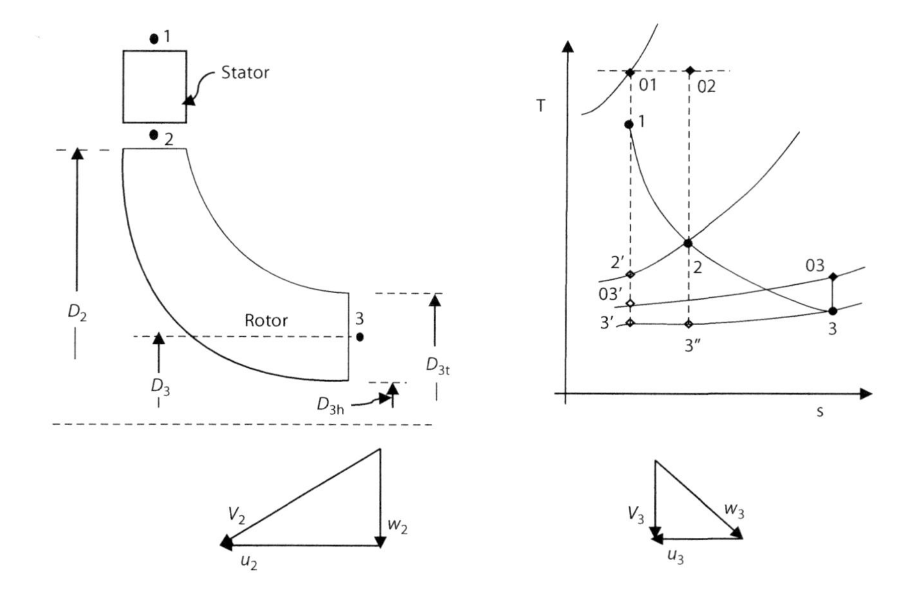

The values of gas constant R and ratio of specific heats k are given as 286.2 J/kg·K and 4/3, respectively. From the definition of $C_p = kR/(k-1)$, its value can be calculated to be 1145 J/kg·K.

The following data are given:

$$p_{01} = 240\,\text{kPa};\ T_{01} = T_{02} = 1200\,\text{K};\ N = 41{,}000\,\text{rpm} = 4293\,\text{rad/s}$$

$$\eta_{ts} = 81\%;\ C_n = 96.4\%;\ V_3 = V_{3ax} = 150\,\text{m/s}$$

$$D_2 = 0.15\,\text{m};\ D_{3t} = 0.12\,\text{m};\ D_{3h} = 0.03\,\text{m};\ \alpha_2 = 65^0$$

Some of the parameters for the inlet triangle can be calculated using the blade speed and trigonometry:

$$u_2 = \frac{\omega D_2}{2} = \frac{(4293)(0.15)}{2} = 322\,\text{m/s}$$

$$E = u_2 V_{u2} - u_3 V_{u3} = u_2 V_{u2} = u_2^2 = (322)^2 = 103692\,\text{J/kg}$$

$$V_2 = \frac{u_2}{\sin\alpha_2} = \frac{322}{\sin 65} = 355\,\text{m/s};\ w_2 = \frac{u_2}{\tan\alpha_2} = \frac{322}{\tan 65} = 150.2\,\text{m/s}$$

However,

$$E = C_p(T_{02} - T_{03}) \Rightarrow T_{03} = T_{02} - \frac{E}{C_p} = 1200 - \frac{103{,}692}{1145} = 1109\,\text{K}$$

$$T_{02} = T_2 + \frac{V_2^2}{2C_p} \Rightarrow T_2 = T_{02} - \frac{V_2^2}{2C_p} = 1200 - \frac{(355)^2}{2(1145)} = 1145\,\text{K}$$

$$T_{03} = T_3 + \frac{V_3^2}{2C_p} \Rightarrow T_3 = T_{03} - \frac{V_3^2}{2C_p} = 1109 - \frac{(150)^2}{2(1145)} = 1099.6\,\text{K}$$

Assuming that the exit velocity is at the mean diameter,

$$D_3 = \frac{D_{3t} + D_{3h}}{2} = \frac{0.12 + 0.03}{2} = 0.075\,\text{m}$$

$$u_3 = \frac{\omega D_3}{2} = \frac{(4293)(0.075)}{2} = 161\,\text{m/s} \text{ and } w_3 = \sqrt{V_3^2 + u_3^2} = 220\,\text{m/s}$$

From the definition of $C_n = V_2/V_{2'}$,

$$C_n = 0.964 = \frac{V_2}{V_{2'}} \Rightarrow V_{2'} = \frac{V_2}{C_n} = \frac{355}{0.964} = 368.6\,\text{m/s}$$

$$\text{Also, } T_{02} = T_{2'} + \frac{(V_{2'})^2}{2C_p} \Rightarrow T_{2'} = T_{02} - \frac{(V_{2'})^2}{2C_p} = 1140.7\,\text{m/s}$$

$$\eta_{ts} \equiv \frac{T_{01} - T_{02}}{T_{01} - T_{3'}} = \frac{E}{(c_0^2)/2} \Rightarrow c_0 = \sqrt{\frac{2E}{\eta_{ts}}} = \sqrt{\frac{2(103692)}{0.81}} = 506\,\text{m/s}$$

But, $$\frac{c_0^2}{2} = h_{01} - h_{3'} = C_p(T_{01} - T_{3'}) \Rightarrow T_{3'} = T_{01} - \frac{c_0^2}{2C_p} = 1088.2\,\text{K}$$

The process 01–3′ is isentropic and hence

$$\frac{p_{3'}}{p_{01}} = \left(\frac{T_{3'}}{T_{01}}\right)^{\frac{k}{(k-1)}} \Rightarrow p_{3'} = p_{01}\left(\frac{T_{3'}}{T_{01}}\right)^{\frac{k}{(k-1)}} = 162.3\,\text{kPa}$$

$$\frac{T_{03}}{T_3} = \frac{T_{03'}}{T_{3'}} \Rightarrow T_{03'} = T_{3'}\frac{T_{03}}{T_3} = (1088.2)\frac{1109}{1099.6} = 1098\,\text{K}$$

$$\text{Hence, } \eta_{tt} \equiv \frac{T_{01} - T_{03}}{T_{01} - T_{03'}} = 88.7\%$$

The blade angles at the tip, mean diameter, and hub can be calculated:

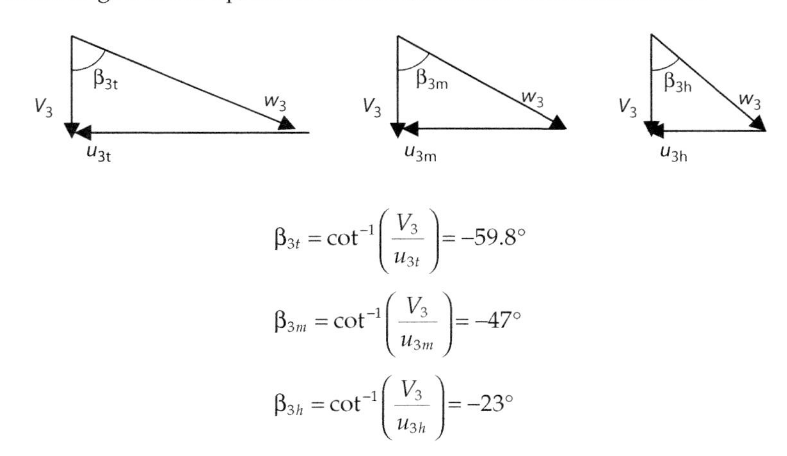

$$\beta_{3t} = \cot^{-1}\left(\frac{V_3}{u_{3t}}\right) = -59.8^\circ$$

$$\beta_{3m} = \cot^{-1}\left(\frac{V_3}{u_{3m}}\right) = -47^\circ$$

$$\beta_{3h} = \cot^{-1}\left(\frac{V_3}{u_{3h}}\right) = -23^\circ$$

Now, the power can be calculated by calculating the mass flow rate:

$$\rho_3 = \frac{p_3}{RT_3} = \frac{(163.2)(1000)}{(286)(1099)} = 0.515\,\text{kg/m}^3$$

$$\dot{m} = \rho_3 (V_{ax})_3 A_3 = \rho_3 V_3 A_3 = \rho_3 V_3 \frac{\pi}{4}(D_{3t}^2 - D_{3h}^2) = 0.82\,\text{kg/s}$$

$$P = \dot{m}E = (0.82)(103{,}692) = 85\,\text{kW}$$

The degree of reaction can now be calculated:

$$R = \frac{\text{Static enthalpy drop across rotor}}{\text{Stagnation (total) enthalpy drop across rotor}}$$

$$= \frac{T_2 - T_3}{T_{02} - T_{03}} = \frac{T_2 - T_3}{T_{01} - T_{03}} = \frac{1145 - 1099.6}{1200 - 1109} = 0.5$$

Alternatively, degree of reaction can also be calculated from the velocities:

$$R = \frac{\left(u_2^2 - u_3^2\right) + \left(w_3^2 - w_2^2\right)}{\left(V_2^2 - V_3^2\right) + \left(u_2^2 - u_3^2\right) + \left(w_3^2 - w_2^2\right)} = 0.5$$

Finally, the nozzle coefficient can be calculated as

$$\lambda_n = \frac{1}{C_n^2} - 1 = 0.076$$

$$\text{Also, } \lambda_n = \frac{h_2 - h_2'}{\frac{1}{2}V_2^2} = \frac{C_p(T_2 - T_2')}{\frac{1}{2}V_2^2} = \frac{(1145)(1145 - 1140.7)}{\frac{1}{2}(355)^2} = 0.076$$

Comment: In the actual design of the blade, the blade angle at the exit changes from the hub to the tip, as shown in the problem, to give a smooth curvature to the blade. Also, note that the blade angles are negative according to the convention, that is, negative when relative velocity opposes the blade velocity.

PROBLEMS

All pressures given are absolute unless specified otherwise. Use air properties unless specified otherwise.

7.1 An IFR turbine with flat blades rotates at 20,000 rpm. The radii at the inlet and mean outlet are 8 in. and 4 in., respectively. If the static conditions at the exit are 14.7 psia and 800°F, calculate the fluid angle (α_3) at the exit and the mass flow rate. The power produced is 140 hp and the blade width (height) is 3 in. at both the inlet and exit.

Ans: 81.6°; 1.7 lbm/s

7.2 The following data refer to an IFR turbine:

Tip diameter = 6 in.

Rate of rotation = 55,000 rpm

Stagnation temperature at rotor inlet = 1900 R

Pressure ratio across the turbine $= p_{01}/p_3 = 2.6$

Mass flow rate = 0.6 lbm/s

Assume standard air properties and that the exit from the turbine has no swirl. Calculate the (a) power produced by the turbine, (b) spouting velocity, and (c) total to static efficiency.

Ans: 70.2 hp; 2335 ft./s; 76%

7.3 Consider the standard inward flow turbine consisting of a ring of stator vanes that act like a nozzle, a flat blade inward flow rotor, and a diffuser. The fuel/air mixture leaves the combustor with a stagnation temperature and pressure of 800°C and 350 kPa, respectively. The exhaust from the diffuser exits into the atmosphere and it can be assumed that the exit kinetic energy is zero. The Mach number at the rotor inlet is 0.95 and the total to total efficiency is 0.85. If the rotor tip diameter is 50 cm, determine the rotational speed and the nozzle angle at the inlet. Assume standard air properties and that the fluid at exit of the rotor has no swirl.

Ans: 19,986 rpm; 65.7°

7.4 An inward flow radial turbine with flat blades has the following specifications:

Speed of rotation = 10,000 rpm

Power = 50 hp

Inlet and mean exit radii = 6 in. and 3 in.

Fluid angle at rotor exit (α_3) = 60°

Static conditions at rotor exit = 14.7 psia and 100 F

Estimate the height of the rotor at the exit.

Ans: 0.256 ft.

7.5 Show that for gas turbines, either axial or radial, $\eta_{tt} > \eta_{ts}$.

7.6 Show that the total to static efficiency is also given by the equation $\eta_{ts} = 2E/c_0^2$.

7.7 Consider a turbine consisting of two stages. The overall pressure ratio is four and the inlet conditions are 500 K and a pressure of four atmospheres. Assume that each stage has the same stage efficiency and pressure ratio. If the isentropic turbine efficiency is 85%, calculate the stage efficiency.

Ans: $\eta_{st=}$ 83.3%; $\eta_t > \eta_{st}$

7.8 An alternative definition of total to static efficiency can be obtained using the so-called power ratio, as $P_R = E/h_{01}$. Show that

$$\eta_{ts} = \frac{P_R}{\left[1 - \left(\frac{p_3}{p_{01}}\right)^{(k-1)/k}\right]}$$

7.9 Combining the rotor loss coefficient λ_r and rotor velocity coefficient C_r, as defined in Equations 7.17 and 7.18, derive Equation 7.19.

7.10 Referring to the enthalpy/entropy diagram for an IFR turbine in Figure 7.7, show that

$$\frac{T_{03}}{T_3} = \frac{T_{03'}}{T_{3'}}$$

7.11 Obtain the relationship between η_{tt}, η_{ts}, and E given by Equation 7.11.

7.12 Obtain the relationship between the specific speed N_s and flow coefficient φ given by Equation 7.23.

7.13 The following data were obtained in connection with estimating the rotor and nozzle loss coefficients for a radial turbine:

Speed of turbine = 20,000 rpm

Inlet diameter/exit mean diameter ($=D_3/D_2$) = 0.5

Stator inlet stagnation pressure = 700 kPa

Stator inlet stagnation temperature = 1200 K

Rotor inlet pressure = 430 kPa

Rotor inlet temperature = 1060 K

Rotor exit pressure = 235 kPa

Rotor exit temperature = 920 K

The exit velocity from the rotor is measured at 245 m/s and the ratio of inlet to exit mean diameters ($=D_2/D_3$) is 2.0. Calculate the (a) nozzle loss coefficient, (b) rotor loss coefficient, (c) total to static efficiency, and (d) nozzle angle at the inlet. The fluid leaves the rotor with no exit swirl.

Ans: $\eta_{ts} = 77.8\%$; $\lambda_r = 0.458$; $\lambda_n = 0.114$

7.14 For an inward flow radial (IRF) turbine, the stagnation pressure and temperature at the inlet are three atmospheres and 350°F, respectively. The total to total and total to static efficiencies (η_{tt} and η_{ts}) are 82% and 73%, respectively. The turbine exhausts to the atmosphere. At the inlet to the rotor, the radius is 5 in. and blade width is 1 in. The rotor loss coefficient is 0.12. The nozzle angle α_2 is 70° and the blade angle at the exit β_3 is 50° (measured with respect to blade speed u_3). If the ratio of inlet diameter to exit mean diameter is 1.8, calculate the following: (a) spouting velocity, (b) rotor loss coefficient, (c) rotor speed, and (d) power produced. Assume that the fluid leaves the rotor axially.

Ans:1619 ft./s; 0.78; 22,417 rpm; 435.8 hp

7.15 Repeat Problem 7.14 assuming that the blades are flat, that is, the relative velocity at the exit is purely axial. Take α_3 to be 68°.

Ans: 1619 ft./s; 0.62; 26,960 rpm; 441.0 hp

7.16 Design a 90° IFR turbine to meet the following specifications:

Required power = 800 kW

Runner speed = 48,000 rpm

Total inlet temperature (T_{01}) = 1000 K

Exhaust pressure = 100 kPa

Assume that the combustion products at the inlet have a ratio of specific heats $k = 1.35$ and $R = 287$ J/kg·K. The fluid leaves the rotor with no whirl.

7.17 The following design problem has the complete inlet conditions specified. Combustion gases from the combustor are available with total pressure and temperature of 60 psia and 2000 R, respectively. Design a 90° IFR turbine that can spin at 40,000 rpm and can produce 750 hp. The rotor exhausts to the atmosphere with no whirl. Assume that the combustion products at the inlet have a ratio of specific heats $k = 1.35$ and $R = 1716$ ft·lbf/slug·°R.

7.18 The following data apply to an IFR turbine with flat radial blades at the inlet. The nozzle ring at the inlet is designed so that $\alpha_2 = 64.5°$. The rotor speed is 30,000 rpm and it receives combustion products from the combustor with stagnation pressure and temperature of $p_{01} = 400$ kPa and $T_{01} = 1200$ K, respectively. At the exit, the tip diameter is 12 cm and the ratio of tip to hub diameter is 4.0, while the inlet diameter is 20 cm. The fluid exits the rotor axially with a speed of 150 m/s. The total to static efficiency is 80% and the nozzle loss coefficient is 0.041. Calculate the following: (a) degree of reaction, (b) total to total efficiency, (c) blade angles at the exit on the hub, tip, and mean diameter, (d) nozzle velocity coefficient, (e) mass flow rate, and (f) rotor loss coefficient. Assume properties for air.

Ans: $\eta_{tt} = 87.9\%$; $R = 0.5$; $\dot{m} = 1.39$ kg / s

7.19 Derive Equation 7.16, which gives the relationship between the nozzle loss coefficient λ_n and nozzle velocity coefficient C_n.

8

Axial Gas Turbines

In the last chapter, the principles of radial flow gas turbines were studied. Another class of gas turbines that plays an important role in aircraft engines belongs to the axial flow type. Such turbines typically produce less power than radial turbines. Hence, they are placed in stages for appreciable power production. As with radial turbines, the working fluid undergoes significant pressure changes and hence temperature is an integral part of the analysis. In this chapter, the features of axial gas turbines will be discussed.

Upon completion of this chapter, the student will be able to

- Analyze the velocity diagrams for axial flow gas turbines
- Predict the stage performance of gas turbines
- Perform analysis and preliminary design of such turbines

8.1 Introduction

Axial flow turbines are most commonly used in gas aircraft turbine engines. Such a turbine could consist of either one or several stages, with each stage consisting of a stationary set of blades (called *stator blades*) that act like nozzles, and a set of moving blades called *rotor blades*. The relative velocities in axial flow turbines are considerably higher than in axial flow compressors, with higher changes in enthalpy per stage. The nozzles or stator blades increase the tangential velocity in the direction of blade rotation with a corresponding drop in pressure. The converse is true in rotor blades, in which the tangential velocity is decreased. Since higher changes in enthalpy take place per stage in an axial flow turbine, the total number of stages in an axial flow turbine is considerably lower than in an axial flow compressor. The reason for this is that in compressors the flow diffuses, with a corresponding rise in pressure. Diffusion allows for only a moderate pressure rise to prevent separation. As opposed to this, the flow in turbines results in an expansion process accompanied by a reduction in pressure and separation is less of an issue. This results in larger pressure drops per stage in axial flow turbines and, consequently, a lower number of stages.

8.2 Stage Velocity Diagrams and Energy Transfer

Unlike radial flow turbines, the flow in axial flow turbines is predominantly in the axial direction. By their very design, they produce significantly lower power than their radial counterparts. Hence, instead of one single turbine, axial flow turbines are usually arranged in stages with alternate rows of stator blades (which act primarily as nozzles) and rotor blades. The flow through one such stage, as shown in Figure 8.1, is considered next.

For simplicity, it is assumed that the flow is uniform across the blade cross section, that is, the flow conditions at one radius are the same as at any other radius. Thus, the velocity triangles at the mean line are considered. This analysis is called *pitch line* or *mean line* analysis. Assuming the flow to be uniform at all radii gives a good approximation for short stubby blades; however, as the length of blades increases, three-dimensional analysis is required. The discussion presented is valid for both axial flow gas turbines and steam turbines, as long as the steam is in the gaseous form (superheated).

The gas enters the nozzle at static pressure p_1, temperature T_1, and velocity V_1, and is expanded to p_2, T_2, and an increased velocity V_2. It then enters the rotor, where it is deflected and is further expanded to p_3, T_3, and exit velocity V_3. For repeating stages (also called *normal stages*), exit velocity V_3 is equal in magnitude and direction to V_1. Thus, $\alpha_1 = \alpha_3$ and the axial velocity is constant. A comment must be made about the sign convention used for blade and nozzle angles: *all the angles are measured with respect to the axial direction* (this is unlike pumps or hydraulic turbines, where the angles were measured with respect to the

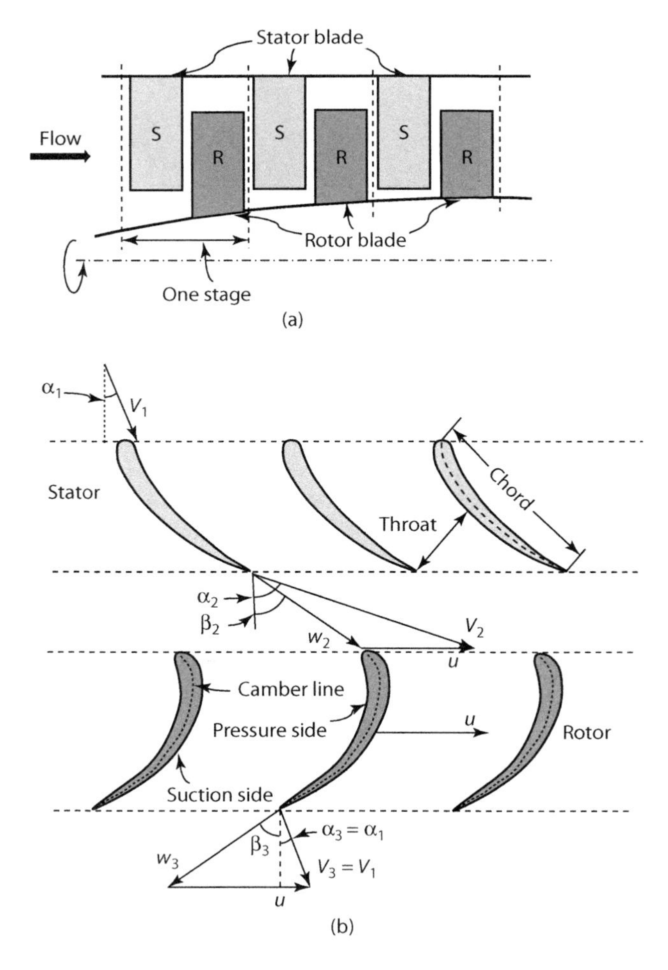

FIGURE 8.1
(a) Arrangement of stator and rotor blades in an axial turbine and (b) single axial turbine repeating stage.

wheel tangent, that is, the blade velocity u). The sign convention used is that the angles will be considered positive when their tangential components are in the same direction as blade velocity, and negative if they are opposite to blade velocity. In Figure 8.1, α_2 and β_2 are positive since V_{u2} and w_{u2} are in the same direction as u; the angle α_3 is positive since V_{u3} is in same direction as u, and β_3 is negative since w_{u3} is in the opposite direction to u. However, the signs of the angles are less relevant if the velocity diagrams are drawn showing the correct directions. Finally, α_3 is called the *swirl angle* (this term has been used earlier), and fluid exiting with no swirl would make the exit velocity triangle a right triangle.

The energy equation can be applied in the same way as was done for radial turbines and yields similar results. For example, across the stator, the total enthalpies would be the same, as there is no exchange of energy, while across the rotor, the difference between inlet and exit total enthalpies would equal E. The temperature versus specific entropy (T–s) diagram looks similar to Figure 7.5. The definitions of total to static efficiency, total to total efficiency, and spouting velocity would also be identical to Equations 7.8, 7.9, and 7.10. The degree of reaction in the context of axial turbines is defined as the ratio of the static enthalpy drop across the stator to the total (stagnation) enthalpy drop across the entire rotor. Thus,

$$R \equiv \frac{\text{Static enthalpy drop in rotor}}{\text{Total (stagnation) enthalpy drop in the stage}} = \frac{h_2 - h_3}{h_{01} - h_{03}} = \frac{\Delta h_{rotor}}{\Delta h_{0,stage}} = \frac{T_2 - T_3}{T_{01} - T_{03}} \tag{8.1}$$

For a normal or repeating stage ($V_3 = V_1$), this equation can be simplified into the usual definition given in Equation 3.13 and 3.14. This is shown in Example 8.7. Also, for repeating stages, another interesting interpretation can be given in terms of static enthalpies as follows (see Problem 8.13):

$$R = \frac{\text{Static enthalpy drop in rotor}}{\text{Static enthalpy drop in the stator and rotor}} = \frac{h_2 - h_3}{(h_1 - h_2) + (h_2 - h_3)} = \frac{h_2 - h_3}{h_1 - h_3} = \frac{T_2 - T_3}{T_1 - T_3} \tag{8.1a}$$

In addition to the degree of reaction, the other variables that influence the performance of axial flow stages are the *stage loading coefficient* (also known as the *blade loading coefficient*) and the *flow coefficient*. These are defined next. The blade loading coefficient is defined as the ratio of the energy extracted from the stage to the square of the blade speed. Thus,

$$\psi = \frac{\Delta h_{0,stage}}{u^2} = \frac{E}{u^2} = \frac{(V_{u2} - V_{u3})}{u} = \frac{\Delta V_u}{u} \tag{8.2}$$

where in the definition of E has been used. The change in angular momentum of the fluid as it flows through the rotor is given by ΔV_u. The flow coefficient is defined as the ratio of the axial velocity to blade velocity.

$$\phi = \frac{V_a}{u} \tag{8.3}$$

High values of the blade loading coefficient would imply blades that have high curvature (impulse blades), indicating highly skewed velocity triangles. As was seen in Chapter 3, impulse stages produce twice as much power as reaction stages for the same speed, but at a lower utilization factor.

It should also be mentioned that higher values of ϕ would imply higher flow rates. It can be shown that ϕ and ψ are related to the traditional flow coefficient Q/ND^3 and head coefficient gH/N^2D^2 introduced in Chapter 2 (see Problem 8.8).

From the velocity triangles shown in Figure 8.1, several relationships can be obtained between R, ϕ, and ψ. All of these can be obtained from trigonometry; hence, the detailed derivations are omitted here and are assigned as exercises (see Problem 8.9).

8.3 Isentropic versus Stage Efficiency

The concepts of isentropic efficiency and stage efficiency have different meanings in the context of compressors or gas turbines, especially since the compression or expansion processes take place over several stages. This is particularly true in the case of axial flow machines. Thus, each stage is affected by the cumulative inefficiency of the preceding stages, since the operating temperature is higher in the preceding stages. This effect can be visualized in the T–s diagram for the expansion process shown in Figure 8.2, in which a gas expands between pressures p_1 and p_2 in three stages. For simplicity, the stages are assumed to be identical, having the same pressure ratio and same isentropic efficiency; that is, ratio of actual to ideal power. It is to be noted that although efficiencies are defined in terms of enthalpies, they are written here and subsequently in terms of temperatures. This is appropriate since the gas is assumed to be an ideal gas with constant specific heats.

Thus, stage efficiency can be written as,

$$\eta_{st} = \frac{T_1 - T_c}{T_1 - T_a} = \frac{T_c - T_f}{T_c - T_d} = \frac{T_f - T_2}{T_f - T_g} = \frac{(T_1 - T_c) + (T_c - T_f) + (T_f - T_2)}{(T_1 - T_a) + (T_c - T_d) + (T_f - T_g)} \tag{8.4}$$

In the last equation, the properties of equality of fractions have been used. After simplification, Equation 8.4 becomes

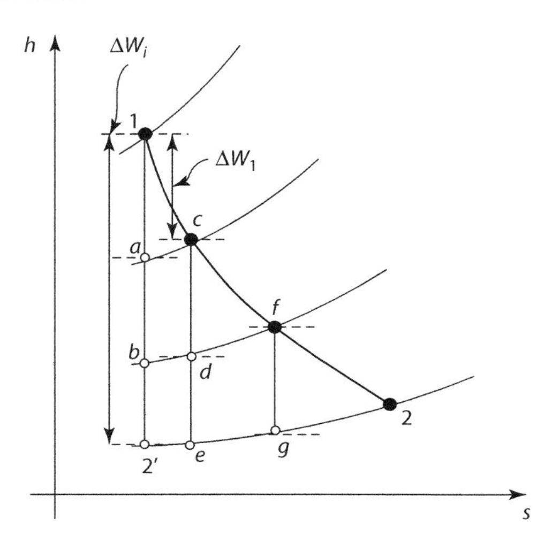

FIGURE 8.2
Isentropic and multistage expansion process in axial flow turbines.

$$\eta_{st} = \frac{(T_1 - T_2)}{(T_1 - T_a) + (T_c - T_d) + (T_f - T_g)} \tag{8.5}$$

The overall turbine efficiency (or simply the turbine efficiency) is given by

$$\eta_t = \frac{(T_1 - T_2)}{(T_1 - T_{2'})} \tag{8.6}$$

However, since the constant pressure lines diverge on the *T–s* diagram, it follows that

$$(T_a - T_b) < (T_c - T_d) \text{ and } (T_b - T_{2'}) < (T_d - T_e) < (T_f - T_g)$$

Thus,

$$\begin{aligned} \eta_t = \frac{(T_1 - T_2)}{(T_1 - T_{2'})} &= \frac{(T_1 - T_2)}{(T_1 - T_a) + (T_a - T_b) + (T_b - T_{2'})} \\ &> \frac{(T_1 - T_2)}{(T_1 - T_a) + (T_c - T_d) + (T_f - T_g)} = \eta_{st} \end{aligned} \tag{8.7}$$

In other words, turbine efficiency η_t is greater than stage efficiency η_{st}, that is, $\eta_t > \eta_{st}$. Stage efficiency is also known as *polytropic efficiency*; similarly, the turbine efficiency is sometimes called *isentropic* or *adiabatic efficiency*. In this book, only the terms *turbine efficiency* η_t (or *compressor efficiency* η_c) and *stage efficiency* η_{st} will be used.

By considering an infinitesimal expansion process, an expression relating the stage efficiency and turbine efficiency can be derived. Consider the incremental expansion process shown in Figure 8.3.

The small stage (polytropic) efficiency is

$$\eta_{st} = \frac{dh}{dh_i} = \frac{C_p dT}{v dp} = \frac{\frac{kR}{k-1} dT}{\frac{RT}{p} dp} = \frac{k}{k-1} \frac{\frac{dT}{T}}{\frac{dp}{p}} \tag{8.8}$$

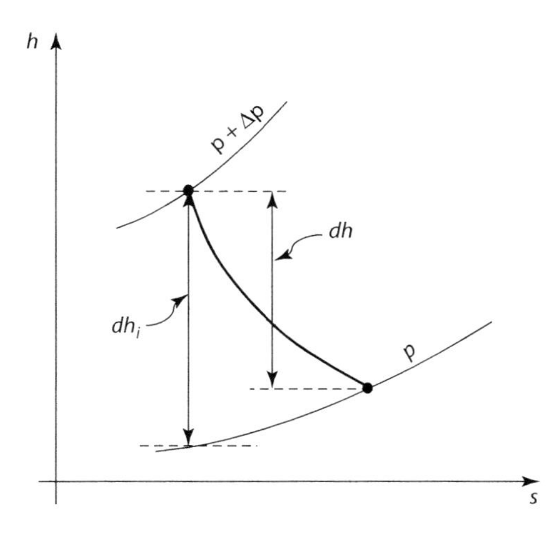

FIGURE 8.3
Incremental changes for expansion process.

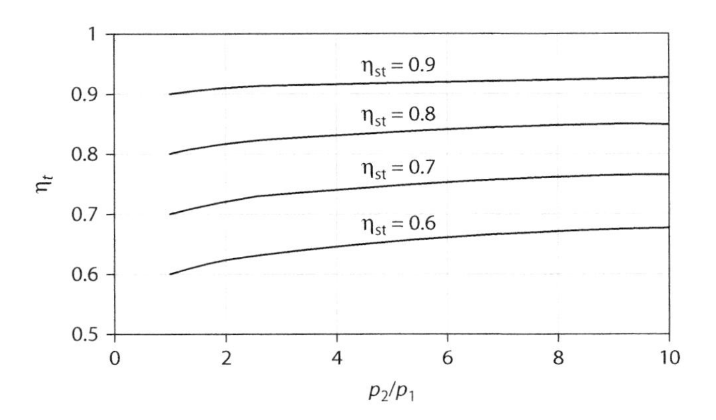

FIGURE 8.4
Effect of pressure ratio and stage efficiency on turbine efficiency.

In Equation 8.8, the relationship $dh_i - vdp = Tds$ has been used. However, $ds = 0$ for isentropic processes. Hence $dh_i = vdp$. Simplification yields

$$\frac{dT}{T} = \frac{(k-1)\eta_{st}}{k}\frac{dp}{p}$$

Integrating for the entire expansion process and assuming that each infinitesimal stage is of equal efficiency yields

$$\frac{T_2}{T_1} = \left(\frac{p_2}{p_1}\right)^{(k-1)\eta_{st}/k_p} \tag{8.9}$$

Using the definition of turbine efficiency as defined in Equation 8.7 gives

$$\eta_t = \frac{\left(1 - \dfrac{T_2}{T_1}\right)}{\left(1 - \dfrac{T_{2'}}{T_1}\right)} = \left[1 - \left(\frac{p_2}{p_1}\right)^{(k-1)\eta_{st}/k}\right] \Bigg/ \left[1 - \left(\frac{p_2}{p_1}\right)^{(k-1)/k}\right] \tag{8.10}$$

A plot of the turbine efficiency η_t versus pressure ratio p_2/p_1 for various values of stage efficiency η_{st} is shown in Figure 8.4. For all pressure ratios, the turbine efficiency is greater than the stage efficiency. Also, $\eta_t \rightarrow \eta_{st}$ as $p_2/p_1 \rightarrow 1$. However, when p_2/p_1 becomes unity, the expansion process becomes meaningless.

A similar set of relationships between η_{st} and η_c can also be obtained for compressors (this is discussed in Chapter 10).

Example 8.1

Consider an axial turbine consisting of two stages. The overall pressure ratio is four. Combustion products enter the turbine at 500 K and a pressure of four atmospheres. Assume that each stage has the same stage efficiency of 85% and that the pressure ratio is also the same for each stage. Calculate the isentropic efficiency of the turbine and compare it with the stage efficiency.

Solution: The turbine efficiency and stage efficiency can be written as (see the figure)

$$\eta_t = \frac{(h_1 - h_3)}{(h_1 - h_{3'})} \text{ and } \eta_{st} = \frac{(h_1 - h_2)}{(h_1 - h_{2'})} = \frac{(h_2 - h_3)}{(h_2 - h_{3''})} = 0.85$$

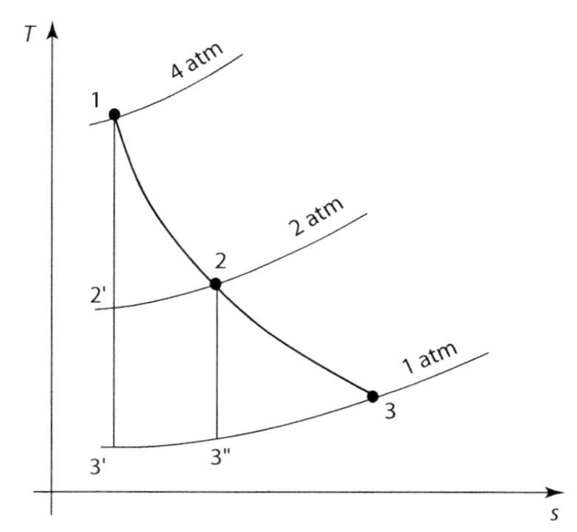

Since 1–2′ and 2–3″ are isentropic processes with equal pressure ratios,

$$\left(\frac{T_1}{T_{2'}}\right) = \left(\frac{p_1}{p_{2'}}\right)^{(k-1)/k} = \left(\frac{p_1}{p_2}\right)^{(k-1)/k} = (2)^{(k-1)/k} = 1.219$$

$$T_{2'} = \frac{T_1}{1.219} = \frac{500}{1.219} = 410.2\,\text{K}$$

From the definition of stage efficiency,

$$(T_1 - T_2) = \eta_{st}(T_1 - T_{2'}) = 0.85(500 - 410.4) = 76.4\,\text{K}$$

$$T_2 = 500 - 76.4 = 423.6\,\text{K}$$

Similarly,

$$\left(\frac{T_2}{T_{3''}}\right) = \left(\frac{p_2}{p_{3''}}\right)^{(k-1)/k} = \left(\frac{p_2}{p_3}\right)^{(k-1)/k} = (2)^{(k-1)/k} = 1.219$$

$$T_{3''} = \frac{T_2}{1.219} = \frac{423.6}{1.219} = 347.5\,\text{K}$$

$$(T_2 - T_3) = \eta_s(T_2 - T_{3''}) = 0.85(423.6 - 347.5) = 64.7\,\text{K}$$

$$T_3 = 423.6 - 64.7 = 358.9\,\text{K}$$

Since process 1–3′ is also isentropic,

$$\left(\frac{T_1}{T_{3'}}\right) = \left(\frac{p_1}{p_{3'}}\right)^{(k-1)/k} = \left(\frac{p_1}{p_3}\right)^{(k-1)/k} = (4)^{(k-1)/k} = 1.486$$

$$T_{3'} = \frac{T_1}{1.486} = \frac{500}{1.486} = 336.8\,K$$

Hence, the turbine isentropic efficiency is

$$\eta_t = \frac{(h_1 - h_3)}{(h_1 - h_{3'})} = \frac{(T_1 - T_3)}{(T_1 - T_{3'})} = \frac{(500 - 358.9)}{(500 - 336.8)} = 0.862$$

The value given is for two stages in the turbine. However, if it is assumed that there is an infinite number of stages, Equation 8.10 can be used to calculate the turbine efficiency.

The value would be

$$\eta_t = \left[1 - (0.25)^{(1.4-1).0.85/1.4}\right] \Big/ \left[1 - (0.25)^{(1.4-1)/1.4}\right] = 0.874$$

Comment: It can be seen that stage efficiency (for both finite and infinite numbers of stages) and overall turbine efficiency have different values; the turbine efficiency is greater than the stage efficiency. It will be seen in Chapter 10 that this is in contrast to compressors, for which the compressor efficiency is less than the stage efficiency.

Example 8.2

Consider the turbine in Example 8.1. If the expansion is from the initial state of 500 K and four atmospheres to a final state of 358.9 K and one atmosphere, calculate the polytropic efficiency (small stage efficiency), assuming that the expansion process takes place in infinitesimal stages.

Solution: For an infinitesimal expansion process as shown in the figure, efficiency can be written as

$$\eta_p = \frac{dh}{dh_i} = \frac{C_p dT}{v dp} = \frac{\frac{kR}{(k-1)} dT}{\frac{RT}{p} dp} = \frac{k}{(k-1)} \frac{\frac{dT}{T}}{\frac{dp}{p}}$$

$$\text{or } \frac{dT}{T} = \eta_p \frac{(k-1)}{k} \frac{dp}{p} \Rightarrow \ln\left(\frac{T_2}{T_1}\right) = \eta_p \frac{(k-1)}{k} \ln\left(\frac{p_2}{p_1}\right)$$

$$\text{or } \eta_p = \frac{k}{(k-1)} \frac{\ln\left(\frac{T_2}{T_1}\right)}{\ln\left(\frac{p_2}{p_1}\right)}$$

Substituting the values,

$$\eta_p = \frac{1.4}{(1.4-1)} \frac{\ln\left(\frac{358.9}{500}\right)}{\ln\left(\frac{1}{4}\right)} = 0.837$$

It is instructive to calculate the isentropic turbine efficiency as follows:

$$\frac{T_1}{T_{2'}} = \left(\frac{p_1}{p_{2'}}\right)^{(k-1)/k} = 4^{(1.4-1)/1.4} = 1.485$$

$$T_{2'} = \frac{500}{1.485} = 336.7\,\text{K}$$

Hence,

$$\eta_t = \frac{(T_1 - T_2)}{(T_1 - T_{2'})} = \frac{(500 - 358.9)}{(500 - 336.7)} = 0.864$$

Thus, the isentropic turbine efficiency is greater than the polytropic (small, even infinitesimal, stage) efficiency.

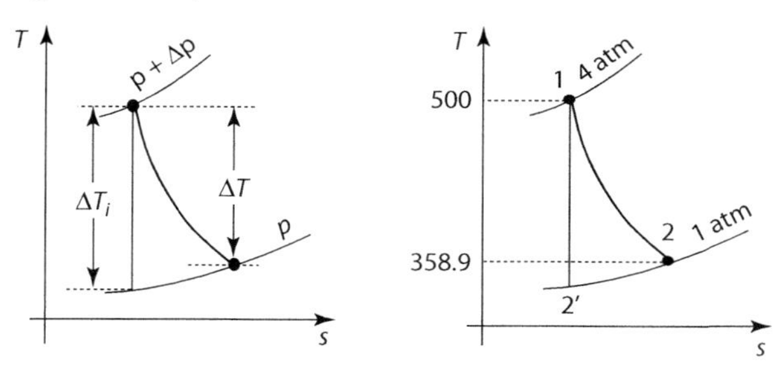

Comment: This is in contrast with compressors, where isentropic compressor efficiency would be less than small stage or infinitesimal stage efficiency. This will be shown in the next chapter.

Example 8.3

A multistage impulse turbine is expected to have a pressure drop from six atmospheres at the inlet to one atmosphere at the exit. The inlet static temperature is 900 K. The stage efficiency of the turbine is 85% and the mean blade speed is 250 m/s. The axial velocity is expected to be constant at 100 m/s. If the nozzle angle is +75° and the blade angles at the inlet and outlet are equal, estimate the number of stages required.

Solution:

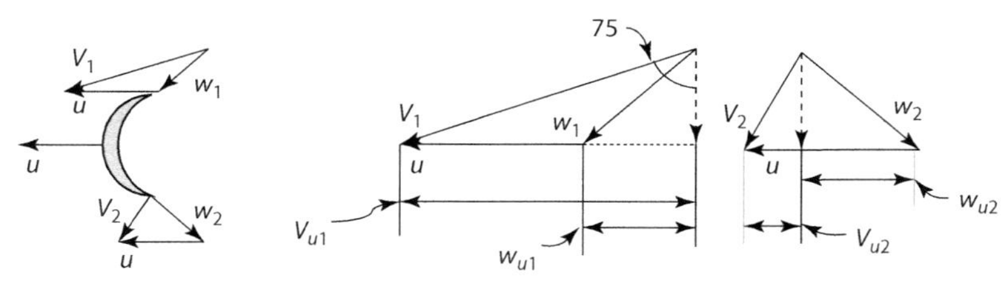

From the given data,

$$u = 250\,\text{m/s};\ \ \alpha_1 = 75°;\ \ V_a = 100\,\text{m/s}$$

From trigonometry, the following results are obtained:

$$V_{u1} = 100\tan 75 = 373.2\,\text{m/s};\ V_1 = \frac{V_a}{\cos 75} = 386.4\,\text{m/s}$$

$$w_{u1} = V_{u1} - u = 123.2\,\text{m/s} = w_{u2}$$

$$w_1 = \sqrt{w_{u1}^2 + V_a^2} = 158.7\,\text{m/s} = w_2$$

For the exit triangle,

$$V_{u2} = u - w_{u2} = 126.8\,\text{m/s}$$

$$V_2 = \sqrt{V_{u2}^2 + V_a^2} = 161.5\,\text{m/s}$$

Now, the energy transfer can be calculated as

$$E = \frac{V_1^2 - V_2^2}{2} + \frac{u_1^2 - u_2^2}{2} + \frac{w_2^2 - w_1^2}{2} = \frac{V_1^2 - V_2^2}{2} = 61600\,\text{J/kg}$$

This is the energy transfer per stage. The stagnation temperature rise per stage is

$$\Delta T_{stage} = \frac{E}{C_p} = \frac{61600}{1006} = 61.3\,\text{K}$$

The overall temperature ratio is given by

$$\frac{p_i}{p_e} = \left(\frac{T_i}{T_e}\right)^{\frac{k}{(k-1)\eta_{st}}} \Rightarrow \frac{T_i}{T_e} = \left(\frac{p_i}{p_e}\right)^{\frac{(k-1)\eta_{st}}{k}}$$

$$\frac{T_e}{T_i} = \left(\frac{p_e}{p_i}\right)^{\frac{(k-1)\eta_{st}}{k}} = \left(\frac{1}{6}\right)^{\frac{(1.4-1)0.85}{1.4}} = 0.65$$

Hence,

$$T_e = (0.65)T_i = (0.65)900 = 582.5\,\text{K}$$

$$\Delta T_{overall} = 900 - 582.5 = 317.5\,\text{K}$$

Thus, the number of stages is given by

$$\#\,\text{of stages} = \frac{\Delta T_{overall}}{\Delta T_{stage}} = \frac{317.5}{61.3} = 5.18$$

The number of stages is rounded to six.

Comment: If the small stage efficiency were not considered, the number would be different.

Example 8.4

An axial flow stage has a stage loading coefficient of 1.5 and a nozzle angle of 70°. The axial velocity is constant throughout the stage and the exit velocity is purely axial. Calculate the (a) degree of reaction, (b) blade angles, and (c) flow coefficient.

Solution: For simplicity, let us take the blade velocity u to be equal to one. The velocity triangles are shown in the following figure:

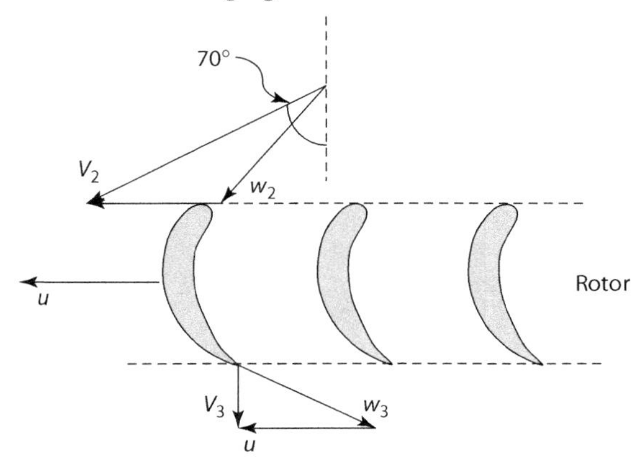

$$\psi = \frac{E}{u^2} = E = 1.5$$

or $E = u_2 V_{u2} - u_3 V_{u3} = u\left(V_{u2} - V_{u3}\right) = u V_{u2} = 1.5$

or $V_{u2} = 1.5$ and $w_{u2} = V_{u2} - u = 1.5 - 1 = 0.5$

Since the nozzle angle at the inlet is known, the axial velocity at the inlet and exit can be calculated (since it constant for the entire stage):

$$\tan 70 = \frac{V_{u2}}{V_a} \text{ or } V_a = \frac{V_{u2}}{\tan 70} = \frac{1.5}{\tan 70} = 0.546 = V_3$$

$$\text{or } \phi = \frac{V_a}{u} = 0.546$$

The blade angles can be calculated from the velocity triangles as follows:

$$\beta_2 = \arctan \frac{w_{u2}}{V_a} = \arctan \frac{0.5}{0.546} = 42.48°$$

$$\beta_3 = \arctan \frac{w_{u3}}{V_a} = \arctan \frac{u}{V_a} = 61.36°$$

The degree of reaction can be calculated from the velocities as follows:

$$w_2 = \sqrt{w_{u2}^2 + V_a^2} = \sqrt{0.5^2 + 0.546^2} = 0.740$$

$$w_3 = \sqrt{w_{u3}^2 + V_a^2} = \sqrt{u^2 + V_a^2} = \sqrt{1^2 + 0.546^2} = 1.139$$

$$R = \frac{\left(u_2^2 - u_3^2\right) + \left(w_3^2 - w_2^2\right)}{2E} = \frac{\left(w_3^2 - w_2^2\right)}{2E} = \frac{\left(1.139^2 - 0.740^2\right)}{2(1.5)} = 0.25$$

Comment: The degree of reaction can be calculated using the longer expression for *E*; this would result in the same answer.

Example 8.5

A single-stage axial flow turbine rotating at 36,000 rpm has the following specifications:

Mean radius of the rotor	= 3.2 in.
Height of the blades	= 2 in.
Axial velocity	= 240 ft./s
Blade angle at inlet of rotor	= −27°
Blade angle at exit of rotor	= −75°
Stagnation pressure at nozzle inlet	= 38 psia
Stagnation temperature at nozzle inlet	= 760°R
Rotor exit static pressure	= 14.7 psia

Calculate the following: (a) total to total efficiency, (b) total to static efficiency, (c) degree of reaction, (d) utilization factor, (e) stage loading factor, (f) power produced and (g) Mach number at entry of rotor, (h) fluid angles at inlet and outlet of rotor (α_2 and α_3).

Solution: The velocity triangles and angles are shown.

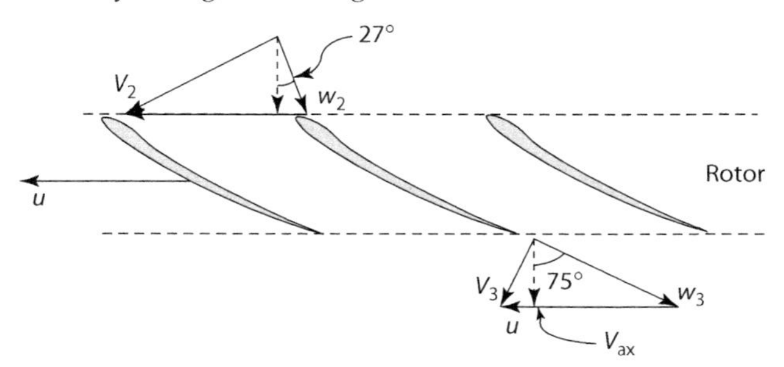

The rotational speed of the rotor is $N = 36000$ rpm $= 3769$ rad/s

Hence, the blade speed is $u =$ (mean radius)(angular speed) $= 1005.3$ ft./s

The flow coefficient can then be calculated as $\phi = \dfrac{V_a}{u} = \dfrac{240}{1005.3} = 0.238$

Since the blade angles at the inlet and exit are known, the following can be calculated from trigonometry:

$$w_2 = \frac{V_a}{\cos\beta_2} = 269.4\,\text{ft./s};\; w_3 = \frac{V_a}{\cos\beta_3} = 927.3\,\text{ft./s}$$

$$w_{u2} = V_a \tan\beta_2 = 122.3\,\text{ft./s};\; w_{u3} = V_a \tan\beta_3 = 895.7\,\text{ft./s}$$

$$V_{u2} = u - w_{u2} = 883.0\,\text{ft./s};\; V_{u3} = u - w_{u3} = 109.6\,\text{ft./s}$$

$$V_2 = \sqrt{(V_a^2 + V_{u2}^2)} = 915\,\text{ft./s};\; V_3 = \sqrt{(V_a^2 + V_{u3}^2)} = 263.8\,\text{ft./s}$$

From the velocity triangles, the energy transferred can be calculated as follows:

$$E = u_2 V_{u2} - u_3 V_{u3} = u(V_{u2} - V_{u3}) = 777500\,\text{ft}\cdot\text{lbf/slug}$$

Energy transfer is also related to the stagnation temperatures at 2 and 3. Thus,

$$T_{03} = T_{02} - \frac{E}{C_p} = 760 - \frac{777500}{6006} = 630.5°\text{R}$$

Also, the process from 01 to 3′is isentropic. Hence,

$$\frac{T_{3'}}{T_{01}} = \left(\frac{p_{3'}}{p_{01}}\right)^{(k-1)/k} = \left(\frac{14.7}{38}\right)^{(k-1)/k} = 0.762$$

Hence, $T_{3'} = (0.762)(760) = 579.4°\text{R}$

The spouting velocity can now be calculated as $c_0 = \sqrt{2C_p(T_{01} - T_{3'})} = 1472.9\,\text{ft./s}$

Now, from the velocities, static temperatures can be calculated at 2 and 3. Thus,

$$T_2 = T_{02} - \frac{V_2^2}{2C_p} = 760 - \frac{(915)^2}{2(6006)} = 690.3\text{R}$$

$$T_3 = T_{03} - \frac{V_3^2}{2C_p} = 630.5 - \frac{(263.8)^2}{2(6006)} = 624.7\text{R}$$

$$\text{Since } \frac{T_{03}}{T_3} = \frac{T_{03'}}{T_{3'}},\text{ we get } T_{03'} = 584.8\text{R}$$

Now, the efficiencies can be calculated as

$$\eta_{ts} = \frac{(T_{02} - T_{03})}{(T_{01} - T_{3'})} = \frac{E}{c_0^2/2} = \frac{777500}{(1472)^2/2} = 0.716$$

$$\eta_{tt} = \frac{(T_{02} - T_{03})}{(T_{01} - T_{03'})} = \frac{(760 - 630)}{(760 - 584)} = 0.738$$

The utilization factor, degree of reaction, and blade loading coefficient can now be calculated:

$$\varepsilon = \frac{E}{E + \frac{V_3^2}{2}} = \frac{777500}{777500 + (263.8)^2 / 2} = 0.957$$

$$R = \frac{(T_2 - T_3)}{(T_{02} - T_{03})} = \frac{(690 - 624)}{(760 - 630)} = 0.506$$

$$\psi = \frac{E}{u^2} = \frac{777500}{(1005)^2} = 0.769$$

The flow angles at the inlet and outlet are

$$\alpha_2 = \text{arccot}\frac{V_a}{V_{u2}} = \text{arccot}\frac{240}{883} = 74.8°$$

$$\alpha_3 = \text{arccot}\frac{V_a}{V_{u3}} = \text{arccot}\frac{240}{109.6} = 24.5°$$

Power can be calculated by first calculating the exit density. Thus,

$$\rho_3 = \frac{p_3}{RT_3} = \frac{(14.7)(144)}{(1716)(624)} = 0.00197\,\text{slugs/ft.}^3$$

$$\text{Area} = 2\pi r_{mean}\, h = 2\pi\left(\frac{3.2}{12}\right)\left(\frac{2}{12}\right) = 0.279\,\text{ft.}^2$$

$$\dot{m} = \rho_3 (V_3)_{ax}\,\text{Area} = (0.00197)(240)(0.279) = 0.132\,\text{slugs/s}$$

$$P = \dot{m}E = \frac{(0.132)(777500)}{550} = 187\,\text{hp}$$

Comment: The energy transferred and utilization factor can also be calculated using alternative formulas involving the velocities.

8.4 Effect of Stage Reaction

The numerical values of degree of reaction greatly influence the performance of axial flow turbines. The definitions of energy transfer, utilization factor, and degree of reaction in terms of kinematical quantities were discussed in Chapter 3. These variables can be redefined in terms of enthalpy as was done in this chapter. The two most important cases for axial flow turbines are $R=0$ and $R=0.5$. These have been considered in detail in Chapter 3 and hence the kinematic aspects are discussed only briefly here, while concentrating on the thermodynamic aspects.

Zero reaction stage: When $R=0$, it can be seen, after simplification, that $w_2=w_3$ and $h_2=h_3$. For maximum utilization, the flow should be axially directed. The Mollier diagram corresponding to such a stage and the corresponding velocity diagram are shown in Figure 8.5. There is a subtle yet very important difference between a zero reaction stage and an impulse stage. An impulse stage is one in which there is no pressure drop, whereas a zero reaction stage is one in which there is no enthalpy drop. For incompressible machines (e.g., hydraulic impulse turbines such as Pelton wheels) constant enthalpy is equivalent to constant pressure, since the flows are isothermal and density is a constant. Consequently, $\Delta h = \Delta p/\rho$. In compressible flow machines, however, isenthalpic is not equivalent to isobaric because

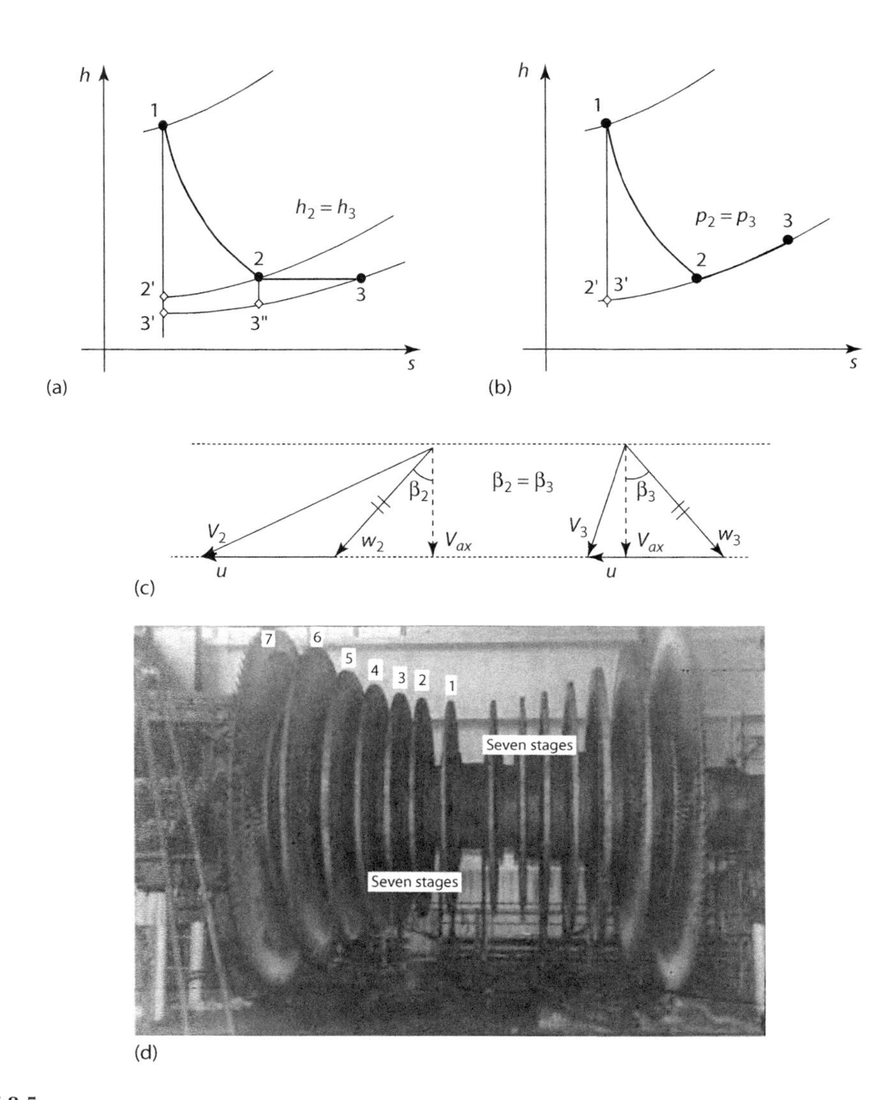

FIGURE 8.5
Mollier diagrams for (a) zero reaction stage, (b) impulse stage, and (c) velocity diagrams, (d) rotor of a fourteen-stage low-pressure steam turbine; steam enters near the center and flows axially through seven stages on each side.

of temperature changes. As can be seen in Figure 8.5a, a zero reaction stage is isenthalpic; however, there is an increase in enthalpy in an impulse stage as can be seen if Figure 8.5b.

50% reaction stage: When $R = 0.5$, from the definition of degree of reaction in terms of enthalpy in Equation 8.1a,

$$h_2 - h_3 = \frac{h_1 - h_3}{2}$$

That is, 50% of the drop in enthalpy for the entire stage takes place in the rotor and the rest occurs in the stator. This is true only for repeating stages. However, the assumption of repeating stages is commonly made in the analysis of axial flow gas turbines. A consequence of this is that $V_2 = w_3$ and $V_3 = w_2$, which results in symmetric velocity diagrams. These, along with the Mollier diagram, are shown in Figure 8.6.

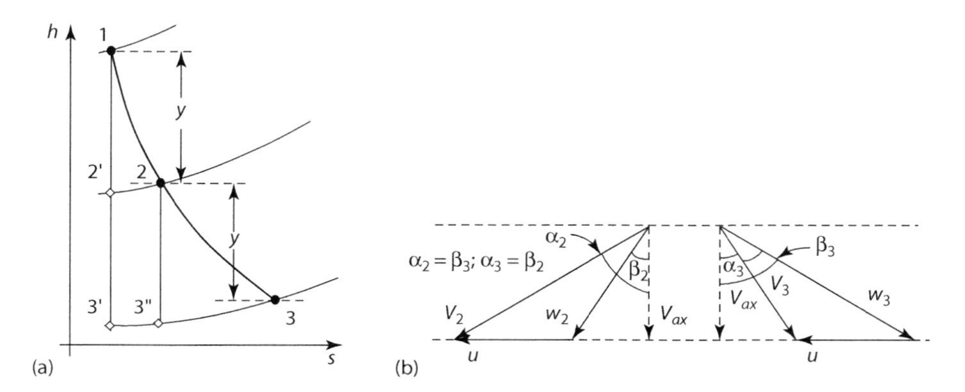

FIGURE 8.6
(a) Mollier diagram and (b) velocity diagram for 50% reaction stage.

In the context of these two types of axial flow turbine stages, it is useful to revisit the ideas that connect utilization factor and degree of reaction that were discussed in Chapter 3. For axial flow stages, these two variables are related by Equations 3.19 and 3.20. It should be noted that the subscripts 1 and 2 that were used in Chapter 3 are replaced by 2 and 3 here as these represent the inlets to the rotor. Also, in line with the angle convention for axial flow turbines, the angle α_1 used in Chapter 3 would be the complement of α_2 used here. Thus,

$$\varepsilon = \frac{V_2^2 - V_3^2}{V_2^2 - RV_3^2}$$

For maximum utilization, the exit velocity V_3 would be as small as possible and would therefore be in the axial direction ($V_3 = V_2\cos\alpha_2$), and the expression for maximum utilization would simplify to

$$\varepsilon_m = \frac{\sin^2\alpha_2}{1 - R\cos^2\alpha_2}$$

The values of ε_m for zero reaction and 50% reaction can be obtained by substituting values of R in the equation. Using the velocity triangles shown in Figure 3.7 and Figure 3.8 for maximum utilization, the energy transferred is $E = 2u^2$ and u^2, respectively. Thus, impulse stages produce twice the amount of power for the same blade speed as a 50% reaction stage, although less efficiently. Thus, the compromise between power and efficiency is to make earlier stages impulse and the later stages 50% reaction. Shown in Figure 8.5d are the 14 stages of a low-pressure steam turbine. The flow enters along the center and moves in either direction to minimize the axial thrust. Thus, there are seven stages on either side.

8.5 Losses in Axial Flow Turbines

The losses occurring in axial flow turbines are quantified in terms of the rotor and stator loss coefficients, which have contributions from profile losses, secondary flow losses, and tip clearance losses. There are three methods that are commonly employed in discussing them. These are due to Soderberg (1949), Ainley and Mathieson (1957), and Craig and

Cox (1970). They are all based on design correlations of numerous axial flow turbines and are empirical in nature. A description of each of the losses follows.

Profile Loss Coefficients: The most common method of obtaining these is to perform cascade tests in wind tunnels in which the total pressure losses are measured. One fundamental difference between compressor cascades and turbine cascades is characterized by the increase or decrease in the mean pressure. Channels that have decreasing mean pressure experience relatively lower mean pressure losses, while channels with increasing mean pressure have higher pressure losses. This difference is reflected in turbine and compressor profile loss coefficients. Sarvanamuttoo et al. (2009) found that the coefficients for rotor and stator depended on blade parameters such as thickness to chord ratio, pitch to chord ratio, inlet and exit blade angles, and lift coefficient. They also obtained correlations for these coefficients for impulse and reaction blading.

Secondary Loss Coefficients: Most of the discussion of flow in turbomachines assumes the flow to be two-dimensional for simplicity. But the actual flow is complicated by its three-dimensional nature and associated thermal aspects such as compressibility and thermal boundary layers resulting in complex secondary flows. Extensive work on this has been done by Ainley and Mathieson (1957), who concluded that the secondary losses can be written in terms of the lift coefficient and pitch to chord ratio. Using Zweifel's criterion (which states that the optimal pitch to chord ratio for minimal secondary losses is obtained when the lift coefficient is 0.8), Ainley obtained a correlation between secondary loss coefficient, pitch chord ratio, and inlet and mean fluid angles. This work was later modified by Dunham and Came (1970), who obtained better correlations to match data.

Tip Clearance Loss Coefficients: The distance between the rotor tip and the casing is called *tip clearance*. A small amount of leakage occurs through this clearance and this results in pressure loss. Dunham and Came (1970) found that the tip clearance coefficient is dependent on the blading loading. They obtained a correlation for the tip clearance coefficient as a function of lift coefficient, exit and mean blade angles, and pitch to chord ratio. Wilson (1984) suggested a simple procedure by which the tip clearance coefficient can be modified from the theoretical "zero clearance" case by multiplying by the ratio of blading annulus area to casing annulus area.

It must be emphasized that all the correlations obtained by gas turbine design practitioners are highly empirical and cover a narrow range of parameters including degrees of reaction, inlet and exit blade and flow angles, inlet and exit Mach numbers, blade geometry, and Reynolds numbers. Obviously, it is impossible to cover the entire range for all these variables. In the absence of accurate theoretical predictions, the correlations mentioned can at best be guidelines on which gas turbine analysis can be based. It is for this reason that no specific formula is presented and the reader is directed to the appropriate references.

8.6 Design Procedure for Axial Flow Turbines

In this section, some of the standard methods used for the design of axial turbine stages are discussed. As has been the practice in the previous chapters, since many of the requirements are interlinked, the procedure outlined in this section is more in the form of design guidelines/practices than the actual design itself. The typical inputs for preliminary design are (a) inlet temperature and pressure (these are from the combustion products of the combustor), (b) exit conditions at the outlet, and (c) either mass flow rate or the power expected. The preliminary design of the turbine stage is then performed using the following parameters (most of these are based on the work of El Sayed (2008), Mattingly (1987), Wilson (1984), and Vavra (1960), among others):

1. Maximum blade speed is restricted to 1200 ft./s. This would be limited by the allowable stress in the blade. Also, the maximum speed would be related to the blade loading and flow coefficients.
2. Temperature drop per stage is restricted to 250°F–350°F. This would influence the total number of stages.
3. Degree of reaction can lie between 0.0 and 0.7. For initial high-pressure stages, it should be closer to zero (indicating impulse) and for later stages it should be closer to 0.5 (reaction stages).
4. Hub to tip ratio should be in the range 0.6–0.88.
5. Blade loading coefficient should be in the range 0.9–2.8.
6. Flow coefficient should be in the range 0.25–1.4.
7. Outlet fluid angle α_3 should be in the range 0°–20°. Note that for repeating stages the inlet and exit fluid angles for the stage are equal, that is, $\alpha_1 = \alpha_3$.
8. Stator and rotor blade loss coefficients are approximately 0.05.
9. Nozzle exit Mach number should near sonic conditions but be less than 1.2, that is, M_2 should be in the range 0.8–1.2.
10. Fluid angle at the rotor inlet should be in the range 60°–75°.

Example 8.6

The following design was specified for a turbine stage:

RPM	= 7000 rpm
Mean radius	= 1.5 ft.
Inlet Mach number	= 1.01
Nozzle inlet angle (α_2)	= 63°
Nozzle exit angle (α_3)	= 0°
Inlet stagnation temperature	= 3400° R
Inlet stagnation pressure	= 300 psia

Assume that the combustion products have $k = 1.3$ and $R = 1718$ ft.·lbf/slug·°R and that this is a repeating stage. Evaluate this design and comment.

Solution: The purpose of this example is to see if these specifications are within the allowable bounds.

The rotational speed of the rotor is $N = 7000$ rpm $= 733$ rad/s

Hence, the blade speed is $u =$ (r_mean)(angular speed) $= 1099.6$ ft./s. This is within the prescribed limits.

$$T_2 = \frac{T_{02}}{1 + \frac{k-1}{2} M_1^2} = 2948°\text{R}$$

$$V_2 = \left(2c_p\left(T_{02} - T_2\right)\right)^{0.5} = 2592\,\text{ft./s}$$

We can now complete the inlet velocity triangle:

$$V_{2a} = V_a = V_2 \cos(\alpha_2) = 1176 \text{ ft./s} = V_3$$

$$V_{u2} = V_2 \sin(\alpha_2) = 2309 \text{ ft./s}$$

$$w_{u2} = V_{u2} - u = 1210 \text{ ft./s}$$

$$w_2 = \sqrt{(V_a^2 + w_{u2}^2)} = 1687 \text{ ft./s}$$

$$\beta_2 = \arctan\left(\frac{w_{u2}}{V_a}\right) = 45.8°$$

$$w_3 = \sqrt{(V_a^2 + u^2)} = 1610 \text{ ft./s}$$

$$\beta_3 = -\arctan\left(\frac{u}{V_a}\right) = -43.1°$$

From the velocity triangles, the energy transferred can be calculated as follows:

$$E = u_2 V_{u2} - u_3 V_{u3} = u V_{u2} = \frac{(2309)(1099)}{550} = 4617 \text{ hp}$$

The flow coefficient and blade loading coefficient can then be calculated as

$$\psi = \frac{E}{u^2} = \frac{(4617)(550)}{1099^2} = 2.10$$

$$\varphi = \frac{V_a}{u} = \frac{1176}{1099} = 1.07$$

These are within the prescribed limits for permissible designs.

The stage temperature drop and degree of reaction can now be calculated:

$$T_{03} = T_{02} - \frac{E}{C_p} = 3058\text{R}$$

$$\Delta T_0 = T_{02} - T_{03} = 341\text{R}$$

$$T_3 = T_{03} - \frac{V_3^2}{2C_p} = 2965\text{R}$$

$$R = \frac{(T_2 - T_3)}{(T_{02} - T_{03})} = \frac{(2948 - 2965)}{341} = -0.05$$

The temperature drop per stage is within limits. The degree of reaction is negative, which is not permissible. However, it is so close to zero that it can be taken to be a reaction stage.

Comment: The specifications are within the prescribed design limits. Since the stage appears to be an impulse stage, it is most likely to be one of the initial high-pressure stages in the turbine.

Example 8.7

Show that for repeating or normal stages, the degree of reaction as given in Equation 8.1 is equivalent to Equations 3.13 and 3.14.

Solution: The definition of degree of reaction is

$$R = \frac{h_2 - h_3}{h_1 - h_3} \quad \text{(a)}$$

From the energy balance across stator and rotor the following equations are obtained:

$$\text{Stator}: \quad h_1 + \frac{V_1^2}{2} = h_1 + \frac{V_1^2}{2} \quad \Rightarrow h_{01} = h_{02}$$

$$\text{Rotor}: \quad h_2 + \frac{V_2^2}{2} = h_3 + \frac{V_3^2}{2} + E \quad \Rightarrow h_{02} = h_{03} + E \tag{b}$$

Combining the two equations, and since $V_1 = V_3$ for repeating stages,

$$h_1 + \frac{V_1^2}{2} = h_3 + \frac{V_3^2}{2} + E \quad \Rightarrow \quad h_1 - h_3 = E \tag{c}$$

Combining Equations (b) and (c),

$$h_1 - h_3 = E = h_{02} - h_{03}$$

Hence, degree of reaction, as given in Equation (a), can be rewritten as

$$R = \frac{h_2 - h_3}{h_{02} - h_{03}} \tag{d}$$

However, using the definition of E in terms of velocities ($u_2 = u_3$ since axial flow),

$$h_2 + \frac{V_2^2}{2} = h_3 + \frac{V_3^2}{2} + E = h_3 + \frac{V_3^2}{2} + \frac{V_2^2 - V_3^2}{2} + \frac{w_3^2 - w_2^2}{2}$$

After simplification, this becomes

$$h_2 - h_3 = \frac{w_3^2 - w_2^2}{2} \tag{e}$$

Also,

$$h_{02} - h_{03} = E = \frac{V_2^2 - V_3^2}{2} + \frac{w_3^2 - w_2^2}{2} \tag{f}$$

From Equations (d), (e), and (f), the following equation is obtained:

$$R = \frac{\left(w_3^2 - w_2^2\right)}{\left(V_2^2 - V_3^2\right) + \left(w_3^2 - w_2^2\right)} = \frac{\left(w_o^2 - w_i^2\right)}{\left(V_i^2 - V_o^2\right) + \left(w_o^2 - w_i^2\right)}$$

which is the usual definition of degree of reaction as given by Equation 3.14. Note that inlet and outlet conditions of the rotor states 2 and 3 have been replaced by states i and o.

Comment: Throughout the derivation, use has been made of the fact that $u_2 = u_3$, since the stage is an axial flow stage. Thus, the definition of degree of reaction for a turbine stage in an axial flow turbine is equivalent to that for a single turbine as defined in Chapter 3.

PROBLEMS

All pressures given are absolute unless specified otherwise. Use air properties unless specified otherwise.

8.1 Show that for an axial flow turbine, E is given by the following expressions:

$$E = u\left(V_{u2} - V_{u3}\right) = u\left(w_{u2} - w_{u3}\right)$$

8.2 Show that for axial flow turbines the degree of reaction R and the flow angles are related by the following expressions:

$$R = 1 - \frac{V_a}{2u}(\tan\alpha_3 + \tan\alpha_2)$$

$$R = \frac{1}{2} - \frac{V_a}{2u}(\tan\beta_3 + \tan\alpha_2)$$

$$R = 1 - \frac{V_a}{2u}(\tan\alpha_1 + \tan\alpha_2)$$

8.3 The initial stages of axial flow turbines are of the impulse type. Consider one such stage. The mean radius of the rotor is 6 in. and the height of the blades is 2 in. The flow leaving the rotor is in the axial direction. The inlet stagnation conditions at the stator are $T_{01} = 1200$ R and $p_{01} =$ five atmospheres. The turbine speed is 10,000 rpm. Calculate the (a) spouting velocity, (b) blade loading coefficient, (c) flow coefficient, (d) inlet and exit Mach numbers for the rotor, (e) power, and (f) inlet nozzle angle. Show the process on a *T–s* diagram. Axial velocity is 450 ft/s.

Ans: $c_0 = 2306$ ft./s; $\psi = 2.00$; $\phi = 0.86$; power = 265 hp; $\alpha_2 = 66.74°$

8.4 The later stages in an axial flow turbine are reaction-type stages. In one such stage, the degree of reaction is 50%. The mean radius and height of the rotor blades are 3 in. and 1 in., respectively. The speed is 40,000 rpm and the flow coefficient $\phi = 0.24$. The rotor exhaust temperature and pressure are 60°F and 14.7 psia, respectively. If $V_{u2} = 1100$ ft./s, calculate the following: (a) nozzle angle, (b) blade loading coefficient, (c) total temperature at the stage inlet and exit, and (d) power. Also, show the process on a *T–s* diagram.

Ans: $\alpha_2 = 78.13°$; $\psi = 1.1$; $T_{02} = 727°$R; $T_{03} = 525.5°$R; power = 171 hp

8.5 a) Show that the head coefficient defined as gH/N^2D^2 for incompressible flow machines (pumps, hydraulic turbines) is proportional to the blade loading coefficient E/u^2 defined for compressible flow machines (gas turbines, compressors, etc.).

b) Show that the flow coefficient defined as Q/ND^3 for incompressible flow machines (pumps, hydraulic turbines) is proportional to the blade loading coefficient V_a/u defined for compressible flow machines (gas turbines, compressors, etc.). Here, V_a is the axial velocity.

8.6 Consider an axial flow turbine in which the flow coefficient and blade loading coefficient are 0.52 and 1.6, respectively. The axial velocity can be assumed to be a constant throughout the stage. The flow at the exit of the rotor is purely axial. Calculate the degree of reaction, blade angles at the inlet and outlet, and inlet nozzle angle.

Ans: $R = 0.2$, $\beta_3 = -62.5°$, $\beta_2 = 49°$, $\alpha_2 = 72°$

8.7 Consider an axial turbine stage for which the blade loading coefficient $\psi = 1.5$. Axial velocity can be assumed to be a constant. The absolute velocity leaving the stage is purely axial, while at the entrance, the absolute velocity makes an angle of 65° with the axial direction. Calculate the (a) flow coefficient ϕ, (b) degree of reaction, and (c) blade angles.

Ans: $\phi = 0.7$; $R = 0.25$; $\beta_2 = 35.6°$; $\beta_3 = -55°$

8.8 Repeat Example 8.3, assuming the stage efficiency to be 100%.

Ans: 6 stages

8.9 Obtain the following relationships between R, ϕ, ψ, β_2, β_3, α_1, α_2, and α_3. Assume that the stages are normal.

$$R = 1 - \frac{\phi^2}{2\psi}\left(\tan^2\alpha_2 - \tan^2\alpha_1\right) = 1 - \frac{\phi^2}{2\psi}\left(\tan^2\alpha_2 - \tan^2\alpha_3\right)$$

$$R = 1 - \frac{\phi}{2}\left(\tan\alpha_2 + \tan\alpha_1\right) = 1 - \frac{\phi}{2}\left(\tan\alpha_2 + \tan\alpha_3\right)$$

$$\psi = 2\left(1 - R - \phi\tan\alpha_1\right) = 2\left(1 - R - \phi\tan\alpha_3\right) \quad \text{aa}$$

$$R = -\frac{\phi}{2}\left(\tan\beta_2 + \tan\beta_3\right)$$

$$\tan\beta_2 = \tan\alpha_2 - \frac{1}{\phi}; \quad \tan\alpha_3 = \tan\beta_3 + \frac{1}{\phi}$$

8.10 An axial flow turbine stage has the following specifications:

Total to static efficiency, η_{ts}	$=0.85$
Flow coefficient, ϕ	$=0.40$
Total temperature at stator inlet	$=2000°$R
Total pressure at stator inlet	$=42$ psi
Exhaust static pressure	$=14.7$ psi

Axial velocity is constant at 550 ft./s and the exit absolute velocity from the rotor is axially directed. Calculate the (a) degree of reaction, (b) utilization factor, (c) Mach number at the exit of the nozzle (or entry of the rotor), (d) blade and flow angles at the inlet and outlet of the rotor, (e) total to total efficiency, (f) blade loading coefficient ψ.

Ans: $R=0.5$; $\varepsilon=0.946$; $M_2=1.0$; $\alpha_2=74.1°$; $\beta_2=45°$; $\beta_3=-68.2°$; $\eta_{tt}=0.892$; $\psi=1.399$

8.11 Verify the equations in Problem 8.10 using the results obtained in Problem 8.9.

8.12 Prove the following relationships for axial flow turbines:

a) If degree of reaction is zero, then $\tan\beta_2+\tan\beta_3=0$

b) If degree of reaction is 0.5, then $\tan\beta_2+\tan\alpha_3=0$

8.13 Show that for a repeating axial flow turbine stage the degree of reaction is given by the ratio of the static enthalpy drop across the rotor to the sum of static enthalpy drops across the stator and the rotor, that is, prove Equation 8.1a.

8.14 An axial flow turbine stage has been designed with the following specifications to produce 2000 hp.

RPM	$=10000$ rpm
Blade speed	$=800$ ft./s
Axial velocity	$=410$ ft./s
Nozzle inlet angle (α_2)	$=70°$
Nozzle exit angle (α_3)	$=10°$
Inlet stagnation temperature	$=1800°$F
Inlet stagnation pressure	$=200$ psia
Stator loss coefficient	$=0.04$

Evaluate the design given and comment.

8.15 The following design has been proposed for an axial turbine stage to produce 500hp: $T_{01} = 1800$ R, $p_{01} = 300$ psia, $\alpha_2 = 61°$, $\alpha_3 = 11°$, blade velocity = 1150 ft./s, $k = 1.3$, and $R = 1718$ ft·lbf/slug·°R. Evaluate the design and comment. The axial velocity is held a constant at 975 ft./s.

8.16 Design an axial flow turbine stage to meet the following expectations. The inlet conditions are $T_{01} = 2000$ R, $p_{01} = 200$ psia, and it is expected to produce 1500 hp. Note that the answer will not be unique.

8.17 A 50% axial flow turbine is being considered as a standby for an oil-fired power plant. The combustion products enter the rotor at a static temperature of 800°C. The stagnation pressure ratio across the turbine is expected to be ten, the nozzle angle at the inlet is 65°; the axial velocity is constant at 130 m/s. Assuming maximum utilization, estimate the number of stages required. Assume that isentropic efficiency of the turbine is 100%. Show the process on an enthalpy–entropy (h–s) diagram.

Ans: 7 stages

8.18 A single-stage axial flow air turbine produces 82 kW at a speed of 50,000 rpm. At the nozzle inlet, the stagnation pressure and temperature are 225 kPa and 425 K, respectively. At the rotor exhaust, the static pressure is 101 kPa. The mean blade speed is 300 m/s. The relative gas angles are −28° and −76° at the rotor inlet and exit, respectively. The flow coefficient is 0.24.

a) Show the process on an h–s diagram

Calculate the following:

b) Spouting velocity
c) Mean blade radius
d) Energy transfer
e) Required mass flow
f) Total to total efficiency
g) Total to static efficiency

Ans: $c_0 = 418$ m/s; $\eta_{ts} = 0.86$; $\eta_{tt} = 0.887$; $\dot{m} = 1.164$ kg/s; $E = 75147$ J/kg; mean blade radius = 5.8 cm

8.19 A single-stage axial flow turbine with a degree of reaction of zero receives air at stagnation conditions of 1000 K and 400 kPa. The inlet velocity to the stage is purely axial and has a value of 275 m/s. At the mean radius, the blade speed is 300 m/s. The stagnation pressure at the exit of the stage is 101 kPa. The mean radius and height of the blades are 0.2 m and 0.05 m, respectively. For maximum utilization, calculate the following: (a) nozzle and rotor exit angles, (b) power, (c) spouting velocity, and (d) utilization factor.

Ans: $P = 3952$ kW; $c_0 = 846$ m/s; $\varepsilon = 0.826$; $\beta_2 = \beta_3 = 47.5°$; $\alpha_2 = 65.38°$

8.20 Consider an axial turbine stage with 50% reaction. There is no exhaust diffuser and the stage is expected to produce 1200 kW when the mass flow rate is 15 kg/s. The inlet stagnation conditions are 1000 K and 400 kPa. The stage can be considered a repeating stage with a mean radius of 0.2 m and constant axial velocity. The mean blade speed is 200 m/s and the inlet velocity is 275 m/s. The exhaust stagnation pressure is one atmosphere. Calculate the (a) nozzle angle and blade angle at inlet of the rotor, (b) spouting velocity, and (c) height of the blades.

Ans: $c_0 = 842$ m/s; blade height = 0.037 m; $\beta_2 = 21.33°$; $\alpha_2 = 49.51°$

8.21 Consider a low-pressure axial flow turbine with a mass flow rate of 40 lbm/s and inlet stagnation conditions of 2200 R and 32 psi. The inlet velocity is 900 ft./s and mean radius is 1.5 ft. It spins at 5600 rpm while producing 12640.6 hp. The axial velocity can be considered to be constant throughout all stages, which are 50% reaction. The blade angles at the inlet and exit of the rotor are 16° and −72°, respectively. Estimate the number of stages, assuming the flow through all of them to be isentropic. Also, calculate the following: (a) blade loading coefficient, (b) flow coefficient, (c) stagnation pressure and temperature at the exit of the first three stages. Show the process on an *h*–*s* diagram.

Ans: Number of stages = 6; $\psi = 1.216$; $\phi = 0.403$

8.22 An axial flow impulse stage has the following conditions:

Nozzle inlet temperature T_1	= 1400 K
Nozzle exit temperature T_2	= 1190 K
Nozzle inlet velocity V_1	= 230 m/s
Stage inlet static pressure p_1	= 230 kPa
Rotor exit pressure p_3	= 101.4 kPa
Nozzle area ratio A_2/A_1	= 1.6

Assume that the gas has a specific heat ratio of 1.3 and gas constant of 287 J/kg•K. Calculate the following: (a) blade angles at the inlet and outlet of the rotor, (b) energy transfer, (c) spouting velocity, (d) total to total and total to static efficiencies, (e) blade speed. Show the process on an *h*–*s* diagram.

Ans: $c_0 = 807$; $\eta_{tt} = 0.867$; $\eta_{ts} = 0.788$

8.23 The following data refers to a 50% reaction turbine stage.

Annular flow area	= 15 sq in.
Mean radius at rotor exit	= 2.2 in.
Rotor speed	= 45,000 rpm
Exhaust conditions: T_3	= 425 R; p_3 = 14.7 psia

If axial velocity is constant at 230 ft/s and V_{u2} = 1150 ft/s, find a) nozzle angle α_2 and b) turbine horsepower.

Ans: 78.7°; 156.9 hp

9

Radial Compressors

In the previous two chapters, the discussion was on power-producing machines involving compressible fluids, namely gas turbines, both radial and axial (although steam turbines could fall under this category, they will be discussed separately). This chapter deals with radial flow compressors, the analogs of centrifugal pumps involving incompressible fluids. Since both density and temperature vary, compressors differ significantly from pumps both in their operation and features.

Upon completion of this chapter, the student will be able to

- Analyze the velocity diagrams for centrifugal compressors
- Predict their performance under on-design and off-design conditions
- Perform their preliminary design

9.1 Introduction

The Brayton cycle requires large volumes of compressed air in order to operate efficiently. This can be achieved using either radial/centrifugal or axial flow compressors. One of the primary difficulties in the design of efficient compressors is the requirement that they work under a variety of operating conditions. For instance, in the field of aviation, the inlet air to the compressor is at atmospheric pressure at takeoff and landing. However, while cruising, the density and pressure are considerably lower. The compressor needs to operate quite efficiently under both these opposing conditions. In addition, its speed must be compatible with the turbine as they are both mounted on the same shaft. The earlier versions of gas turbines developed in the 1940s and 1950s used radial flow types; however, modern gas turbines use axial flow stages or a combination of axial and centrifugal types (usually several axial flow stages followed by a radial stage).

9.2 Radial/Centrifugal Compressors

A schematic diagram of a centrifugal compressor along with the geometry of the blades is shown in Figure 9.1. The flow is somewhat similar to that in a centrifugal pump and opposite to that of a radial flow gas turbine. At the exit of the rotor there may be additional stationary blades called *diffuser* or *stator vanes*. The fluid enters the rotor axially at the eye of the compressor and moves radially outward. After leaving the rotor, it enters the diffuser, which contains a set of stationary blades, where the pressure increases further before it enters the scroll case. The advantages of centrifugal compressors are their higher

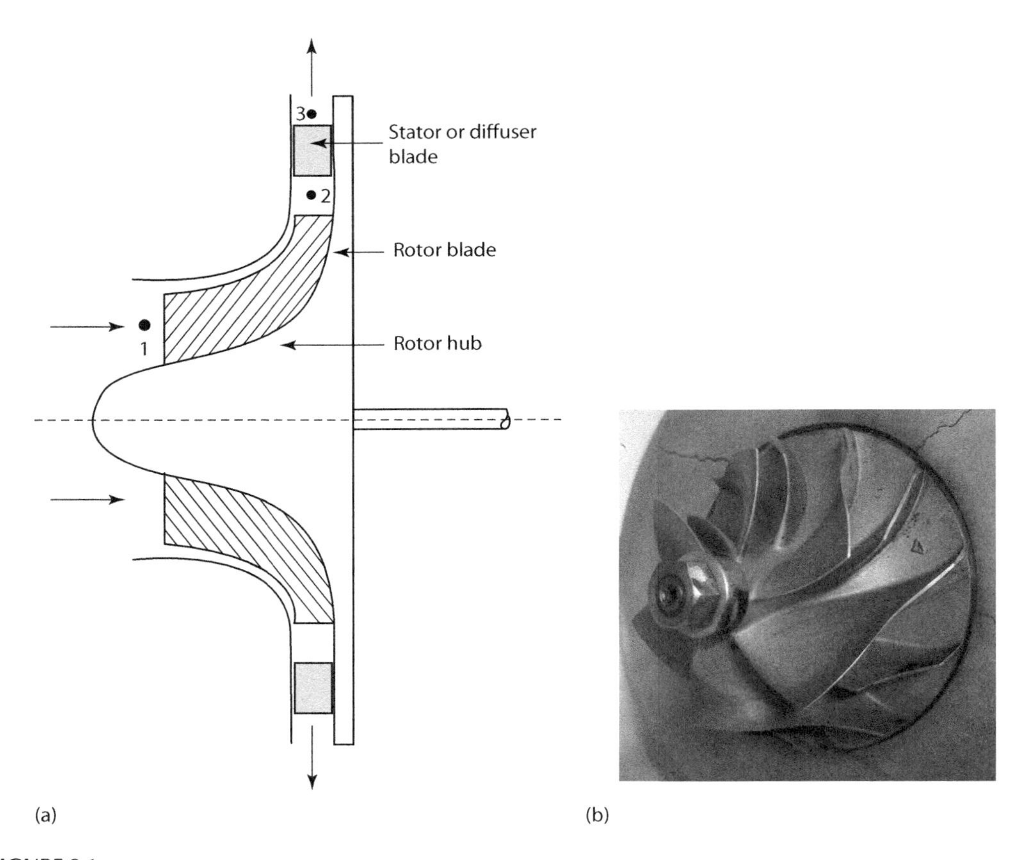

FIGURE 9.1
(a) Schematic diagram and (b) prototype of a centrifugal compressor.

pressure ratios, smaller size (length), ease of adding intercoolers if needed, simplicity of design and construction, and efficient operation over a wider range of mass flows. One major disadvantage is the difficulty of adding additional stages (this may not be required in view of the higher pressure ratios that can be achieved). Typical pressure ratios that can be obtained are 4:1. However, this ratio can be increased to as high as 6:1 with the use of higher-strength materials such as titanium.

9.3 Velocity Triangles and Energy Transfer

The nomenclature of the compressor and the velocity triangles at the inlet and exit of the impeller are shown in Figure 9.2.

From the Euler equation, the energy E becomes

$$E = u_1 V_{u1} - u_2 V_{u2} \tag{9.1}$$

Sometimes, guide vanes are provided at the inlet to impart a limited amount of whirl to the fluid. These are called *inlet guide vanes*. In such situations, the value of V_{u1} would not be

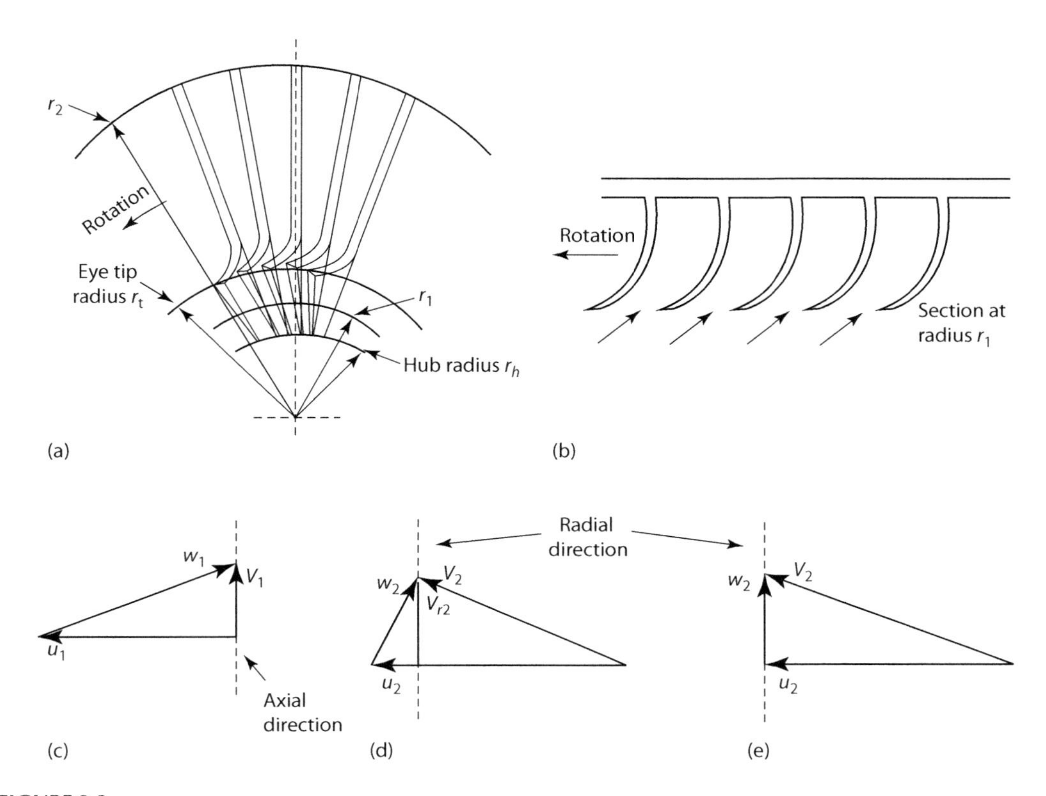

FIGURE 9.2
Nomenclature and velocity triangles for the compressor: (a) schematic diagram of the compressor viewed axially, (b) section through the eye at radius r_1, (c) inlet velocity triangle (viewed radially from top), (d) actual exit velocity triangle (viewed axially), and (e) ideal exit velocity triangle (viewed axially).

zero. The fluid entering the rotor shown in Figure 9.2 has no inlet momentum, as shown in the inlet velocity triangle. It is customary to measure the angles with respect to the axial direction at the inlet and radial direction at the exit. Without inlet guide vanes, the fluid inlet angular momentum ($\alpha_1 = 0$) is zero. Thus,

$$|E| = u_2 V_{u2} \tag{9.2}$$

For convenience, the negative sign for E has been dropped. For typical exit velocities, both tangential and small radial velocities exist, although the radial velocity is quite small. Ideally, the exit velocity would be such that its whirl component is exactly equal to u_2, as shown in the figure. Then, $\beta_2 = 90°$ and

$$|E| = u_2^2 \tag{9.3}$$

As was the case with radial flow turbines, the flow coefficient φ and blade loading coefficient ψ are defined as follows:

$$\phi = \frac{V_{r2}}{u_2}; \quad \psi = \frac{E}{u_2^2} \tag{9.4}$$

Several relationships can be derived involving the degree of reaction, flow coefficient, blade loading coefficient, and the blade angles. Some of these have been included in the problems at the end of the chapter.

9.4 Compressor Enthalpy–Entropy Diagram

The enthalpy–entropy (h–s) diagram for the flow process is shown in Figure 9.3. As in the previous chapters, in order not to clutter up the diagram, relative stagnation states have not been shown. Only stagnation states based on absolute velocities have been shown. Process i-1 on the diagram represents the flow through the inlet casing containing the inlet guide vanes. The fluid expands from pressure p_i to p_1 and accelerates from velocity V_i to V_1. However, in the absence of any energy transfer, the total enthalpy remains the same. Thus,

$$h_i + \frac{V_i^2}{2} = h_1 + \frac{V_1^2}{2} \quad \Rightarrow \quad h_{0i} = h_{01} \tag{9.5}$$

Process 1–2 represents the flow through the impeller or rotor. The energy equation becomes

$$\begin{aligned} h_1 + \frac{V_1^2}{2} &= h_2 + \frac{V_2^2}{2} + E \quad \Rightarrow \quad h_{01} = h_{02} + E \\ \text{or} \quad E &= h_{01} - h_{02} = u_1 V_{u1} - u_2 V_{u2} \end{aligned} \tag{9.6}$$

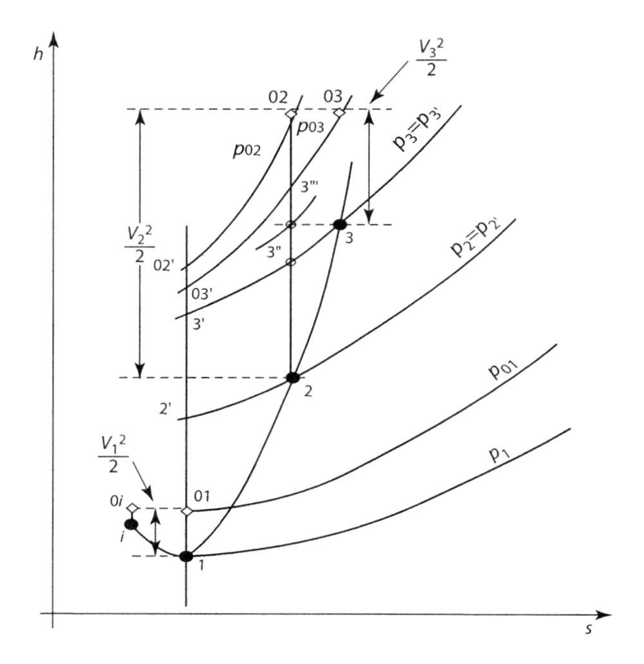

FIGURE 9.3
Enthalpy–entropy diagram for a centrifugal compressor.

It should be recalled that Equation 9.6 can be simplified and written in terms of rothalpy, as shown in Equation 3.11b in Chapter 3.

Flow through the diffuser or the stator is represented by process 2–3. Here, the fluid is decelerated from a speed V_2 to V_3 accompanied by a rise in the static pressure from p_2 to p_3. Since there is no energy transfer in the form of either heat or work, the energy balance equation across the stator becomes

$$h_2 + \frac{V_2^2}{2} = h_3 + \frac{V_3^2}{2} \Rightarrow h_{02} = h_{03} \tag{9.7}$$

In the context of compressors, degree of reaction can be defined as the ratio of enthalpy rise in the rotor to enthalpy rise in the rotor and stator (diffuser). Hence,

$$R = \frac{h_2 - h_1}{h_3 - h_1} \tag{9.7a}$$

A few comments on this definition are in order. In the context of reaction of a stage, the usual assumption is that the stages are repeating or normal stages. In such cases, the definition of degree of reaction becomes

$$R = \frac{h_2 - h_1}{h_{03} - h_{01}} = \frac{h_2 - h_1}{h_{02} - h_{01}} \tag{9.7b}$$

While staging in axial flow compressors is quite simple and perhaps logical, it is difficult to do multiple staging in centrifugal compressors, primarily due their geometry. Hence, for centrifugal compressors, Equation 9.7a is more commonly used to define degree of reaction, while Equation 9.7b is used for axial compressors. In the subsequent discussion, which of the definitions should be used will either be clear from the context or will be explicitly specified. However, it should be noted that, the concept of degree of reaction is most relevant for turbines and used less frequently in the discussion of compressors, both centrifugal and axial.

The pressure rise through the compressor can be evaluated from the h–s diagram. It is usual to consider the stagnation pressures, since stagnation states are good reference states as the fluid is at rest. From Figure 9.3, the pressure ratio is defined with respect to the inlet to the rotor 01 and the exit of the diffuser 03. Due to the isentropic process between 01 and 03′, the pressure ratio becomes

$$P_r = \frac{p_{03}}{p_{01}} = \frac{p_{03'}}{p_{01}} = \left(\frac{T_{03'}}{T_{01}}\right)^{k/(k-1)} \tag{9.8}$$

As with turbines, the total to total efficiency can be defined as

$$\begin{aligned} \eta_{tt} &= \frac{h_{03'} - h_{01}}{h_{03} - h_{01}} = \frac{h_{03'} - h_{01}}{h_{02} - h_{01}} = \frac{C_p\left(T_{03'} - T_{01}\right)}{h_{02} - h_{01}} \\ &= \frac{C_p\left(T_{03'} - T_{01}\right)}{E} = \frac{C_p T_{01}\left(\dfrac{T_{03'}}{T_{01}} - 1\right)}{E} \end{aligned} \tag{9.9}$$

If the fluid enters without swirl, a simple expression can be obtained for E $(=u_2V_{u2})$. Introducing the equation for pressure ratio from Equation 9.8 into 9.9, an expression relating the pressure ratio and efficiency can be obtained. Also, if the velocities are written in terms of the inlet and exit blade velocities using the fluid and blade angles at the exit (β_2 and β_2), other expressions can be obtained. It is quite easy to obtain them when necessary and hence they are not derived here. Some of these are left as exercises at the end of the chapter, while some are derived in the examples. It suffices to say that the overall pressure ratio P_r can be written in terms of the total to total efficiency, exit blade velocity, inlet temperature, and the angles. The total to total efficiency in Equation 9.9 is often called the *compressor efficiency*.

In the context of centrifugal compressors, another efficiency called *impeller efficiency* η_i is defined as follows:

$$\eta_i = \frac{h_{02'} - h_{01}}{h_{02} - h_{01}} = \frac{T_{02'} - T_{01}}{T_{02} - T_{01}} \tag{9.9a}$$

Typical values of impeller efficiency are in the range 0.87–0.93.

9.5 Diffuser

The concept of draft tubes was introduced in the chapter on hydraulic turbines. It was seen that since reaction turbines operated under lower heads, their overall efficiency can be increased by using draft tubes at the exit. Thus, part of the exit kinetic energy that would otherwise be wasted is used to create a partial vacuum at the exit of the turbine, thereby increasing the overall head. Diffusers in centrifugal compressors serve a similar purpose.

The absolute velocity V_2 at the exit of the impeller is quite high, and it is reduced to a lower velocity V_3 in the diffuser as shown by an h–s diagram. This deceleration is dependent on the amount and efficiency of diffusion that takes place in the diffuser. In this context, diffuser efficiency and diffuser pressure recovery coefficient are introduced next. It is customary to assume that the fluid is incompressible in these definitions. Diffuser efficiency is the ratio of the actual static pressure rise to the ideal (isentropic) static pressure rise in the diffuser. Thus,

$$\eta_D = \frac{p_3 - p_2}{p_{3'''} - p_2} = \frac{p_{3''} - p_2}{p_{3'''} - p_2} \tag{9.10}$$

The pressure recovery coefficient C_{pr} is quite analogous to pressure coefficient, which is most useful in studying flow over air foils. Again assuming incompressible flows and density at the inlet,

$$C_{pr} = \frac{p_{3'''} - p_2}{\frac{1}{2}\rho V_2^2} \tag{9.11}$$

Several simplifications are possible using conservation of mass.

9.5.1 Vaneless Diffuser

This uses the simplest concept of the diffusion process in radial flow machines, according to which the swirl velocity is inversely proportional to the radius, a result of conservation of angular momentum. The variation of radial velocity is governed by the mass flow rate and change of area along the length of the diffuser. Thus, the increase in pressure in the vaneless diffuser is due to the diffusion process and the corresponding decrease in velocity. From conservation of mass (with a small pressure rise across the diffuser, that is, $\rho_2 = \rho_3$ and constant width of the diffuser $b_2 = b_3$):

$$\rho_2 V_{r2}(2\pi r_2 b_2) = \rho_3 V_{r3}(2\pi r_3 b_3) \quad \text{or} \quad r_2 V_{r2} = r_3 V_{r3} \tag{9.12}$$

For frictionless flows, the angular momentum is conserved in the diffuser. Hence,

$$r_2 V_{u2} = r_3 V_{u3} \tag{9.13}$$

Also,

$$\begin{aligned} V_2^2 &= V_{r2}^2 + V_{u2}^2 = V_{r2}^2\left(1+\frac{V_{u2}^2}{V_{r2}^2}\right) = \left(\frac{r_3^2}{r_2^2}\right)V_{r3}^2\left(1+\frac{V_{u3}^2}{V_{r3}^2}\right) \\ &= \left(\frac{r_3^2}{r_2^2}\right)\left(V_{r3}^2 + V_{u3}^2\right) = \left(\frac{r_3^2}{r_2^2}\right)V_3^2 \end{aligned} \tag{9.14}$$

After simplifying, the equation becomes

$$r_2 V_2 = r_3 V_3 \tag{9.15}$$

Combining Equations 9.12, 9.13, and 9.15, the following useful relationship between the velocities and the radii of the compressor is obtained:

$$\frac{V_2}{V_3} = \frac{V_{u2}}{V_{u3}} = \frac{V_{r2}}{V_{r3}} = \frac{r_3}{r_2} = \frac{d_3}{d_2} \tag{9.16}$$

In the diffuser, the velocity is expected to decrease to give rise to an increase in static pressure. Thus, inlet velocity V_2 would be greater than V_3. The amount of diffusion can be loosely defined in terms of the ratio of the velocities. As can be seen from Equation 9.16, the amount of diffusion is proportional to the diameter ratio. Hence, large-sized diffusers are required to obtain appreciable pressure increases. This is one of the main reasons that vaneless diffusers have found little application in aircraft propulsion. Most aviation applications use axial flow compressors. Hence, this might be a moot point. Nevertheless, size and the accompanying lower efficiency for vaneless diffusers makes them unsuitable for aviation. However, when size is not a factor, as with industrial compressors, vaneless diffusers are acceptable, since they are economical and have a wider range of operation. Also, such diffusers do not face the problems of choking and blade stall.

9.5.2 Vaned Diffuser

When significant pressure rises are needed over short distances, vaneless diffusers are not useful. In such situations, it is necessary to guide the flow through the diffuser passages. This is accomplished by blade rings of various shapes. Three such shapes are shown in Figure 9.4. The blades can be straight, cambered, or airfoil-shaped. One design challenge in vaned diffusers is to see that flow separation does not take place. This is accomplished by making the diffusion angle small (10°–12°). Increasing the number of diffuser blades also minimizes the possibility of separation since the divergence of the blades is small. However, an increased number of blades leads to increased friction losses. Hence, an optimum number of blades is used. One disadvantage of vaned diffusers is that they have a narrow range of operation.

9.6 Slip Coefficient

As was the case in centrifugal pumps, because of the pressure difference between the suction and pressure sides of the rotor blades, there is a difference between the actual and ideal tangential velocity components at the exit of the rotor (see Figure 5.5). This is quantified by means of the slip coefficient μ_s, defined as $\mu_s = \frac{V'_{u2}}{V_{u2}}$, where V'_{u2} and V_{u2} are the actual and ideal velocity components. The difference between V'_{u2} and V_{u2} is called the *slip velocity* ΔV_{u2}.

An accurate estimate of the slip coefficient is required to enable good design of the compressor. Several correlations, based on inviscid flow theory, are available in the literature for prediction of μ_s. All of these are based on the behavior of the relative eddy, which is a vortex flow formed due to the pressure difference between the pressure and suction sides

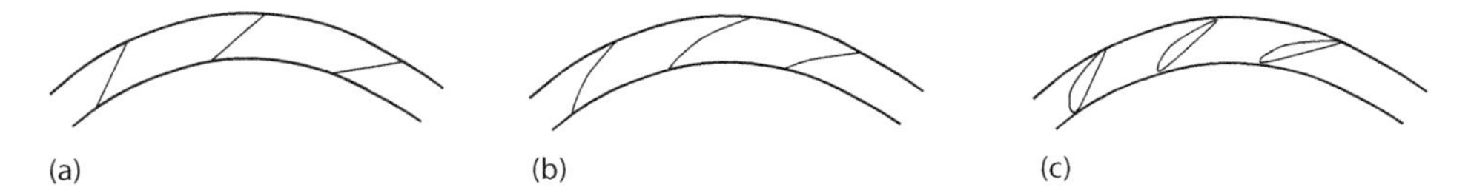

FIGURE 9.4
Diffuser rings with (a) straight blades, (b) cambered blades, and (c) airfoil blades.

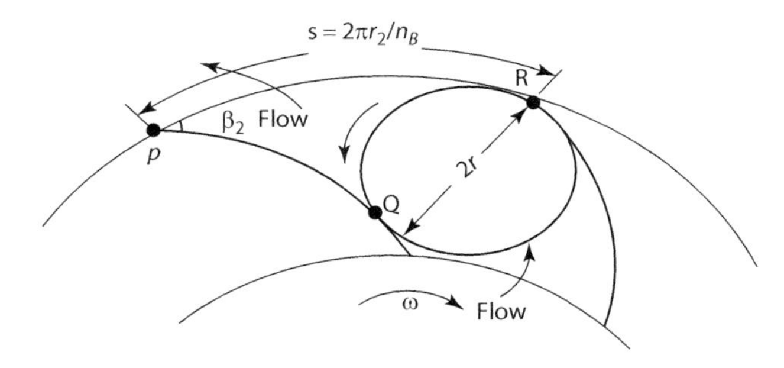

FIGURE 9.5
Flow model at the end of the rotor for slip factor.

of the blade. Three such correlations are most commonly used in the study of compressors. All of these are useful for the range $45° < \beta_2 < 90°$.

1. Stodola's (1927) correlation: Here, the assumption made is that the *eddy that is formed rotates at the same speed as the runner.* Thus, if the eddy is treated as a cylinder of radius $2r$, then its speed of rotation can be determined using the following arguments. The blade pitch is related to the number of blades as

$$s = \frac{2\pi r_2}{n_B} \tag{9.17}$$

Now, *approximating* the shape PQR in Figure 5.5 to be a right triangle,

$$\sin\beta_2 = \frac{2r}{s} \Rightarrow 2r = s\sin\beta_2 = \frac{2\pi r_2}{n_B}\sin\beta_2 \tag{9.18}$$

According to Stodola's theory, the slip velocity is assumed to be the same as the speed of the relative eddy. Thus,

$$V_{u2} - V_{u2}' = \Delta V_{u2} = Nr = \frac{N\pi r_2}{n_B}\sin\beta_2 = \frac{\pi u_2}{n_B}\sin\beta_2 \tag{9.19a}$$

After simplifying,

$$(1-\mu_s)V_{u2} = \frac{\pi u_2}{n_B}\sin\beta_2 \tag{9.19b}$$

Thus,

$$\begin{aligned} V_{u2} - V_{u2}' = \Delta V_{u2} = Nr = \frac{N\pi r_2}{n_B}\sin\beta_2 = \frac{\pi u_2}{n_B}\sin\beta_2 \\ \text{or } (1-\mu_s)V_{u2} = \frac{\pi u_2}{n_B}\sin\beta_2 \end{aligned} \tag{9.20}$$

$$\mu_s = 1 - \frac{\pi u_2}{n_B V_{u2}}\sin\beta_2 \tag{9.21}$$

2. Stanitz's (1952) correlation: Stanitz made an extensive theoretical analysis of flow between compressor blades, assuming inviscid flows. The results he obtained are valid for β_2 in the range 45°–90°. In contrast to Stodola, he found that the slip velocity depended neither on compressibility nor the blade angle and obtained the following simple expression for slip velocity:

$$\begin{aligned} \Delta V_{u2} = \frac{0.63\pi u_2}{n_B} \Rightarrow (1-\mu_s)V_{u2} = \frac{0.63\pi u_2}{n_B} \\ \text{or } \mu_s = 1 - \frac{0.63\,\pi u_2}{n_B V_{u2}} \end{aligned} \tag{9.22}$$

For radial blades (when $V_{u2}=u_2$), this equation simplifies to

$$\mu_s = 1 - \frac{0.63\pi}{n_B} = 1 - \frac{1.98}{n_B} \approx 1 - \frac{2}{n_B} \tag{9.23}$$

which is most commonly used for preliminary design purposes.

3. Balje's (1981) correlation: This is more restrictive, since it is valid only for $\beta_2=90°$.

$$\mu_s = \left(1 + \frac{6.2}{n_B n^{2/3}}\right)^{-1} \tag{9.24}$$

where n is the ratio of the impellor tip to eye diameter. This equation is sometimes approximated as

$$\mu_s = \left(1 - \frac{6.2}{n_B n^{2/3}}\right) \tag{9.25}$$

There are several other theories with varying degrees of mathematical sophistication. These have been documented by Ferguson (1963), who compared the slip factors from various correlations with known experimental values.

Another variable that is of interest in centrifugal compressors is the so-called *power input factor,* which accounts for the fact that the actual work input needs to be higher due to friction between the casing and the gas carried, and also disk friction or windage losses. Thus, the actual power would be the product of theoretical power and the power input factor. Typical values of this factor are between 1.035 and 1.04, indicating that the windage and disk friction losses are quite small. This topic will not be discussed further, since some of these losses are included in the mechanical efficiency of the compressor.

Example 9.1

A centrifugal compressor is expected to produce a stagnation pressure ratio (p_{03}/p_{01}) of 1.5 while rotating at 8000 rpm. The compressor draws air from the atmosphere at a temperature of 15°C. The mass flow rate is 1.5 kg/s. For zero swirl on entry and radial blades at the exit, what size should the compressor be? Also calculate the power input. Assume isentropic flow throughout.

Solution: The velocity triangles and the *T–s* diagram are provided.

$$N = 8000\,\text{rpm} = 837.8\,\text{rad/s}; \quad T_{01} = 15 + 273 = 288\,\text{K}$$

From the velocity triangles and Euler's equation, the energy E can be written as

$$|E| = u_1 V u_1 - u_2 V u_2 = u_2 V u_2 = u_2^2$$
$$= C_p\left(T_{02} - T_{01}\right) = C_p\left(T_{03} - T_{01}\right)$$

From the *h–s* diagram, the process 01 to 03 is isentropic. Thus,

$$1.5 = \frac{p_{03}}{p_{01}} = \left(\frac{T_{03}}{T_{01}}\right)^{k/(k-1)}$$

$$\text{or} \quad \frac{T_{03}}{T_{01}} = \left(\frac{p_{03}}{p_{01}}\right)^{(k-1)/k} = (1.5)^{(1.4-1)/1.4} = 1.12$$

Hence,

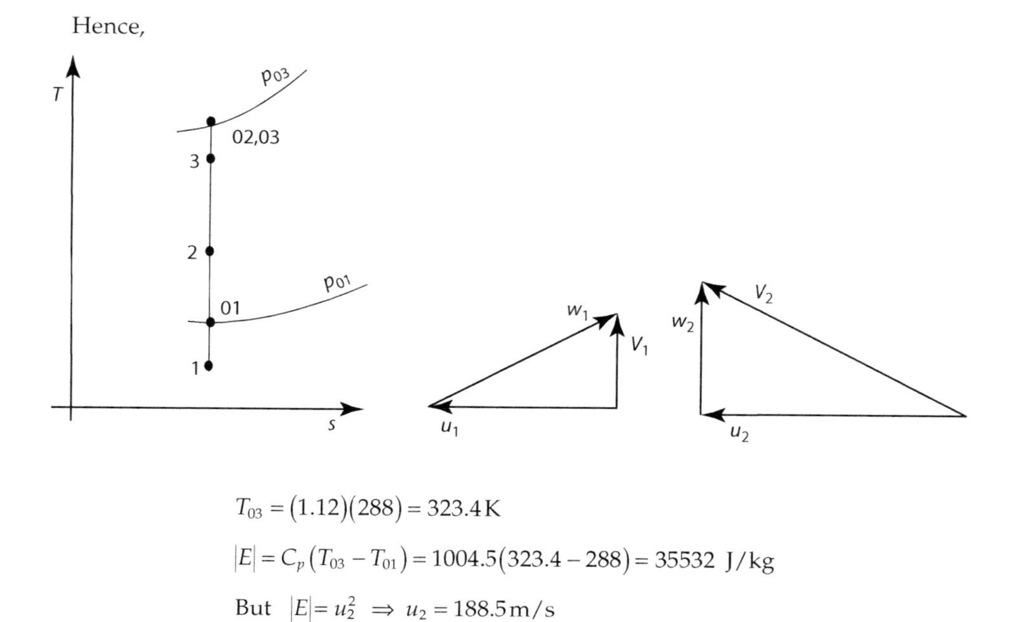

$$T_{03} = (1.12)(288) = 323.4\,\text{K}$$

$$|E| = C_p(T_{03} - T_{01}) = 1004.5(323.4 - 288) = 35532 \text{ J/kg}$$

$$\text{But } |E| = u_2^2 \Rightarrow u_2 = 188.5\,\text{m/s}$$

The diameter of the compressor and the power input can now be calculated:

$$r = \frac{188.5}{837.8} = 0.225 \text{ m} = 22.5 \text{ cm}$$

$$P = E\dot{m} = (35532)(1.5) = 53298 \text{ W} = 53.30 \text{ kW}$$

Comment: If the process is not isentropic, the *T–s* diagram will not be vertical, and information about the efficiency of the compressor will also be needed.

Example 9.2

The impeller radius of a centrifugal compressor is 0.25 m. At the inlet, the stagnation conditions are 15°C and 100 kPa. When rotating at 20,000 rpm, the mass flow rate is 2.1 kg/s. The total to total efficiency is 85% and the number of blades in the rotor is 19. The fluid enters the rotor with no swirl and the rotor blades are radially tipped at the exit. Using Stanitz's equation for slip coefficient, calculate the following: (a) the power input in kW, (b) the total pressure rise that can be expected, (c) the static pressure at the exit of the diffuser if the diffuser exit velocity is 100 m/s.
Solution: The *h–s* diagram is given in Figure 9.3. The velocity triangles are provided.

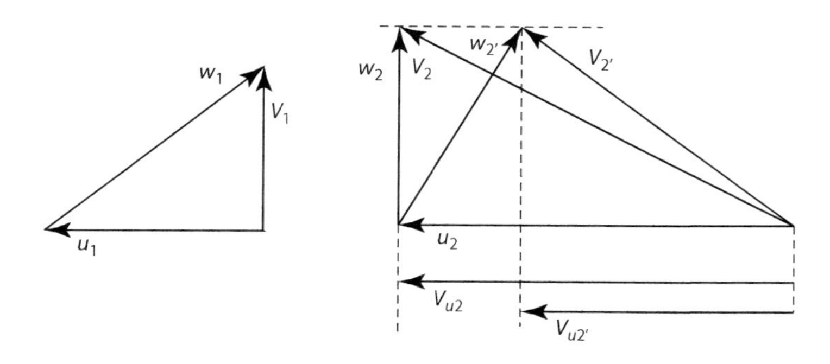

The blade velocity is given by

$$N = 20{,}000 \text{ rpm} = 2094.4 \text{ rad / s}; \ T_{01} = 15 + 273 = 288 \text{ K}$$

$$u_2 = N \times r_2 = (2094)(0.25) = 523.6 \text{m/s}$$

For the case of no slip, $V_{u2}=u_2=523.6$ m/s. However, due to slip, the tangential component of V_2 is reduced. For this, the slip coefficient needs to be calculated. From Stanitz's formula,

$$\mu_s = 1 - \frac{2}{n_B} = 1 - \frac{2}{19} = 0.89$$

$$\text{Hence, } V_{u2'} = \mu_s V_{u2} = (0.89)(523.6) = 468.5 \text{m/s}$$

Thus,

$$|E| = |u_1 V u_1 - u_2 V_{u2'}| = u_2 V_{u2'} = (523.6)(468.5) = (2.45)10^5 \text{ J/kg}$$

$$P = E\dot{m} = ((2.45)10^5)(2.1) = 515 \text{kW}$$

The temperatures can now be calculated using the total to total efficiency. Thus,

$$|E| = C_p (T_{02} - T_{01}) = C_p (T_{03} - T_{01})$$

$$\text{or} \quad T_{03} - T_{01} = \frac{|E|}{C_p} = \frac{(2.45)10^5 \text{ J/kg}}{1004 \text{ J/kg} - \text{K}} = 244.2 \text{K}$$

$$T_{03} = 244.2 + 288 = 532.2 \text{K}$$

Using the definition of total to total efficiency,

$$\eta_{tt} = \frac{T_{03'} - T_{01}}{T_{03} - T_{01}} \Rightarrow T_{03'} - T_{01} = \eta_{tt} (T_{03} - T_{01}) = (0.85)(244.2) = 207.6 \text{K}$$

$$T_{03'} = T_{01} + 207.6 = 495.6 \text{ K}$$

Since process 01–03′ is isentropic,

$$\frac{p_{03'}}{p_{01}} = \frac{p_{03}}{p_{01}} = \left(\frac{T_{03'}}{T_{01}}\right)^{k/(k-1)} = \left(\frac{495.6}{288}\right)^{1.4/(1.4-1)} = 6.68$$

$$\text{Hence, } p_{03} = (6.68)(100) = 668 \text{kPa}$$

The exit static pressure can be calculated by noting the relationship between the stagnation and static states at 3 and that process 3–03 is isentropic. Thus,

$$h_{03} = h_3 + \frac{V_3^2}{2} \Rightarrow T_{03} = T_3 + \frac{V_3^2}{2C_p}$$

$$\text{or } T_3 = T_{03} - \frac{V_3^2}{2C_p} = 532.2 - \frac{100^2}{2(1004.5)} = 527.2 \text{K}$$

$$\frac{p_{03}}{p_3} = \left(\frac{T_{03}}{T_3}\right)^{k/(k-1)} = \left(\frac{532.2}{527.2}\right)^{1.4/(1.4-1)} = 1.033$$

$$\text{Hence, } p_3 = \frac{p_{03}}{1.033} = \frac{668}{1.033} = 646 \text{kPa}$$

Comment: If the effect of slip were ignored, that is, $\mu_s = 1$, then the exit pressures, both stagnation and static, would be considerably higher (see Problem 9.4).

Example 9.3

Derive the following formula for pressure rise through a radial compressor (with no inlet swirl).

$$\frac{p_{03}}{p_{01}} = \left[1 + \frac{(k-1)\eta_{tt}u_2^2 \cot\alpha_2}{kRT_{01}(\cot\alpha_2 + \cot\beta_2)}\right]^{k/(k-1)}$$

Solution: The velocity triangles for zero inlet swirl and arbitrary exit angle are shown. Total to total efficiency is given by the equation

$$\eta_{tt} = \frac{h_{03'} - h_{01}}{h_{03} - h_{01}} = \frac{T_{03'} - T_{01}}{h_{03} - h_{01}} = \frac{C_p T_{01}\left(\frac{T_{03'}}{T_{01}} - 1\right)}{E} = \frac{\frac{kR}{(k-1)} T_{01}\left(\frac{T_{03'}}{T_{01}} - 1\right)}{u_2 V_{u2}}$$

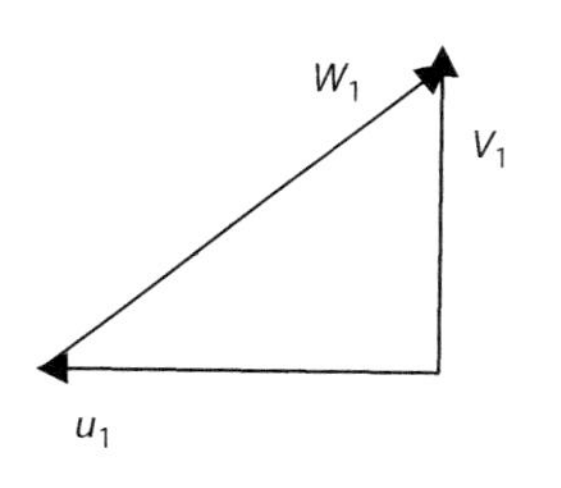

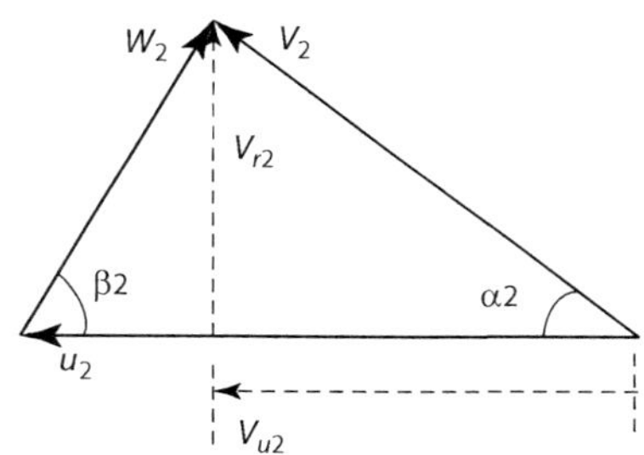

This equation can be simplified to obtain

$$\left(\frac{T_{03'}}{T_{01}} - 1\right) = \frac{(k-1)\eta_{tt}u_2 V_{u2}}{kRT_{01}} \Rightarrow \frac{T_{03'}}{T_{01}} = 1 + \frac{(k-1)\eta_{tt}u_2 V_{u2}}{kRT_{01}}$$

Since process 01–03′ is isentropic, the pressure ratio can be obtained as

$$\frac{p_{03'}}{p_{01}} = \frac{p_{03}}{p_{01}} = \left(\frac{T_{03'}}{T_{01}}\right)^{k/(k-1)} = \left(1 + \frac{(k-1)\eta_{tt}u_2 V_{u2}}{kRT_{01}}\right)^{k/(k-1)} \tag{a}$$

Now, from the exit triangle, the following relationship can be written:

$$\begin{aligned} u_2 &= V_{u2} + w_{u2} = V_{r2}\cot\alpha_2 + V_{r2}\cot\beta_2 = V_{r2}(\cot\alpha_2 + \cot\beta_2) \\ &= \frac{V_{u2}}{\cot\alpha_2}(\cot\alpha_2 + \cot\beta_2) \end{aligned} \tag{b}$$

Hence,

$$V_{u2} = \frac{u_2 \cot\alpha_2}{(\cot\alpha_2 + \cot\beta_2)} \tag{c}$$

Substituting Equations (b) and (c) into (a), the pressure ratio can be written as

$$\frac{p_{03}}{p_{01}} = \left[1 + \frac{(k-1)\eta_{tt}u_2^2 \cot\alpha_2}{kRT_{01}(\cot\alpha_2 + \cot\beta_2)}\right]^{k/(k-1)}$$

Comment: The pressure ratio can be further simplified. See Problems 9.5 and 9.6.

Example 9.4

A centrifugal compressor draws air from the atmosphere at a temperature of 75°F. While rotating at 15,000 rpm, the pressure ratio is 4. The mean radius at the inlet is 0.4 ft. and the radius of the impeller is 1 ft. At the inlet, the velocity is 580 ft./s and the blades are curved, such that the inlet fluid angle measured with respect to the wheel tangent is 60°. Estimate the slip coefficient and the number of blades according to the Stanitz equation. The total to total efficiency is 0.8.

Solution: From the information given, the inlet velocity triangle can be constructed.

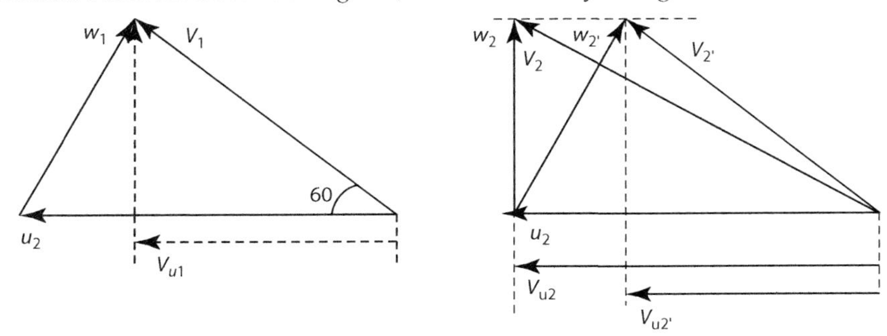

From the speed of rotation, the blade speed at the inlet and exit can be computed:

$$N = 15{,}000\,\text{rpm} = 1570.8\,\text{rad/s}$$

$$T_{0i} = 75 + 460 = 535\text{R}$$

$$u_1 = Nr_1 = (1570.8)(0.4) = 628.3\,\text{ft./s}$$

$$u_2 = Nr_2 = (1570.8)(1.0) = 1570.8\,\text{ft./s}$$

$$V_{u1} = V_1 \cos 60 = (580)\cos 60 = 290\,\text{ft./s}$$

Since the pressure ratio and total to total efficiency is known, the stagnation temperatures can be calculated. Thus,

$$\frac{p_{03}}{p_{01}} = \frac{p_{03'}}{p_{01}} = 4$$

$$\frac{T_{03'}}{T_{01}} = \left(\frac{p_{03'}}{p_{01}}\right)^{(k-1)/k} = (4)^{(1.4-1)/1.4} = 1.486$$

$$\text{Hence, } T_{03}' = (1.406)(T_{01}) = (1.406)(535) = 795\text{R}$$

The definition of total to total efficiency can now be used to get T_{03}:

$$\eta_{tt} = \frac{T_{03'} - T_{01}}{T_{03} - T_{01}} \Rightarrow (T_{03} - T_{01}) = \frac{(T_{03'} - T_{01})}{\eta_{tt}} = \frac{(795.0 - 535)}{0.8} = 325\text{R}$$

$$T_{03} = T_{01} + 325 = 860.0\text{R} = T_{02}$$

Energy per unit mass E can now be calculated using the stagnation temperature rise in the compressor:

$$|E| = C_p(T_{02} - T_{01}) = C_p(T_{03} - T_{01}) = (6006)(325) = 1952002\,\text{ft} - \text{lbf/slug}$$

But, from Euler's equation,

$$|E| = u_2V_{u2'} - u_1V_{u1} \Rightarrow 1952002 = u_2V_{u2} - (628.3)(290)$$

$$\text{Hence, } u_2V_{u2'} = 1952002 + (628.3)(290) = 2134215\,\text{ft.}^2/\text{s}^2$$

$$\text{or } V_{u2'} = \frac{2134215}{u_2} = \frac{2134215}{1570.8} = 1358.7\,\text{ft./s}$$

Next, the slip coefficient and the number of blades are calculated from the Stanitz equation:

$$\mu_s = \frac{V_{u2'}}{u_2} = \frac{1358.7}{1570.8} = 0.865$$

$$\mu_s = 1 - \frac{2}{n_B} = 0.865 \Rightarrow n_B = 14.81 \text{ or } 15\,\text{blades.}$$

Comment: Note that V_{u2} has been used in place of $V_{u2'}$; this should not make any difference in the context of this problem since total to total efficiency was used to calculate the actual temperature difference T_{02}–T_{01}.

Example 9.5

A radially tipped centrifugal air compressor running at 15,000 rpm has an impeller tip radius of 0.8 ft. The slip coefficient and impeller total to total efficiency are 0.9 and 0.8, respectively. The inlet stagnation conditions are 75°F and 15.23 psia. The tip and hub radii at the inlet are 0.4 and 0.2 ft., respectively. If the mass flow rate is 0.25 slugs/s, calculate the total pressure ratio, power required to drive the compressor, the inlet (at mean radius) blade and fluid angles, and inlet Mach number, under the following conditions: (a) without inlet guide vanes and (b) with inlet guide vanes (so that the inlet relative velocity is axial).

Solution: (a) Without inlet guide vanes. The velocity triangles are provided. From the rotational rate of the rotor and the inlet radius, the blade speed at the inlet can be calculated.

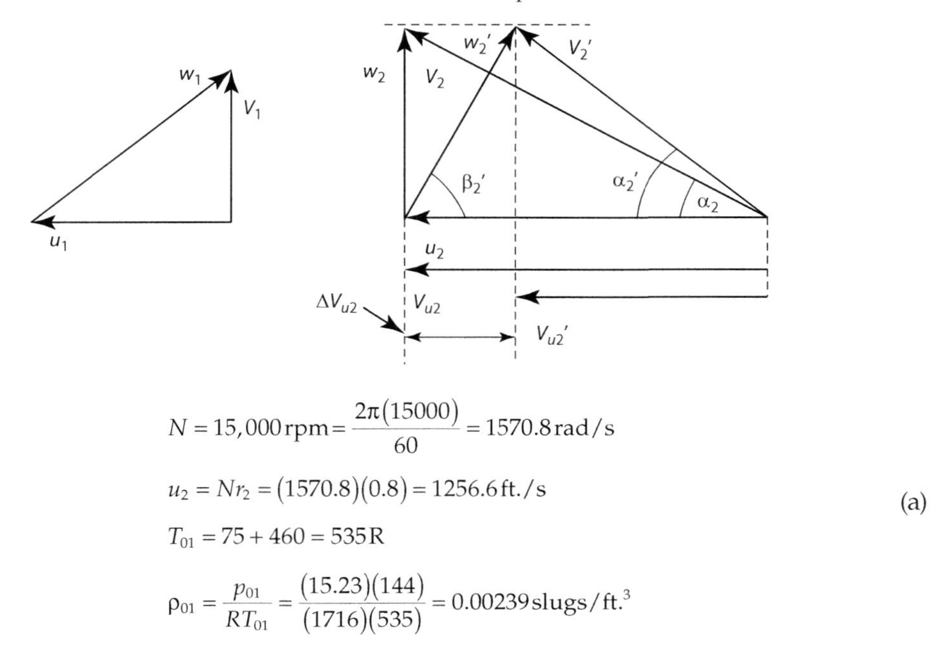

$$N = 15{,}000\,\text{rpm} = \frac{2\pi(15000)}{60} = 1570.8\,\text{rad/s}$$

$$u_2 = Nr_2 = (1570.8)(0.8) = 1256.6\,\text{ft./s} \quad \text{(a)}$$

$$T_{01} = 75 + 460 = 535\,\text{R}$$

$$\rho_{01} = \frac{p_{01}}{RT_{01}} = \frac{(15.23)(144)}{(1716)(535)} = 0.00239\,\text{slugs/ft.}^3$$

The area of the inlet and the mass flow rate are given by

$$A_1 = \pi\left(r_{tip}^2 - r_{hub}^2\right) = \pi\left(0.4^2 - 0.2^2\right) = 0.377\,\text{ft.}^2 \tag{b}$$

$$\dot{m} = \rho_1 V_1 A_1 = \frac{p_1}{RT_1} V_1 A \tag{c}$$

In Equation (b), the pressure and temperature needed are the static values, whereas only the stagnation values are available. However, since 1 to 01 is isentropic, the following relationships can be written:

$$T_1 = T_{01} - \frac{V_1^2}{2C_p}$$
$$p_1 = p_{01}\left(\frac{T_1}{T_{01}}\right)^{k/(k-1)} \tag{d}$$

If Equation (d) is substituted into Equation (c), a highly nonlinear equation is obtained for V_1 that needs to be solved by trial and error. This is illustrated as follows:

V_1	T_1	p_1	ρ_1	$\dot{m}$
ft./s	**R**	**psia**	**slug/ft.³**	**slug/s**
250	529.8	14.72	0.00233	0.220
260	529.4	14.68	0.00233	0.228
280	528.5	14.59	0.00232	0.245
285	528.2	14.57	0.00231	0.249
286.6	528.2	14.56	0.00231	0.250

Thus, the inlet static conditions corresponding to the mass flow of 0.25 slugs/s are

$$T_1 = 528.2\,\text{R};\; p_1 = 14.56\,\text{psi};\; \text{and } V_1 = 286.6\,\text{ft./s}$$

The energy transfer per unit mass is

$$|E| = \left|u_1 V u_1 - u_2 V_{u2}'\right| = u_2 V_{u2}' = (u_2)(\mu_s u_2)$$
$$= (0.9)(1256.6)^2 = 1421223\,\text{ft}\cdot\text{lbf/slug} \tag{e}$$
$$P = E\dot{m} = \frac{(1421223)(0.25)}{550} = 645.6\,\text{hp}$$

The corresponding temperature rise can now be calculated:

$$|E| = C_p\left(T_{02} - T_{01}\right) \Rightarrow T_{02} - T_{01} = \frac{|E|}{C_p} = \frac{1421223}{6006} = 236.6\text{R}$$

$$T_{02} = T_{01} + \frac{|E|}{C_p} = 535 + 236.6 = 771.6\,\text{R} \tag{f}$$

$$\eta_{tt} = \frac{T_{02'} - T_{01}}{T_{02} - T_{01}} = 0.8 \Rightarrow T_{02'} = \eta_{tt}\left(T_{02} - T_{01}\right) + T_{01} = 724.3\text{R}$$

$$\frac{T_{02'}}{T_{01}} = \frac{724.3}{535} = 1.353$$

Since process 01–02′ is isentropic, the pressure ratio can be calculated:

$$\frac{p_{02'}}{p_{01}} = \left(\frac{T_{02'}}{T_{01}}\right)^{k/(k-1)} = \left(\frac{724.3}{535}\right)^{3.5} = 2.887 = \frac{p_{02}}{p_{01}}$$

The blade angle and inlet Mach number at the inlet mean radius can now be calculated (it will be assumed that the velocity is uniform at the inlet):

$$r_{mean} = \frac{0.4 + 0.2}{2} = 0.3\,\text{ft.}$$

$$u_{mean} = N\,r_{mean} = (1570.8)(0.3) = 471.2\,\text{ft./s}$$

$$\beta_{mean} = \arctan\frac{V_1}{u_{mean}} = \arctan\frac{286.6}{471.2} = 31.3^\circ \tag{g}$$

$$\alpha_{mean} = 90^\circ$$

$$M_1 = \frac{V_1}{\sqrt{kRT_1}} = \frac{286.6}{\sqrt{(1.4)(1716)(528.2)}} = \frac{286.6}{1126.4} = 0.254$$

Now, the calculations are repeated with inlet guide vanes. The velocity triangle at the inlet at the mean radius is as shown.

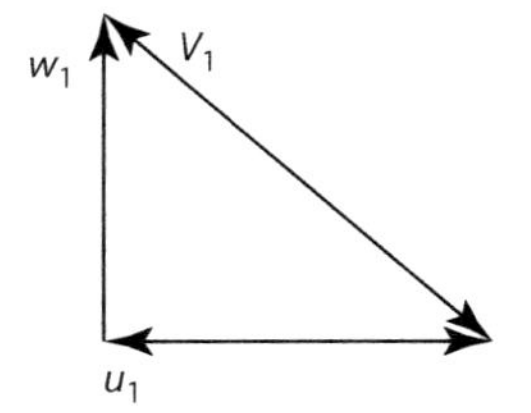

$$|E| = |u_1 V u_1 - u_2 V_{u2'}| = u_{mean} u_{mean} - (u_2)(\mu_s u_2)$$

$$= \left|(471.2)(471.2) - (0.9)(1256.6)^2\right| = 1199157\ \text{ft} \cdot \text{lbf/slug} \tag{h}$$

$$P = E\dot{m} = \frac{(1199157)(0.25)}{550} = 544\,\text{hp}$$

Now, the temperatures and pressure ratios can be calculated:

$$|E| = C_p(T_{02} - T_{01}) \Rightarrow T_{02} = T_{01} + \frac{|E|}{C_p} = 535 + \frac{1199157}{6006} = 734.7\text{R}$$

$$\eta_{tt} = \frac{T_{02'} - T_{01}}{T_{02} - T_{01}} = 0.8 \Rightarrow T_{02'} = \eta_{tt}(T_{02} - T_{01}) + T_{01} \tag{i}$$

$$= (0.8)(199.7) + 535 = 694.7\text{R}$$

$$\frac{T_{02'}}{T_{01}} = \frac{694.7}{535} = 1.299$$

The pressure ratio can now be calculated:

$$\frac{p'_{02}}{p_{01}} = \left(\frac{T'_{02}}{T_{01}}\right)^{k/(k-1)} = \left(\frac{694.7}{535}\right)^{3.5} = 2.495 = \frac{p_{02}}{p_{01}} \tag{j}$$

The inlet Mach number and inlet fluid angle at the mean radius are given by

$$\alpha_{mean} = \arctan\frac{w_1}{u_{mean}} = \arctan\frac{286.6}{471.2} = 31.3°$$

$$\beta_{mean} = 90°$$

$$V_1 = \sqrt{w_1^2 + u_{mean}^2} = 551.5\,\text{ft./s} \tag{k}$$

$$T_1 = T_{01} - \frac{V_1^2}{2C_p} = 535 - \frac{(551.5)^2}{2(6006)} = 509.7\,\text{ft./s}$$

$$M_1 = \frac{V_1}{\sqrt{kRT_1}} = \frac{551.5}{\sqrt{(1.4)(1716)(509.7)}} = \frac{551.5}{1106.5} = 0.499$$

Comment: By having inlet guide vanes, the power input and pressure ratio are reduced.

9.7 Design of Centrifugal Compressors

The basic design features of centrifugal compressors are presented in this section. As has been the practice in the previous chapters, only factors affecting thermodynamics and fluid flow will be presented. Mechanical details regarding stress analysis, vibrations, and other considerations are beyond the scope of the text. The input parameters that are essential are the inlet conditions (pressure and temperature), stagnation pressure ratio for the compressor, and mass flow rate. The rotational speed N is either specified or picked to ensure that the specific speed falls in the region of centrifugal compressors. According to Biederman et al. (2004), optimum efficiency of centrifugal compressors is obtained when the dimensionless specific speed is between 0.6 and 0.9. This corresponds to a dimensionless specific diameter range between 3 and 5. A more detailed analysis of effect of efficiency on specific speed and specific diameters can be found in Balje (1981). The design procedure can be summarized in the following steps.

1. The impeller inlet configuration needs to be designed first. Here, it is assumed that the inlet fluid has no angular momentum, that is, the velocity triangle is a right triangle and the absolute flow is axially directed. The velocity diagram is as viewed from above radially. Since the radius changes from the hub to the tip, so too does the blade speed. If inlet absolute velocity V_1 is assumed to be a constant throughout the section (this is the usual practice), the relative velocity and blade angles change from the hub to the tip. Therefore, the design is performed as follows (since the notation is quite cumbersome, the subscript t will be dropped with the understanding that the tip conditions are calculated first; the subscripts t and h refer to the tip and hub of the impeller at the inlet, respectively).
   a. Assume the relative Mach number at the tip ($=M_{1R}$) is between 0.7 and 0.8. The inlet flow will be subsonic since it is usually drawn from the atmosphere.
   b. Assume a value for tip blade angle β_1 between 20° and 32°.
   c. Calculate M_1 from the equation $M_1 = M_{1R}\sin\beta_1$ (see Problem 9.15).

d. Calculate the static temperature at the inlet from the relationship

$$T_{01} = T_1\left(1+\frac{k-1}{2}M_1^2\right)$$

e. Calculate the speed of sound a_1 $(=\sqrt{kRT_1}\,)$.

f. Now, calculate $V_1 = M_1 a_1$ and $w_1 = M_{1R} a_1$; complete the inlet triangle by calculating $u_1 = \sqrt{w_1^2 - V_1^2}$

g. Using the blade speed and the angular velocity, tip radius can be calculated.

h. The hub radius can be calculated from the mass flow rate equation given as

$$\dot{m} = \rho_1 V_1 \pi\left(r_t^2 - r_h^2\right)\;;\quad \frac{\rho_1}{\rho_{01}} = \left(\frac{T_1}{T_{01}}\right)^{1/(k-1)}\;;\ \text{and}\ \rho_{01} = \frac{p_{01}}{RT_{01}}$$

i. The blade angle β_1 changes gradually by about 2°–3° from the hub to the tip. This completes the inlet portion of the impeller.

2. Calculate the head, also called the *adiabatic head*, from the inlet conditions and pressure ratio, assuming the flow from 01 to 03 to be isentropic, using the following equation:

$$H = \frac{C_p T_{01}\left[\left(\frac{p_{03}}{p_{01}}\right)^{(k-1)/k} - 1\right]}{g}$$

3. Estimate the volumetric flow using the mass flow rate and density at stagnation conditions, $Q = \frac{\dot{m}}{\rho_{01}}$
4. Calculate the specific speed N_s and ensure that it falls in the range of centrifugal compressors. From Figure 9.7, calculate the specific diameter D_s and a suitable value for efficiency.
5. From the specific diameter D_s, calculate the outer diameter D_2; now the tip speed or blade speed u_2 can be calculated since N is known. The impeller tip speeds are limited to 2000–2200 ft./s (Bathie, 1984). As a check, the ratio of tip diameter at the inlet to exit diameter should be between 0.5 and 0.7 (Whitfield and Baines, 1990).
6. From the stagnation pressure ratio p_{03}/p_{01} and total to total efficiency η_{tt}, calculate temperature $T_{03}(T_{02})$.
7. Using the relationship $E = u_2 V_{u2} = C_p(T_{02} - T_{01})$, calculate V_{u2}. If necessary, mechanical efficiency can be introduced into this equation; however, for preliminary design calculations, the efficiency can be taken to be 100%. It is usually quite high anyway.
8. Assume a value of flow coefficient $\phi_2 = V_{r2}/u_2$ between 0.23 and 0.35 (Ferguson (1963)) and calculate V_{r2}.

9. Calculate $w_2 = \sqrt{V_{r2}^2 + w_{u2}^2}$; it is to be noted that $w_{u2} = u_2 - V_{u2}$; the diffusion ratio defined as w_1/w_2 should be less than 1.9 (Whitfield and Baines, 1990).
10. Assume a value for impeller efficiency (Equation 9.9a) between 0.87 and 0.93. Using this value, calculate $T_{02'}$.
11. Since process 01 to 02′ is isentropic, calculate p_{02} using the isentropic relationship.
12. Calculate $V_2 = \sqrt{V_{r2}^2 + V_{u2}^2}$ and using this value calculate $T_2 = T_{02} - \frac{V_2^2}{2C_p}$
13. Since process 2 to 02 is isentropic, calculate p_2 using the isentropic relationship $\frac{p_2}{p_{02}} = \left(\frac{T_2}{T_{02}}\right)^{k/(k-1)}$; now, using the ideal gas law, calculate $\rho_2 = p_2/RT_2$
14. Using the mass flow equation, $\dot{m} = \rho_2 V_{r2} 2\pi r_2 b_2$, the height of the blades at the exit can be estimated.
15. The number of blades is somewhat arbitrary and is selected to give a slip coefficient between 0.85 and 0.94.

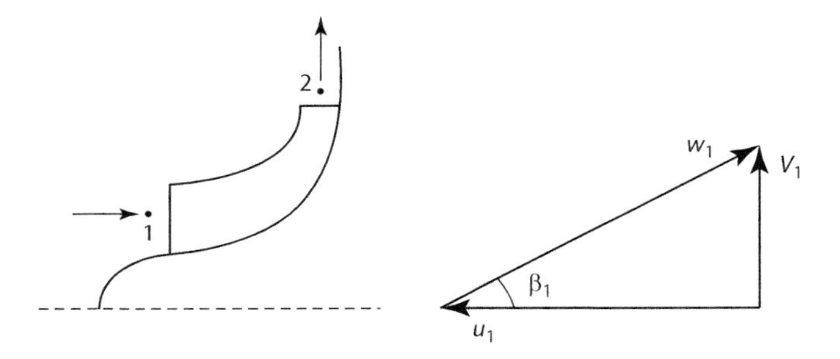

FIGURE 9.6
Inlet configuration and velocity triangle (as viewed radially from above) for centrifugal compressors.

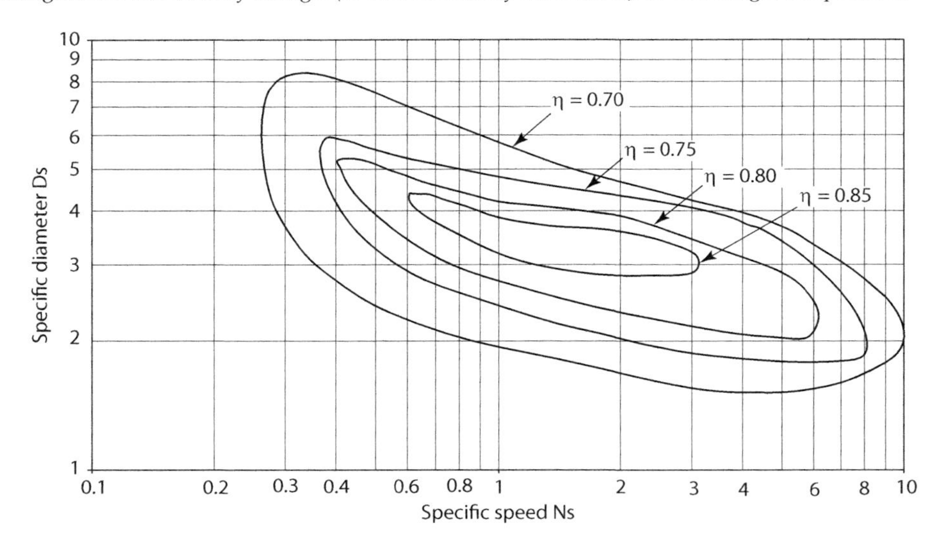

FIGURE 9.7
Variation of specific diameter versus specific speed for design of centrifugal compressors. (Modified from Balje, O.E., *Turbomachines: A Guide to Design, Selection, and Theory*, Wiley, New York, 1981.)

Some comments on the design procedure outlined are in order. Strictly speaking, the value of slip coefficient needs to be included in the design process. However, it can be ignored during the preliminary stages. The thickness of the blades also affects the mass flow rate since it decreases the effective area available for flow. This effect is also expected to be minimal. Although the procedure appears to be cumbersome, the only design decisions that need to be made are inlet relative Mach number, inlet tip blade angle, exit flow coefficient, and impeller efficiency. The rest of the calculations follow quite logically from the *h–s* diagram shown in Figure 9.3. These steps are illustrated in the following example.

Example 9.6

A centrifugal compressor is to be designed to produce a pressure ratio of two. The inlet stagnation conditions are 20°C and one atmosphere. The available mass flow rate of air is 5 kg/s. Obtain the basic v of the impeller.

Solution: The *h–s* diagram is given in Figure 9.3; the inlet hub tip configuration is shown in Figure 9.6. The inlet conditions are $p_{01} = 1$ atm = 101 kPa; $T_{01} = 20°C = 293$ K.

Inlet features of the impeller will be designed first. Since the speed of rotation is not given, let it be assumed to be 15,000 rpm. For the *tip* conditions, values for relative Mach number and blade angle are assumed. Thus, let

$$N = 15000\,\text{rpm} = 1570.8\,\text{rad/s};\; M_{1R} = 0.75;\; \beta_1 = 25°$$

The relationship between relative and absolute Mach numbers is given by

$$M_1 = M_{1R}\sin\beta_1 = (0.75)(\sin\beta_1) = 0.317$$

The inlet static temperature can now be calculated. Hence,

$$T_1 = \frac{T_{01}}{1+\frac{k-1}{2}M_1^2} = \frac{293}{1+\frac{1.4-1}{2}(0.317)^2} = 287.2\,\text{K}$$

$$\text{Hence, } a_1 = \sqrt{kRT_1} = \sqrt{(1.4)(287)(287.2)} = 339.7\,\text{m/s}$$

$$M_{1R} = \frac{w_1}{a_1} \Rightarrow w_1 = (339.7)(0.75) = 254.8\,\text{m/s}$$

$$M_1 = \frac{V_1}{a_1} \Rightarrow V_1 = (339.7)(0.317) = 107.7\,\text{m/s}$$

$$u_1 = \sqrt{w_1^2 - V_1^2} = \sqrt{(254.8)^2 - (107.7)^2} = 230.9\,\text{m/s from velocity triangle}$$

The tip radius can now be calculated since the blade velocity and rotation are known. Hence,

$$u_1 = Nr_1 \Rightarrow r_1 = \frac{u_1}{N} = \frac{230.9}{1579.8} = 0.147\,\text{m} = 14.7\,\text{cm}$$

To calculate the hub radius, the mass flow rate and hence the inlet density are needed. Thus,

$$\rho_{01} = \frac{p_{01}}{RT_{01}} = \frac{101000}{(287)(293)} = 1.21\,\text{kg/m}^3$$

The process 1 to 01 is isentropic. Hence,

$$\frac{\rho_1}{\rho_{01}} = \left(\frac{T_1}{T_{01}}\right)^{1/(k-1)} \Rightarrow \rho_1 = \rho_{01}\left(\frac{T_1}{T_{01}}\right)^{1/(k-1)}$$

$$= (1.21)\left(\frac{287.2}{293}\right)^{1/(1.4-1)} = 1.15\,\text{kg/m}^3$$

From the mass flow rate, the hub radius can be calculated. Thus,

$$\dot{m} = \rho_1 V_1 \pi\left(r_t^2 - r_h^2\right) = (1.15)(107.1)\pi\left[(0.147)^2 - r_h^2\right]$$

$$\Rightarrow r_h = 0.093\,\text{m} = 9.3\,\text{cm}$$

This completes the inlet configuration of the impeller. The process 01 to 03′ is isentropic. Hence,

$$\frac{p_{03}}{p_{01}} = 2 = \frac{p_{03'}}{p_{01}} = \left(\frac{T_{03'}}{T_{01}}\right)^{k/(k-1)} \Rightarrow T_{03'} = T_{01}\left(\frac{p_{03'}}{p_{01}}\right)^{(k-1)/k}$$

$$\text{or} \quad T_{03'} = (293)(2)^{(1.4-1)/1.4} = 357.2\,\text{K}$$

The ideal or adiabatic head can now be calculated. Thus,

$$H_{ad} = \frac{E_{ad}}{g} = \frac{C_p\left(T_{03'} - T_{01}\right)}{g} = \frac{(1004.5)(357.2 - 293)}{9.81} = 6570.8\,\text{m}$$

The flow rate at the inlet can is given by

$$Q = \frac{\dot{m}}{\rho_1} = \frac{5}{1.15} = 4.36\,\text{m}^3/\text{s}$$

The dimensionless form of specific speed is given by

$$N_s = \frac{N\sqrt{Q}}{\left(gH_{ad}\right)^{0.75}} = \frac{(1570.8)\sqrt{4.36}}{\left((9.81)(6570.8)\right)^{0.75}} = 0.81$$

From Figure 9.7,

$$\eta_{tt} = 0.87 \text{ and}$$

$$D_s = 3.57 = \frac{D_2\left(gH_{ad}\right)^{0.25}}{\sqrt{Q}} = \frac{D_2\left((9.81)(6570.8)\right)^{0.25}}{\sqrt{4.36}}$$

$$\text{or} \quad D_2 = \frac{D_s\sqrt{Q}}{\left(gH_{ad}\right)^{0.25}} = \frac{(3.57)\sqrt{4.36}}{\left((9.81)(6570.8)\right)^{0.25}} = 0.467\,\text{m} = 46.7\,\text{cm}$$

$$u_2 = N\frac{D_2}{2} = (1570.8)\frac{0.467}{2} = 367.4\,\text{m/s}$$

From the definition of total to total efficiency, $T_{03} = T_{02}$ can be calculated. Thus,

$$\eta_{tt} = 0.87 = \frac{T_{03'} - T_{01}}{T_{03} - T_{01}} = \frac{357.2 - 293}{T_{03} - 293} \Rightarrow T_{03} = 366.8\,\text{K} = T_{02}$$

Also,

$$E = C_p(T_{03} - T_{01}) = (1004.5)(366.2 - 293) = u_2 V_{u2} - u_1 V_{u1} = u_2 V_{u2}$$

$$\text{or } V_{u2} = 201.6\,\text{m/s}$$

Pick $\varphi_2 = 0.3$, and the radial velocity at the exit can be calculated:

$$\varphi_2 = \frac{V_{r2}}{u_2} = 0.3 \Rightarrow V_{r2} = (0.3)(367.4) = 110.2\,\text{m/s}$$

The exit triangle can now be completed:

$$V_2 = \sqrt{V_{r2}^2 + V_{u2}^2} = \sqrt{110.2^2 + 201.6^2} = 229.8\,\text{m/s}$$

$$w_{u2} = u_2 - V_{u2} = 165.8\,\text{m/s}$$

$$w_2 = \sqrt{V_{r2}^2 + w_{u2}^2} = \sqrt{110.2^2 + 165.8^2} = 199.1\,\text{m/s}$$

Check the diffusion coefficient w_1/w_2, which can be calculated to be 1.28. This is within acceptable bounds. Assume that the impeller efficiency is 0.9. The rest of the calculations for the exit conditions can now be performed:

$$T_2 = T_{02} - \frac{V_2^2}{2C_p} = 366.8 - \frac{229.8^2}{2(1004.5)} = 340.5\,\text{K}$$

$$\eta_{imp} \equiv \frac{T_{02'} - T_{01}}{T_{02} - T_{01}} \Rightarrow 0.9 = \frac{T_{02'} - 293}{366.8 - 293}$$

$$\text{or } T_{02'} = 293 + (0.9)(366.8 - 293) = 359.4\,\text{K}$$

The processes 01 to 02′ and 2 to 02 are isentropic. Hence,

$$\frac{p_{02'}}{p_{01}} = \frac{p_{02}}{p_{01}} = \left(\frac{T_{02'}}{T_{01}}\right)^{k/(k-1)} \Rightarrow p_{02} = p_{01}\left(\frac{T_{02'}}{T_{01}}\right)^{k/(k-1)}$$

$$\text{or } p_{02} = (101)\left(\frac{359.4}{293}\right)^{1.4/(1.4-1)} = 207.1\,\text{kPa}$$

$$\frac{p_2}{p_{02}} = \left(\frac{T_2}{T_{02}}\right)^{k/(k-1)} \Rightarrow p_2 = p_{02}\left(\frac{T_2}{T_{02}}\right)^{k/(k-1)}$$

$$\text{or } p_2 = (207.1)\left(\frac{340.5}{366.8}\right)^{1.4/(1.4-1)} = 159.7\,\text{kPa}$$

$$\rho_2 = \frac{p_2}{RT_2} = \frac{(159.7)(1000)}{(287)(340.5)} = 1.63\,\text{kg/m}^3$$

Now, from the mass flow, the width of the blades at the exit of the impeller can be calculated:

$$\dot{m} = 5 = \rho_2 V_{r2} 2\pi r_2 b_2 = (1.63)(110.2)2\pi\left(\frac{0.467}{2}\right)b_2$$

$$\Rightarrow b_2 = 0.019\,\text{m} = 1.9\,\text{cm}$$

Set the number of blades to be 19; the slip coefficient would be 0.894, well within the acceptable range.

Also, as a final check, the ratio of inlet tip to exit radii is calculated. This ratio is $0.63\left(=\dfrac{0.147}{0.233}\right)$, within acceptable limits.

Comments: The design presented in this example is by no means unique, since several variations are possible, starting with the rotational speed. However, the rudiments of the design process can be seen. Also, the slip coefficient and mechanical efficiency need to be included in the example. These have been left out for simplicity, as they would neither dramatically alter the results nor enhance the understanding of the design process.

9.8 Performance Characteristics

To obtain the performance characteristics of centrifugal compressors, the first step is to perform a dimensional analysis of the variables involved (see Problems 9.17 and 9.18). For simplicity, the subscripts 2 and 02 will refer to exit conditions (without making a distinction between the exit of the impeller or diffuser) and 1 and 01 will denote the inlet conditions. The expression relating the dimensionless variables is given as (Problem 9.18)

$$\hat{f}\left(\frac{p_{02}}{p_{01}},\eta,\frac{E}{kRT_{01}},\frac{ND}{\sqrt{kRT_{01}}},\frac{\dot{m}\sqrt{kRT_{01}}}{p_{01}D^2},\frac{ND^2}{\nu}\right)=0 \tag{9.26}$$

The last variable, ND^2/v, can be shown to be a variation of the Reynolds number, which is the ratio of inertial to viscous forces. For turbomachines, viscous forces are very difficult to quantify and are confined to the boundary layers. For large Reynolds numbers, the viscous effects are small. In any case, they are included in the efficiency. Hence, this variable is ignored and the relationship between the dimensionless numbers therefore becomes

$$\bar{f}\left(\frac{p_{02}}{p_{01}},\eta,\frac{E}{kRT_{01}},\frac{ND}{\sqrt{kRT_{01}}},\frac{\dot{m}\sqrt{kRT_{01}}}{p_{01}D^2}\right)=0 \tag{9.27}$$

Expressing E in terms of stagnation temperature rise, Equation 9.27 reduces to

$$\bar{f}\left(\frac{p_{02}}{p_{01}},\eta,\frac{C_p\Delta T_0}{kRT_{01}},\frac{ND}{\sqrt{kRT_{01}}},\frac{\dot{m}\sqrt{kRT_{01}}}{p_{01}D^2}\right)=0 \tag{9.27a}$$

The variables k, D, Cp, and R are usually omitted from Equation 9.27a, since the analysis would be for a specific machine and specific gas. Hence, the equation can be further simplified into

$$f\left(\frac{p_{02}}{p_{01}},\eta,\frac{\Delta T_0}{T_{01}},\frac{N}{\sqrt{T_{01}}},\frac{\dot{m}\sqrt{T_{01}}}{p_{01}}\right)=0 \tag{9.28}$$

The variables $\dfrac{ND}{\sqrt{kRT_{01}}}$ and $\dfrac{\dot{m}\sqrt{kRT_{01}}}{p_{01}D^2}$, or alternatively, $\dfrac{N}{\sqrt{T_{01}}}$ and $\dfrac{\dot{m}\sqrt{T_{01}}}{p_{01}}$, are called *non-dimensional rotational speed* and *non-dimensional mass flow rate*, respectively.

It is possible to give a physical interpretation to the non-dimensional rotational speed and non-dimensional mass flow rate. The former can be written as

$$\frac{ND}{\sqrt{kRT_{01}}} \propto \frac{u}{a_{01}} \tag{9.29}$$

where a_{01} is the speed of sound under inlet stagnation conditions. Thus, the non-dimensional rotational speed is a form of Mach number related to blade speed. Similarly, the non-dimensional mass flow rate can be written as

$$\frac{\dot{m}\sqrt{kRT_{01}}}{p_{01}D^2} \propto \frac{\rho AV\sqrt{kRT_{01}}}{p_{01}D^2} \propto \frac{AV\sqrt{kRT_{01}}}{RT_{01}D^2} \propto \frac{V\sqrt{kRT_{01}}}{kRT_{01}} = \frac{V}{\sqrt{kRT_{01}}} = \frac{V}{a_{01}} \tag{9.30}$$

which is a form of Mach number related to the flow. Thus, if the non-dimensional rotational speed and mass flow rate are proportional under different operating conditions for a compressor, then their velocity triangles will be similar, and the performance characteristics in terms of the pressure ratio and efficiency will be the same.

The expression given by Equation 9.28 can be expressed graphically by plotting one group against the other while holding the third constant. Typically, the variables of interest are the pressure ratio p_{02}/p_{01} and efficiency η, which can be plotted as a function of $\dot{m}\sqrt{T_{01}}/p_{01}$ with $N/\sqrt{T_{01}}$ as a parameter. These are shown in Figure 9.8. Such plots, which are obtained either experimentally or numerically, are called *compressor maps*. These maps are of several forms: one where the actual values of mass flow and speed are used; another where non-dimensional mass flow and speed are scaled against the design values; and a third, where they are scaled to standard temperature and pressure. It should be noted in passing that the temperature ratio $\Delta T_0/\sqrt{T_{01}}$ is a simple function of the pressure ratio p_{02}/p_{01} and efficiency η, and that its shape is similar to Figure 9.8a. Hence, it is not shown separately here.

As a closing comment, the similarities between the pump characteristics and compressor maps should be noted. It should be realized that the delivery pressure and mass flow rate for compressors are analogous to the "head" and "discharge" of pumps. Also, the

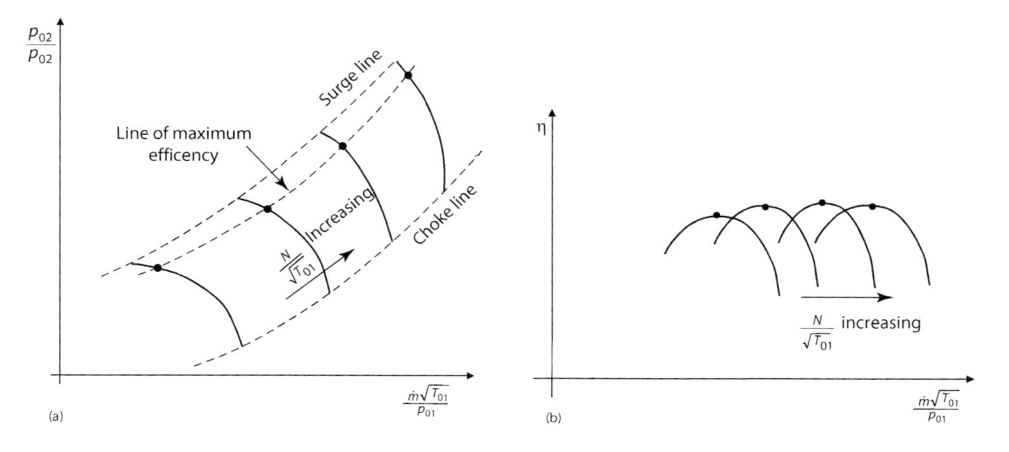

FIGURE 9.8
A typical compressor map: (a) variation of pressure ratio with mass flow and (b) variation of efficiency with mass flow.

shape of the efficiency curve for compressors is similar to that of pumps, although only one curve pertaining to a single rotational speed is provided for pumps.

Finally, the two lines enveloping the compressor pressure maps on the left and right are the surge line and choke line. Points on the choke line represent the maximum mass flow rate possible for the given speed. Points on the surge line represent the instability associated with drops in delivery pressure and decreasing mass flow rate. Surging and choking are discussed in the following section. The line in the middle is the locus of points of peak efficiency.

9.9 Surging, Choking, and Stalling

There are certain limits to the mass flow rates in compressors beyond which the flow becomes unstable. These limits are due to surging and choking, each of them occurring for different reasons. In both these conditions, the machine is operating at off-design conditions that are aerodynamically and mechanically undesirable.

Surging in compressors is quite similar to surging in pumps (see Chapter 5) and occurs for the same reasons. The fundamental difference is that while the head and flow rate were the variables under discussion for pumps, those for compressors are the pressure ratio and mass flow rate. To fix our ideas, consider the situation shown in Figure 9.9, which shows a compressor characteristic curve at a particular speed. Consider an operating point to the right of C, point A for example. Any decrease in mass flow rate due to an increase in resistance would tend to move the operating point to the left. This corresponds to a higher pressure being produced by the compressor that increases the mass flow rate and moves the operating point to A, thus stabilizing the system. Similarly, an increase in mass flow rate would result in the compressor producing lower pressure, which in turn decreases the mass flow, and the system would move back to A. The case would be similar for point B or, for that matter, any point lying to the right of C.

This is in contrast with a point to the left of C, for example D. Any decrease of mass flow rate now would correspond to a decrease in delivery pressure to the compressor, which in turn produces even less flow, eventually leading to shut-off. Thus, the system moves away from D, leading to instability. But, since the compressor is still producing pressure, the system restarts and mass flow increases. This phenomenon is called *surging*. Typically, the portion to the left of C is not obtainable due to this phenomenon. Surging is usually associated with a sudden drop in delivery pressure, with violent pulsation and eventual mechanical failure of the compressor.

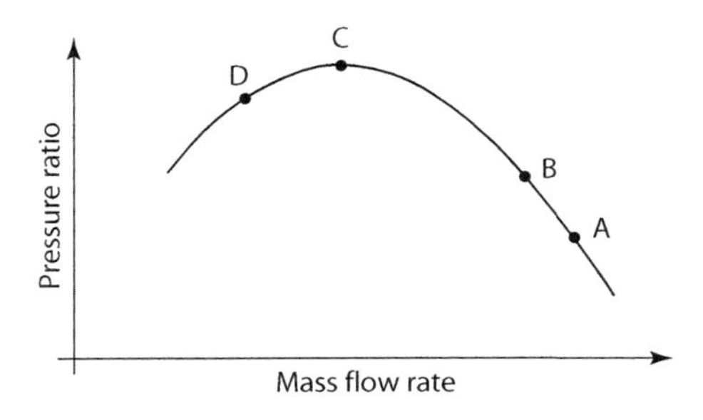

FIGURE 9.9
Surging in compressors.

Choking corresponds to the compressor delivering the maximum mass flow rate for a given value of rotational speed. This occurs when the flow at any section of the compressor reaches sonic conditions. Choking needs to be analyzed separately for stationary sections of the compressor (e.g., the stator or diffuser) and moving sections such as the rotor. These can be subjected to an extremely simplified analysis. Choking in the stator is left as an exercise (see Problem 9.19).

Flow through the rotor is more complex since the vanes are rotating. A simplified analysis for the same is presented here. Choking in a rotor occurs when the *relative Mach number* becomes equal to one at any section, that is,

$$M_{1R} = \frac{w}{a} = \frac{w}{\sqrt{kRT}} = 1 \tag{9.31}$$

Across the rotor, the energy equation becomes

$$\begin{aligned} h_2 + \frac{V_2^2}{2} &= h_3 + \frac{V_3^2}{2} + E = h_3 + \frac{V_3^2}{2} + u_2V_{u2} - u_3V_{u3} \\ \text{or } h_2 + \frac{V_2^2}{2} - u_2V_{u2} &= h_3 + \frac{V_3^2}{2} - u_3V_{u3} \end{aligned} \tag{9.32}$$

which can be further simplified as

$$h_2 + \frac{w_2^2}{2} - \frac{u_2^2}{2} = h_3 + \frac{w_3^2}{2} - \frac{u_3^2}{2} = h + \frac{w^2}{2} - \frac{u^2}{2} \tag{9.33}$$

It may be recalled that the quantity on either side of the equality sign in Equation 9.33 is called the *rothalpy*. From Equation 9.32,

$$u_3V_{u3} = \frac{V_3^2}{2} + \frac{u_3^2}{2} - \frac{w_3^2}{2}$$

$$\text{and hence } h_2 + \frac{V_2^2}{2} - u_2V_{u2} = h_3 + \frac{V_3^2}{2} - \left(\frac{V_3^2}{2} + \frac{u_3^2}{2} - \frac{w_3^2}{2}\right) \tag{9.34}$$

$$h_2 + \frac{V_2^2}{2} = h_3 + \frac{w_3^2}{2} - \frac{u_3^2}{2} = h + \frac{w^2}{2} - \frac{u^2}{2}$$

In Equation 9.34, it has been assumed that the inlet to the impeller has no angular momentum in the ideal case; also, the left side of Equation 9.34 is the stagnation enthalpy at the inlet of the impeller which is equal to $h_{02} = h_{01}$. Thus,

$$\begin{aligned} &h_{02} = h + \frac{w^2}{2} - \frac{u^2}{2} \quad \text{or} \quad T_{02} = T_{01} = T + \frac{w^2}{2C_p} - \frac{u^2}{2C_p} \\ &\text{or} \quad T_{01} = T + \frac{kRT}{2C_p} - \frac{u^2}{2C_p} \Rightarrow 1 = \frac{T}{T_{01}} + \frac{kRT}{2C_pT_{01}} - \frac{u^2}{2C_pT_{01}} \\ &\text{or} \quad \frac{T}{T_{01}} = \frac{2}{k+1}\left(1 + \frac{u^2}{2C_pT_{01}}\right) \end{aligned} \tag{9.35}$$

Assuming the process to be isentropic, densities and temperatures can be related as

$$\frac{\rho}{\rho_{01}} = \left(\frac{T}{T_{01}}\right)^{1/(k-1)} \Rightarrow \rho = \rho_{01}\left(\frac{T}{T_{01}}\right)^{1/(k-1)} \tag{9.36}$$

The mass flow rate is given by (for choked flow in the impeller $w = a = \sqrt{kRT}$)

$$\begin{aligned} \dot{m} = \rho A w &= \rho_{01}\left(\frac{T}{T_{01}}\right)^{1/(k-1)} A\sqrt{kRT} = \rho_{01}\left(\frac{T}{T_{01}}\right)^{1/(k-1)} A\sqrt{kRT}\frac{\sqrt{kRT_{01}}}{\sqrt{kRT_{01}}} \\ &= \rho_{01}\left(\frac{T}{T_{01}}\right)^{1/(k-1)} A\sqrt{\frac{T}{T_{01}}}\sqrt{kRT_{01}} = \rho_{01} A a_{01}\left(\frac{T}{T_{01}}\right)^{(k+1)/2(k-1)} \end{aligned} \tag{9.37}$$

Now, by substituting Equation 9.35 into 9.37, mass flow rate can be obtained. Thus,

$$\begin{aligned} \dot{m} &= \rho_{01} A a_{01}\left[\frac{2}{k+1}\left(1+\frac{u^2}{2C_p T_{01}}\right)\right]^{(k+1)/2(k-1)} \\ &= \rho_{01} A a_{01}\left[\frac{2}{k+1}\left(1+\frac{(k-1)}{2}\frac{u^2}{a_{01}^2}\right)\right]^{(k+1)/2(k-1)} \end{aligned} \tag{9.38}$$

It can therefore be seen that the mass flow rate increases continuously with blade speed until choking occurs at some point in the compressor. It may be noted in passing that the limiting point for flow rate in pumps is set by cavitation, whereas in the case of compressors it is choking.

PROBLEMS

All pressures given are absolute unless specified otherwise. Use air properties unless specified otherwise. Assume that the values of slip factor and power input factor are one unless specified otherwise.

9.1 A centrifugal compressor is spinning at 9700 rpm and takes in air at 20°C. The rotor tip diameter is 50 cm and the mass flow is 0.8 kg/s. Assume no inlet swirl and that the blades are radial at the exit. What stagnation pressure ratio and power input can be expected if the flow is considered to be isentropic?

Ans: 51.5 kW; Pressure ratio = 2.0

9.2 The inlet stagnation temperature of a centrifugal compressor rotating at 10,000 rpm is 20°C. The inlet flow is axial and the blades are radially tipped at the exit. It is required to produce a stagnation pressure ratio of 2.8 when the total to total efficiency is 0.85. It has 19 blades and a mass flow rate of 3.0 kg/s. Through measurements, the density of air at the exit of the impeller is found to be 1.5 kg/m³. The width of the blades at the exit is 1 cm. Calculate the following: (a) power required to drive the compressor, (b) diameter of the impeller, and (c) absolute Mach number at the exit. Use Stanitz's equation for the slip coefficient.

Ans: 355 kW; 0.694 m; 0.897

9.3 A centrifugal compressor has a mass flow rate of 1.6 kg/s while rotating at 11,000 rpm. The diameter of the impeller is 0.6 m and the inlet stagnation temperature at the impeller is 15°C. Width of the blades at the exit is 1 cm and exit density is 1.5 kg/m^3. If the total to total efficiency is 83%, calculate the following: (a) input power, (b) the pressure ratio that can be expected, (c) fluid angle at the exit of the rotor, and (d) exit Mach number. Assume that the slip coefficient is unity. Also assume axial flow at inlet and radial blades at exit.

Ans: 191 kW; 2.8; 9.3°; 0.94

9.4 Repeat Example 9.2 when slip is ignored, that is, $\mu_s = 1$.

9.5 The pressure rise through radial compressors is given by the following formula:

$$\frac{p_{03}}{p_{01}} = \left[1 + \frac{(k-1)\eta_{tt}u_2^2(1-\phi_2\cot\beta_2)}{kRT_{01}}\right]^{k/(k-1)}$$

where $\phi_2 = \left[\frac{V_{r2}}{u_2}\right]$. Derive it.

9.6 For radial blades, the pressure ratio equations given in Problem 9.5 simplify to

$$\frac{p_{03}}{p_{01}} = \left[1 + \frac{(k-1)\eta_{tt}u_2^2\mu_s}{kRT_{01}}\right]^{k/(k-1)}$$

where μ_s is the slip coefficient. Derive this formula.

9.7 (a) Show that the blade loading coefficient ψ and flow coefficient ϕ are related by the following expression:

$$\psi = 1 - \phi_2\cot\beta_2$$

(b) Show that for radial tipped blades and isentropic flow $\psi = 1$.

(c) Show that the pressure ratio for radial tipped blades and isentropic flow is given by the following expression:

$$\text{Pressure ratio} = \left(1 + \frac{u_2}{C_pT_{01}}\right)^{k/(k-1)}$$

(d) Show that the degree of reaction can be expressed as follows:

$$R = \frac{1}{2} + \frac{1}{2}\phi_2\cot\beta_2$$

9.8 A centrifugal compressor rotating at 16,000 rpm produces a stagnation pressure ratio of 5.3. Inlet air is drawn from the atmosphere, where the temperature is 25°C. The mean radius at the inlet is 0.125 m and at the exit is 0.3 m. The inlet velocity is 180 m/s and there is no swirl. If the total to total efficiency is 0.81, calculate the slip coefficient and estimate the number of blades from Stanitz's equation.

Ans: 0.892; 19 blades

9.9 The rotor diameter of a centrifugal compressor is 80 cm and the average inlet diameter is 40 cm. The inlet absolute velocity is 175 m/s and the blade angle at the inlet is 65° with respect to the wheel tangent. If the compressor has 19 blades and a total to total efficiency of 82%, how fast should it spin to produce a pressure ratio of 5.1? Use Stanitz's correlation for slip coefficient.

Ans: 12,146 rpm

9.10 The following data refer to a radially tipped centrifugal compressor:

Calculate the power, pressure ratio, fluid, and blade angles for the following situations: (a) without inlet guide vanes and (b) with inlet guide vanes.

Entry conditions (stagnation)	$T_{01}=25°C$, $p_{01}=105$ kPa
Entry conditions (static)	$T_1=271.85$ K, $p_1=69.4$ kPa
Mass flow rate	6 kg/s
Speed	15,000 rpm
Number of blades	17
Total to total efficiency	84%
Impeller tip radius	25 cm
Inlet mean radius	8 cm

9.11 A single-stage centrifugal compressor, while running at 50,000 rpm, draws air at a rate of 3 lbm/s from the atmosphere when the stagnation conditions are 14.7 psia and 65°F. The delivery pressure is 73.5 psia. The impeller tip radius is 3 in. and the number of vanes is 21. The exit vane angle is 90°. Calculate the total to total efficiency, β_2, β_2, β_2', and β_2'.

Ans: $\eta_{tt}=78.2\%$

9.12 Compare the slip coefficients of centrifugal compressors using the three formulas of Balje, Stodola, and Stanitz. All of the compressors have an impeller diameter of 50 cm and a tip diameter of the eye of 30 cm. The number of blades is (a) 15, (b) 17, and (c) 19.

Ans: Balje's formula: a) 0.773; b) 0.794; c) 0.812

9.13 A centrifugal compressor with 21 blades has an impeller tip speed of 350 m/s and total to total efficiency of 85%. It takes air from the atmosphere at 20°C and 100 kPa. The radial component of velocity at the exit is 25 m/s and the exit area of the impeller is 0.1 kg/s. The vanes are radial at the exit and the inlet flow is axial. Calculate the power to drive the compressor if the mechanical efficiency of the compressor is 95%.

Ans: 477.6 kW

9.14 Estimate the slip coefficient for a centrifugal compressor spinning at 14,000 rpm under the following conditions:

Pressure ratio	5
Inlet stagnation temperature	72°F
Mean diameter at inlet	1 ft.
Impeller diameter	2 ft.
Total to total efficiency	85%
Absolute velocity at inlet	40 ft./s
Inlet blade angle	65°

Ans: 0.91

9.15 Show that the absolute and relative Mach numbers at the inlet (M_1 and M_{1R}) of centrifugal compressors under ideal conditions are related by the equation $M_1 = M_{1R}$ $\text{Sin}\beta_1$, where β_1 is the blade angle at the inlet.

9.16 A centrifugal compressor is required to meet the following specifications:

Mass flow	0.5 slugs/s
Pressure ratio	3
Inlet stagnation temperature	72°F
Inlet stagnation pressure	14.7 psia

Perform the preliminary design and obtain the dimensions.

9.17 For centrifugal compressors such as pumps, dimensional analysis can be performed to obtain the relevant dimensional variables. The variables can be expressed as $\tilde{f}(\dot{m}, p_{02}, p_{01}, T_{01}, k, R, D, N, \nu, \eta, E) = 0$

where

p_{02}, p_{01}	Exit and inlet stagnation pressures, respectively
T_{01}	Inlet stagnation temperature
D	Diameter of impeller
N	Revolutions per minute
η	Impeller efficiency
E	Energy per unit mass
k, R	Specific heat ratio and gas constant, respectively
v	Kinematic viscosity

There are eleven variables and four dimensions. Using Buckingham's Pi theorem and p_{01}, T_{01}, D, and N as repeating variables, obtain the six dimensionless groups.

$$\text{Ans}: \quad \pi_1 = k; \ \pi_2 = \eta; \ \pi_3 = \frac{p_{02}}{p_{01}}; \pi_4 = \frac{\dot{m}N}{p_{01}D}; \ \pi_5 = \frac{RT_{01}}{N^2D^2}$$

$$\pi_6 = \frac{\nu}{ND^2}; \ \pi_7 = \frac{E}{N^2D^2}$$

9.18 (a) By suitably combining the dimensionless parameters, show the relationship between them can be expressed as

$$\tilde{f}\left(\frac{p_{02}}{p_{01}}, \eta, \frac{E}{kRT_{01}}, \frac{ND}{\sqrt{kRT_{01}}}, \frac{\dot{m}\sqrt{kRT_{01}}}{p_{01}D^2}, \frac{ND^2}{\nu}\right) = 0$$

(b) By letting $E = C_p \Delta T_0$, show a new functional form for the equation in (a) can be written as

$$\tilde{f}\left(\frac{p_{02}}{p_{01}}, \eta, \frac{C_p \Delta T_0}{kRT_{01}}, \frac{ND}{\sqrt{kRT_{01}}}, \frac{\dot{m}\sqrt{kRT_{01}}}{p_{01}D^2}, \frac{ND^2}{\nu}\right) = 0$$

9.19 Using a simplified one-dimensional, isentropic analysis of flow through the stator, show that the condition for choking can be expressed as follows:

$$\dot{m} = \rho_0 a_0 A \left(\frac{2}{k+1} \right)^{(k+1)/2(k-1)}$$

where the subscript 0 refers to stagnation conditions at the inlet of the stator. A similar expression for flow through the diffuser can also be derived with stagnation conditions referring to the exit of the diffuser.

9.20 The following data refer to a centrifugal compressor with a radial inlet and exit for the impeller:

Inlet stagnation conditions p_{01}, T_{01}	100 kPa, 25°C
Speed of rotation	10,000 rpm
Impeller inlet mean radius	12.5 cm
Impeller exit radius	25 cm
Mean blade height at inlet	3 cm
Inlet fluid angle (with respect to axial direction)	53.13°

If inlet axial velocity is the same as exit radial velocity, calculate the power and exit fluid angle.

Ans: 124 kW, 69.4°

9.21 Draw the velocity triangles and shape of blades at the inlet and exit of an impeller of a centrifugal compressor for which the energy transfer is given by

$$E = u_1^2 - u_2^2$$

For one such impeller, the axial velocity component at the inlet and radial velocity component at the exit of the impeller are equal. If $u_1/u_2 = 0.5$ and the inlet nozzle angle measured with respect to the axial direction is 60°, calculate the ratio of absolute velocities at the exit and inlet; that is, V_2/V_1.

Ans: 1.80

10

Axial Compressors

In the previous chapter, the discussion was on centrifugal compressors. This chapter deals with axial compressors, the analogs of axial pumps/fans. Since these also use compressible fluids, the flow is heavily influenced by thermodynamics. As both density and temperature vary, along with velocities and pressures, there will be significant differences in their operation when compared with blowers and fans, wherein the flows are isothermal.

Upon completion of this chapter, the student will be able to

- Analyze the velocity diagrams for axial compressors
- Predict the performance of a single stage
- Distinguish between small stage and compressor efficiency
- Perform the preliminary design of a compressor stage

10.1 Introduction

Axial flow compressors are most suitable for high flow rates, as required in the operation of aircraft engines. To accommodate the high flow rates, atmospheric air enters at a large radius. As the compression process proceeds in stages, density increases and the area reduces, with the axial velocity being approximately constant. Although the pressure rise in a single stage is small, the use of several stages results in a significant pressure rise through the compressor. Modern aircraft engines achieve this through the use of multiple shafts, solid and hollow, typically called the *low-pressure* (LP), *intermediate-pressure* (IP), and *high-pressure* (HP) *spools*. Such an arrangement is common in axial flow turbines also. The typical practice is to match the pressure levels of compressors and turbines with the appropriate spool. For example, low-pressure compressor stages (initial) and low-pressure turbine stages (latter) are mounted on a LP spool. Similarly, high-pressure compressor stages (latter) and high-pressure turbine stages (the initial stages of which are impulse) are mounted on a HP spool, and so on. The typical pressure ratio per stage in modern compressors is 1.1–1.3.

10.2 Analysis of a Single Stage

A schematic diagram of an axial compressor with three stages is shown in Figure 10.1. Inlet guide vanes are optional, although they are often provided for smooth air entry into the first stage. The fluid enters the rotor in the axial direction and flows axially from stage

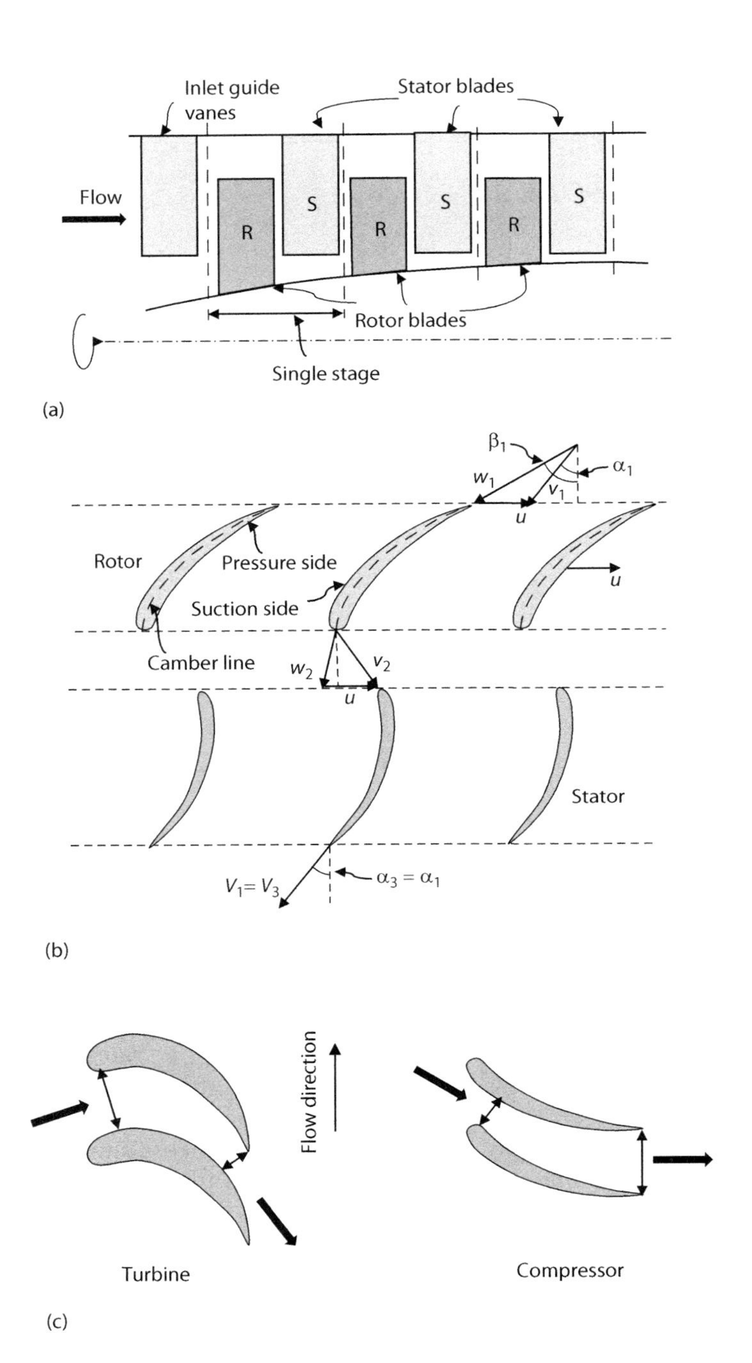

FIGURE 10.1
(a) Schematic diagram with arrangement of stator and rotor blades and (b) single axial compressor repeating stage. (c) Comparison of typical turbine and compressor blades.

to stage. Each row of rotor blades is followed by a row of stationary blades called *stator blades* that act as nozzles to direct the fluid on to the rotor of the succeeding stage. A single stage consists of a rotor and a stator. The flow through the compressor is subjected to an adverse pressure gradient since the pressure increases along the axis. Thus, a series of diffusions take place in both the rotor and stator. Since excessive diffusion (a result of large increases in area) causes problems with separation, the increase in area and the corresponding increase in pressure need to be moderate. This is in contrast to a turbine stage that sees a favorable pressure gradient along the direction of flow. The limit on the amount of diffusion results in a moderate increase in pressure for compressor stages as opposed to the larger pressure drops in turbine stages. Thus, to achieve the same pressure ratio, a larger number of compressor stages is needed (during compression) in comparison with axial flow turbines (expansion). Consequently, there is a striking contrast between turbine blades and compressor blades, as shown in Figure 10.1c. As with any turbomachine, the flow is three-dimensional and turbulent, making a general solution of the flow problem almost impossible. Several simplifying assumptions are made, the most important being that the flow is uniform along the annulus, and analysis is performed along the mean line. Such analysis is called *mean line flow analysis*.

The sign convention for the angles used is the same as for gas turbines, that is, all angles are measured with respect to the axial direction and are considered positive if the component of the velocity is the direction of blade velocity. Three compressor stages are shown in Figure 10.1a. There are remarkable similarities between this axial flow compressor and the axial flow turbine shown in Figure 8.1. However, the difference is that a single stage of the compressor consists of a rotor followed by a stator. The flow enters the rotor at low pressure and temperature p_1 and T_1, and speed V_1, and is compressed to the state p_2, T_2, and V_2. It then enters the stator and leaves at the state p_3, T_3, and V_3. Also, as for turbines, exit velocity V_3 is equal in magnitude and direction to V_1 for repeating stages (called *normal* stages). Thus, $\alpha_1 = \alpha_3$ and the axial velocity is constant. The analysis proceeds in the same manner as for axial flow turbines, and many of the equations and definitions are identical to those discussed in the previous chapters. Hence, they are mentioned only briefly here.

The enthalpy–entropy (h–s) diagram for the flow process is shown in Figure 10.2. The energy equations across the rotor and stator will respectively yield

$$\begin{aligned} E &= h_{02} - h_{01} \quad \text{for the rotor} \\ h_{02} &= h_{03} \quad \text{for the stator} \end{aligned} \tag{10.1}$$

In terms of the velocities, the energy transfer E per stage can be written as

$$E = \frac{V_2^2 - V_1^2}{2} + \frac{w_1^2 - w_2^2}{2} \tag{10.2}$$

The degree of reaction is written as

$$\begin{aligned} R &\equiv \frac{\text{Static enthalpy change in rotor}}{\text{Total (stagnation) enthalpy change in the stage}} \\ &= \frac{h_2 - h_1}{h_{03} - h_{01}} = \frac{\Delta h_{rotor}}{\Delta h_{0,stage}} = \frac{T_2 - T_1}{T_{03} - T_{01}} \end{aligned} \tag{10.3}$$

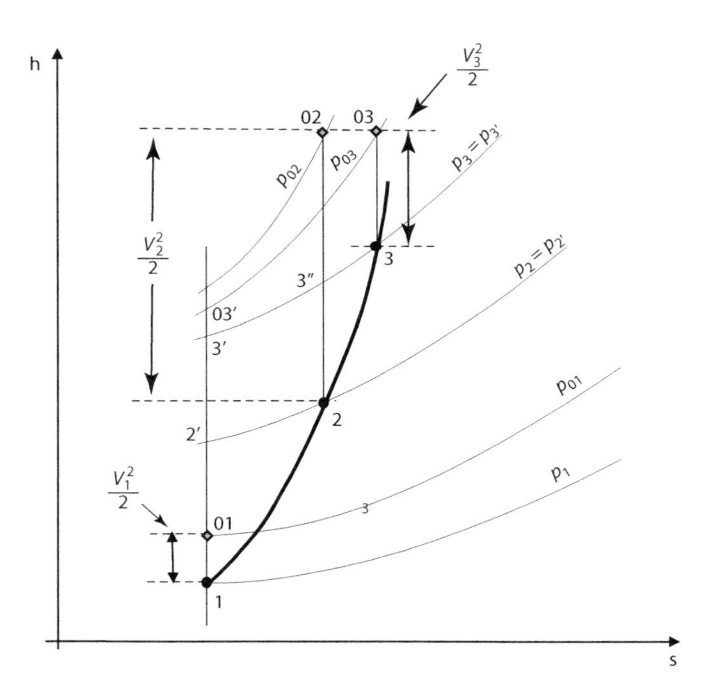

FIGURE 10.2
Enthalpy–entropy diagram for axial compressor stage.

Since the stages are assumed to be repeating stages, the following forms of degree of reaction can be obtained:

$$\begin{aligned} R &= \frac{h_2 - h_1}{h_{03} - h_{01}} = \frac{h_2 - h_1}{h_3 - h_1} \\ &= \frac{\left(w_1^2 - w_2^2\right)}{\left(V_2^2 - V_1^2\right) + \left(w_1^2 - w_2^2\right)} = \frac{\left(w_i^2 - w_o^2\right)}{\left(V_o^2 - V_i^2\right) + \left(w_i^2 - w_o^2\right)} \end{aligned} \tag{10.4}$$

In the preceding equations, the inlet and outlet states of the rotor, 1 and 2, have been replaced by the subscripts i and 0, respectively. The other definitions for stage/blade loading coefficient and flow coefficient are similar to Equations 8.2 and 8.3, respectively. Thus,

$$\psi = \frac{\Delta h_{0,stage}}{u^2} = \frac{E}{u^2} = \frac{\left(V_{u2} - V_{u1}\right)}{u} = \frac{\Delta V_u}{u} \tag{10.4a}$$

$$\phi = \frac{V_a}{u} \tag{10.4b}$$

10.3 Small Stage Efficiency for Axial Compressors

It was found in Chapter 8, during the discussion of axial flow turbines, that the concepts of isentropic efficiency and stage efficiency have different meanings. Unlike in centrifugal compressors, wherein the compression process usually takes place in a single stage, several stages need to be used in axial flow compressors to get appreciable pressure rises. Thus, each stage is affected by the cumulative inefficiency of the preceding stages since it operates at a higher temperature than the preceding stages. This effect can be visualized in the temperature versus specific entropy (*T–s*) diagram for the compression process shown in Figure 10.3, in which a gas is compressed between pressures p_1 and p_2 in three stages.

For simplicity, the stages are assumed to be identical, having the same pressure ratio and isentropic efficiency. It is to be noted that although efficiencies are defined in terms of enthalpies, they are written here and subsequently in terms of temperatures. This is appropriate since the gas is assumed to be an ideal gas with constant specific heats. Thus,

$$\eta_{st} = \frac{T_a - T_1}{T_c - T_1} = \frac{T_d - T_c}{T_f - T_c} = \frac{T_g - T_f}{T_2 - T_f} = \frac{(T_a - T_1) + (T_d - T_c) + (T_g - T_f)}{(T_c - T_1) + (T_f - T_c) + (T_2 - T_f)} \tag{10.5}$$

In the last equation, the properties of equality of fractions has been used. Thus, after simplification, Equation 10.5 becomes

$$\eta_{st} = \frac{(T_a - T_1) + (T_d - T_c) + (T_g - T_f)}{(T_2 - T_1)} \tag{10.6}$$

The compressor efficiency is defined as

$$\eta_c = \frac{(T_{2'} - T_1)}{(T_2 - T_1)} \tag{10.7}$$

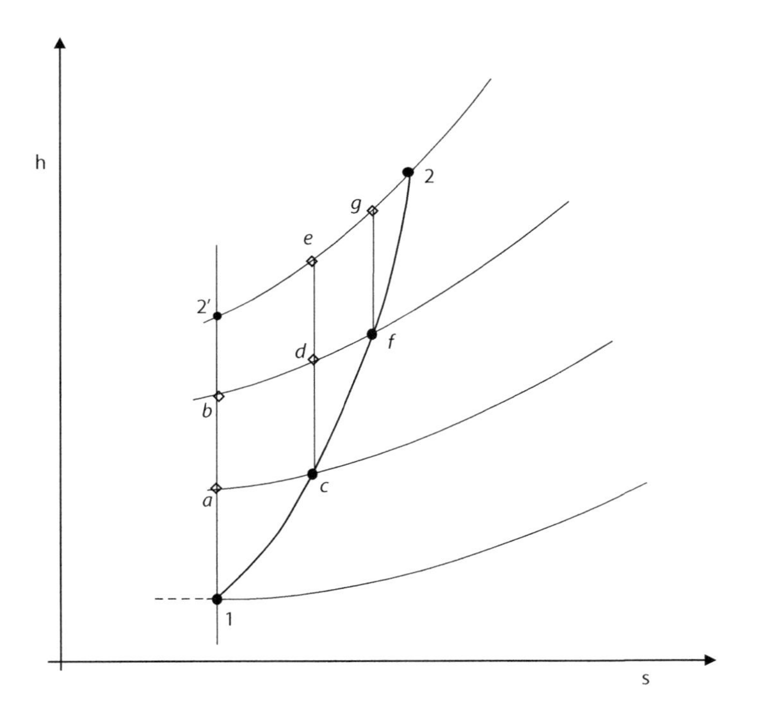

FIGURE 10.3
Isentropic and multistage compression process in axial compressors.

However, since the constant pressure lines diverge on the T–s diagram, it follows that

$$(T_b - T_a) < (T_d - T_c) \text{ and } (T_{2'} - T_b) < (T_e - T_d) < (T_g - T_f)$$

Thus,

$$\eta_c = \frac{(T_{2'} - T_1)}{(T_2 - T_1)} = \frac{(T_a - T_1) + (T_b - T_a) + (T_{2'} - T_b)}{(T_2 - T_1)}$$
$$< \frac{(T_a - T_1) + (T_d - T_c) + (T_g - T_f)}{(T_2 - T_1)} = \eta_{st} \tag{10.8}$$

In other words, stage efficiency η_{st} is greater than compressor isentropic efficiency η_c, that is, $\eta_{st} > \eta_c$. Stage efficiency is also known as *polytropic efficiency*, denoted by η_p; similarly, the compressor efficiency is sometimes called *isentropic* or *adiabatic efficiency*. In this book, the terms used will be *stage efficiency* and *compressor efficiency*.

An expression relating the stage and compressor efficiencies can be derived by considering an infinitesimal compression process. Consider the incremental compression process shown in Figure 10.4.

Since $dh_i = Tds - vdp$ and for isentropic process $ds = 0$, $|dh_i| = vdp$. The small stage (polytropic) efficiency is

$$\eta_{st} = \frac{dh_i}{dh} = \frac{vdp}{C_p dT} = \frac{\frac{RT}{p} dp}{\frac{kR}{k-1} dT} = \frac{k-1}{k} \frac{dp}{p} \frac{T}{dT}$$
$$\text{or} \quad \frac{dT}{T} = \frac{k-1}{k\eta_{st}} \frac{dp}{p} \tag{10.9}$$

Integrating for the entire compression process and assuming that each infinitesimal stage is of equal efficiency yields

$$\frac{T_2}{T_1} = \left(\frac{p_2}{p_1}\right)^{(k-1)/k\eta_{st}} \tag{10.10}$$

Using the definition of compressor efficiency from Equation 10.7 gives

$$\eta_c = \frac{(T_{2'} - T_1)}{(T_2 - T_1)} = \frac{\left(\frac{T_{2'}}{T_1} - 1\right)}{\left(\frac{T_2}{T_1} - 1\right)}$$
$$= \left[\left(\frac{p_2}{p_1}\right)^{(k-1)/k} - 1\right] \Big/ \left[\left(\frac{p_2}{p_1}\right)^{(k-1)/k\eta_{st}} - 1\right] \tag{10.11}$$

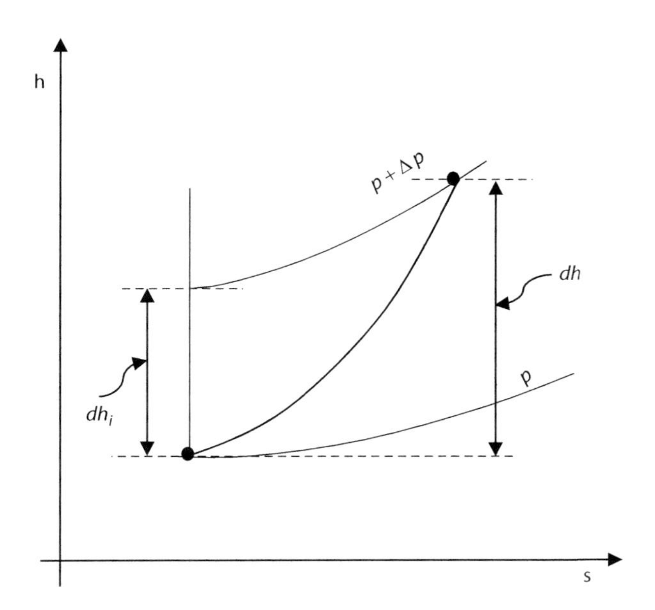

FIGURE 10.4
Incremental compression process.

A plot of the compressor efficiency η_c versus pressure ratio p_2/p_1 for various values of stage efficiency η_s is shown in Figure 10.5. For all pressure ratios, the compressor efficiency is lower than the stage efficiency. Also, $\eta_c \to \eta_s$ as $p_2/p_1 \to 1$. However, when p_2/p_1 becomes unity, the compression process becomes meaningless.

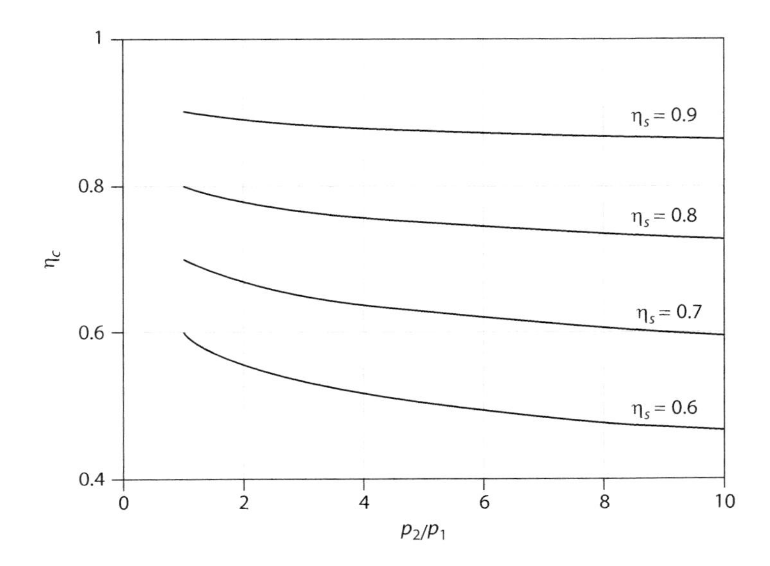

FIGURE 10.5
Effect of pressure ratio and stage efficiency on compressor efficiency.

Example 10.1

Show for axial flow compressors (or turbines) with constant axial velocity, the degree of reaction is given by the following expression. The angles are measured with respect to the axial direction and 1 and 2 are the inlet and exit states, respectively:

$$R = \frac{1}{2} + \frac{V_a}{2u}(\tan\beta_2 - \tan\alpha_1)$$

Solution: The degree of reaction for axial flow machines is given as follows ($u_1 = u_2 = u$):

$$R = \frac{(w_2^2 - w_1^2)}{2E} = \frac{(w_2^2 - w_1^2)}{2u(V_{u1} - V_{u2})} \tag{a}$$

From the velocity triangles shown,

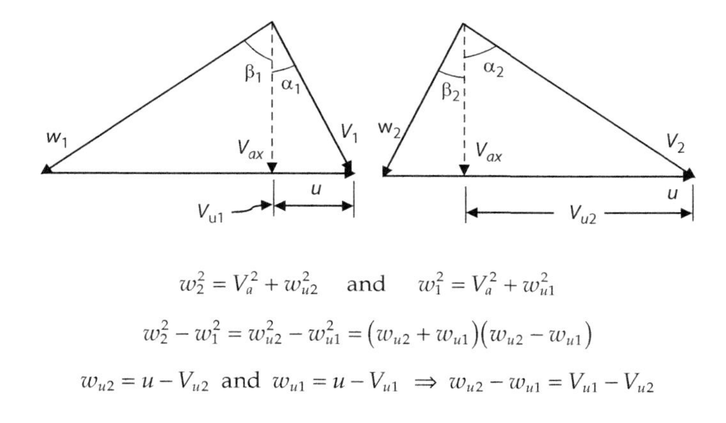

$$w_2^2 = V_a^2 + w_{u2}^2 \quad \text{and} \quad w_1^2 = V_a^2 + w_{u1}^2$$

$$w_2^2 - w_1^2 = w_{u2}^2 - w_{u1}^2 = (w_{u2} + w_{u1})(w_{u2} - w_{u1})$$

$$w_{u2} = u - V_{u2} \text{ and } w_{u1} = u - V_{u1} \Rightarrow w_{u2} - w_{u1} = V_{u1} - V_{u2}$$

Hence,

$$R = \frac{(w_2^2 - w_1^2)}{2u(V_{u1} - V_{u2})} = \frac{(w_{u2} + w_{u1})(V_{u1} - V_{u2})}{2u(V_{u1} - V_{u2})} = \frac{(w_{u2} + w_{u1})}{2u}$$

$$= \frac{(V_a \tan\beta_2 + u - V_a \tan\alpha_1)}{2u} = \frac{1}{2} + \frac{V_a}{2u}(\tan\beta_2 - \tan\alpha_1)$$

Comment: According to the convention used, the blade angles β_1 and β_2 are negative while angles α_1 and α_2 are positive.

Example 10.2

An axial flow air compressor has an overall compression ratio of six. The inlet conditions are $T_{01} = 300$ K and $p_{01} = 100$ kPa. The degree of reaction is 0.5 and the fluid angle at the stator exit is +30°. The stage efficiency is 85% and the axial velocity, which is constant in each stage, is 130 m/s. The hub and tip diameters are 0.4 and 0.7 m, respectively, and the speed of rotation is 9000 rpm. (a) Estimate the number of stages, (b) mass flow rate, and (c) blade angles. Assume all stages are identical.

Solution:

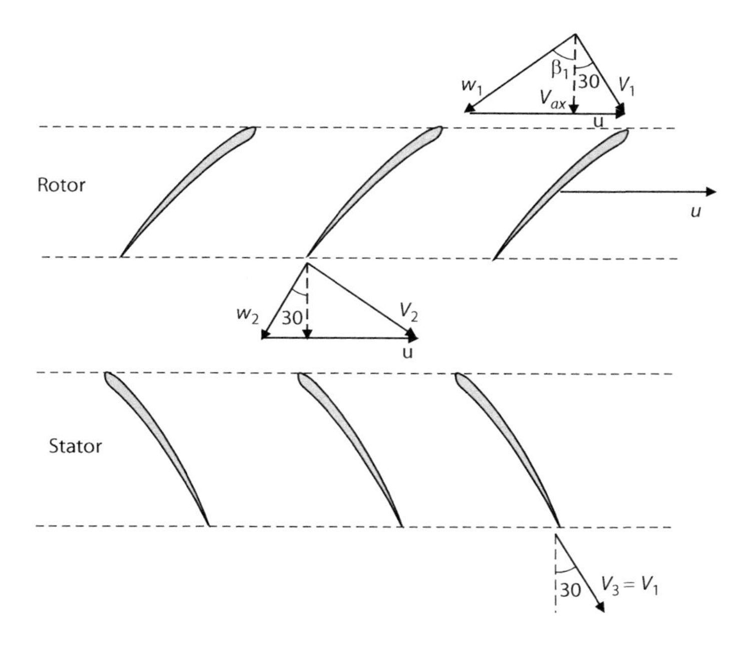

Using the overall pressure ratio and the stage efficiency, let us calculate the overall temperature rise through the compressor. Using the subscripts i and e for the inlet and exit conditions, respectively,

$$\frac{T_{0e}}{T_{0i}} = \left(\frac{p_{0e}}{p_{0i}}\right)^{\frac{(k-1)}{k\eta_{st}}} = (6)^{\frac{1.4-1}{(1.4)(0.85)}} = 1.826$$

$$\text{Hence, } T_{0e} = 300 \times 1.826 = 547.88\,\text{K}$$

The overall temperature rise is the difference between the stagnation temperatures at the inlet and the exit. Thus,

$$\Delta T_{\text{overall}} = 547.88 - 300 = 247.88\,\text{K}$$

The velocity diagram for a single stage is shown.

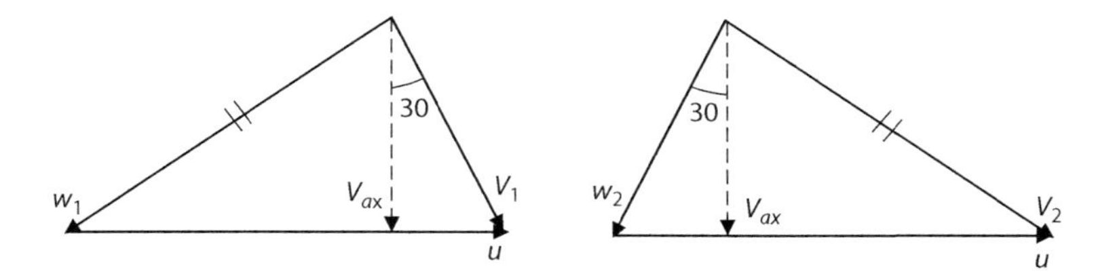

Thus,

$$V_1 = \frac{V_{ax}}{\cos 30} = 150.1\,\text{m/s} = w_2; \quad V_{u1} = V_{ax}\tan 30 = 75.05\,\text{m/s}$$

$$u = \frac{2\pi N}{60}\frac{D_{mean}}{2} = 259.2\,\text{m/s}; \quad w_{u1} = u - V_{u1} = 184.1\,\text{m/s}$$

$$w_1 = \sqrt{V_{ax}^2 + w_{u1}^2} = 225.4\,\text{m/s} = V_2$$

$$E = \frac{V_1^2 - V_2^2}{2} + \frac{u_1^2 - u_2^2}{2} + \frac{w_2^2 - w_1^2}{2} = V_1^2 - V_2^2 = -28269\,\text{J/kg}$$

The temperature rise per stage can be calculated as

$$\Delta T_{\text{stage}} = \frac{|E|}{C_p} = \frac{28,269}{1004} = 28.14\,\text{K}$$

$$\text{Number of stages} = \frac{\Delta T_{\text{overall}}}{\Delta T_{\text{stage}}} = \frac{247.9}{28.14} = 8.8\,\text{stages}$$

Hence, the number of stages is 9.
The blade angles can be calculated from the velocity triangles as

$$\alpha_1 = \beta_2 = \arctan\left(\frac{V_{u1}}{V_{ax}}\right) = \arctan\left(\frac{75.05}{130.0}\right) = 30°$$

$$\alpha_2 = \beta_1 = \arctan\left(\frac{w_{u1}}{V_{ax}}\right) = \arctan\left(\frac{184.1}{130.0}\right) = 54.8°$$

To calculate the mass flow rate, the density at the inlet needs to be calculated. Thus,

$$T_1 = T_{01} - \frac{V_1^2}{2C_p} = 300 - \frac{(150.1)^2}{2(1004)} = 288.8\,\text{K}$$

$$p_1 = p_{01}\left[\frac{T_1}{T_{01}}\right]^{k/(k-1)} = 100\left[\frac{288.8}{300}\right]^{3.5} = 87.5\,\text{kPa}$$

$$\rho_1 = \frac{p_1}{RT_1} = \frac{(87.5)(1000)}{(287)(288.9)} = 1.056\,\text{kg/m}^3$$

$$A_{\text{annulus}} = \frac{\pi\left(d_{\text{tip}}^2 - d_{\text{hub}}^2\right)}{4} = 0.259\,\text{m}^2$$

$$\dot{m} = \rho_1 V_{ax} A_{\text{annulus}} = (1.055)(130)(0.259) = 35.57\,\text{kg/s}$$

As an aside, let us calculate the static temperatures at all three points in the stage, namely, rotor inlet, rotor outlet, and stator outlet.

$$T_{02} = T_{01} - \frac{E}{C_p} = 300 - \frac{-28,269}{1004} = 328.14$$

$$T_2 = T_{02} - \frac{V_2^2}{2C_p} = 328.14 - \frac{(225.4)^2}{2(1004)} = 302.9\,\text{K}$$

$$T_3 = T_{03} - \frac{V_3^2}{2C_p} = 328.14 - \frac{(150.1)^2}{2(1004)} = 316.9\,\text{K}$$

It has been assumed that the stage is repeating; hence $V_3 = V_1$. Also, an energy balance across the stator will imply that $h_{02} = h_{03}$ or $T_{02} = T_{03}$.

The static temperature changes across the rotor and stator can now be calculated:

$$T_3 - T_2 = 316.9 - 302.9 = 14\,\text{K}$$

$$T_2 - T_1 = 302.9 - 288.9 = 14\,\text{K}$$

Thus, the temperature changes across the stator and rotor are equal, a requirement for 50% reaction repeating stages using ideal gases. Also, the static enthalpy change across the rotor is exactly half of the stagnation change.

Comment: Degree of reaction can also be calculated from the velocities, and it would be found to be the same.

Example 10.3

The high-pressure stages of an axial flow air compressor have inlet stagnation conditions of 431 K and 200 kPa. The number of stages is 15, and the compressor rotates at 12,000 rpm. For each stage, the stage efficiency is 89%. The mean radius of all the stages is 25 cm and the stage loading factor for each stage is 0.32. The inlet Mach number is 0.5 and mass flow rate through the compressor is 10 kg/s. Calculate the pressure ratio, compressor isentropic efficiency, and the height of the blades at the inlet. Assume all stages are identical, with no inlet swirl.

Solution: Let subscripts *i* and *e* denote the inlet and exit conditions. Thus,

$$T_{0i} = 431\,\text{K and } p_{0i} = 200\,\text{kPa}$$

$$N = 12,000\,\text{rpm and } \omega = 1256\,\text{rad/s}$$

$$u = \omega r = (1256\,\text{rad/s})(0.25\,\text{m}) = 314\,\text{m/s}$$

The *T*–*s* diagram for the compressor is shown.

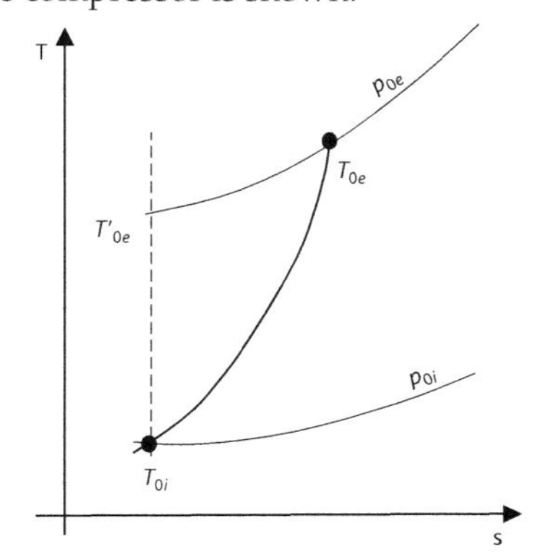

We can now use the definition of blade coefficient to calculate the energy transfer per stage. Thus,

$$\psi = \frac{E}{u^2} = 0.32 \Rightarrow E_{\text{stage}} = (0.32)u^2 = 31,582\,\text{J/kg}$$

$$(\Delta T_0)_{\text{stage}} = \frac{E}{C_p} = \frac{31,582}{1006} = 31.44\,\text{K}$$

The overall temperature rise and the exit temperature for the compressor can now be calculated:

$$(\Delta T_0)_{\text{overall}} = n\,(\Delta T_0)_{\text{stage}} = (15)(31.44\,\text{K}) = 471.6\,\text{K}$$

$$T_{0e} = T_{0i} + (\Delta T_0)_{\text{overall}} = 431 + 471.6 = 902.6\,\text{K}$$

The temperature ratio, and hence the pressure ratio, can now be calculated:

$$\frac{T_{0e}}{T_{0i}} = \frac{902.6}{431.0} = 2.094 = \left(\frac{p_{0e}}{p_{0i}}\right)^{\frac{k-1}{k\eta_p}} \Rightarrow \frac{p_{0e}}{p_{0i}} = \left(\frac{T_{0e}}{T_{0i}}\right)^{\frac{k\eta_p}{k-1}} = 10$$

$$p_{0e} = (10)(200) = 2000\,\text{kPa} = p'_{0e}$$

The compressor efficiency can be calculated as

$$\eta = \frac{T_{oe'} - T_{0i}}{T_{0e} - T_{0i}} = \frac{\dfrac{T_{oe'}}{T_{0i}} - 1}{\dfrac{T_{0e}}{T_{0i}} - 1} = \frac{\left(\dfrac{p_{0e}}{p_{0i}}\right)^{\frac{k-1}{k}} - 1}{\left(\dfrac{p_{0e}}{p_{0i}}\right)^{\frac{k-1}{k\eta_p}} - 1} = 0.8504$$

The velocity diagram for a single stage is shown in the following figure. The entrance to the first stage is under the inlet conditions T_{0i} and p_{0i}. Denoting the conditions for the first-stage inlet with subscript 1 and for the exit with subscript 2, the following relations are obtained:

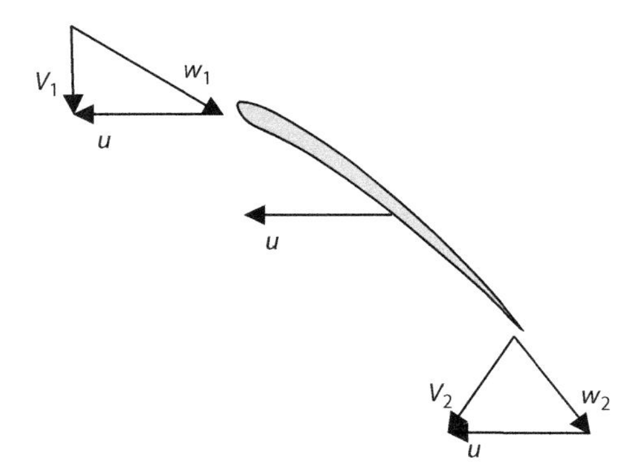

$$\frac{T_{01}}{T_1} = 1 + \frac{k-1}{2} M_1^2 = 1 + \frac{k-1}{2}(0.5)^2 = 1.05$$

$$\text{Hence } T_1 = \frac{T_{01}}{1.05} = 410.5\text{K} \Rightarrow c_1 = \sqrt{kRT_1} = 406\,\text{m/s}$$

$$V_1 = V_a = M_1 c_1 = 203\,\text{m/s}$$

The flow coefficient and other variables can now be calculated as follows:

$$\varphi = \frac{V_a}{u} = \frac{203\,\text{m/s}}{314\,\text{m/s}} = 0.646$$

$$\frac{T_1}{T_{01}} = \frac{410.5}{431} = 0.952$$

Since the process 1 to 01 is isentropic,

$$\frac{p_1}{p_{01}} = \left(\frac{T_1}{T_{01}}\right)^{\frac{k}{k-1}} = 0.843 \Rightarrow p_1 = 0.843 p_{01} = 168.6\,\text{kPa}$$

$$\rho_1 = \frac{p_1}{RT_1} = 1.43\,\text{kg/m}^3$$

$$\dot{m} = 10\,\text{kg/s} = \rho_1 V_1 A_1 \Rightarrow A_1 = 0.0344\ \text{m}^2$$

$$A_1 = 2\pi r_m h \Rightarrow h = 0.0219\,\text{m} = 2.19\,\text{cm}$$

Comment: Since the blade angles are not known at the exit, the kinematics at the exit cannot be determined.

10.4 Work Done Factor

Because the pressure rise in each stage of the compressor is quite small, several stages are stacked in axial flow compressors, as shown in Figure 10.1a. As the pressure increases in each subsequent stage, the density of fluid increases, thus decreasing the annular area (the axial velocity does not change appreciably). However, the boundary layer tends to grow along the axial direction. Thus, although the flow in the first stage is close to being uniform across the entire height of the blade, it becomes quite non-uniform in subsequent stages due to boundary layer effects and secondary flows. This gives rise to another variable that needs to be introduced in the study of axial flow compressors, which is the *work done factor*. Interestingly, this results from the three-dimensional nature of the flow in the annulus. However, a very qualitative explanation of the work done factor can be given as follows. Although the flow is assumed to be uniform in the first stage across the rotor, the boundary layer on the outer wall and on the shaft thicken due to the adverse pressure gradient in the axial direction. This factor, along with the secondary leakage losses at the tip of the blade, affects the axial velocity profile, which will no longer be uniform. The situation is shown in Figure 10.6, where the axial flow profiles are shown across the initial stages.

It can be seen that the axial velocity increases near the center of the blade and decreases near the wall and the shaft (a consequence of constant mass flow across the annulus). Thus, the design axial velocity is not the same along the entire height of the rotor blade. It decreases near the tip and the hub and increases near the center. Since the axial velocity differs along the height, the energy exchange is not uniform and is not equal to the design value. How this affects the flow and energy transfer can be seen from the velocity triangles at the inlet and exit of a typical rotor blade, as shown in Figure 10.7.

From Euler's equation,

$$\begin{aligned} E &= u_2 V_{u2} - u_1 V_{u1} = u\left(V_{u2} - V_{u1}\right) = u\left[\left(u - V_a \tan\beta_2\right) - V_a \tan\alpha_1\right] \\ &= u\left[u - V_a\left(\tan\alpha_1 + \tan\beta_2\right)\right] \end{aligned} \tag{10.12}$$

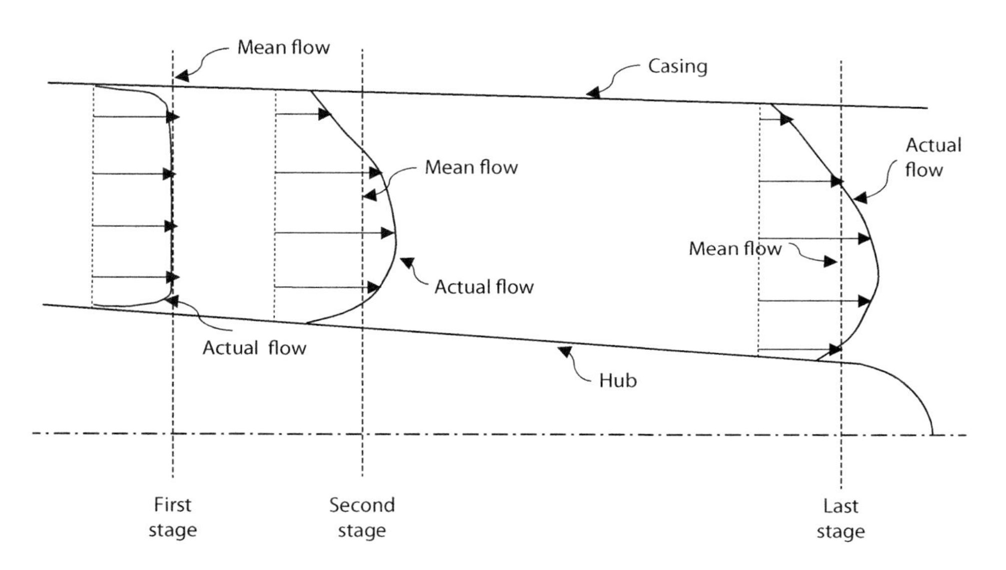

FIGURE 10.6
Developing axial velocity profiles in the rotor annulus.

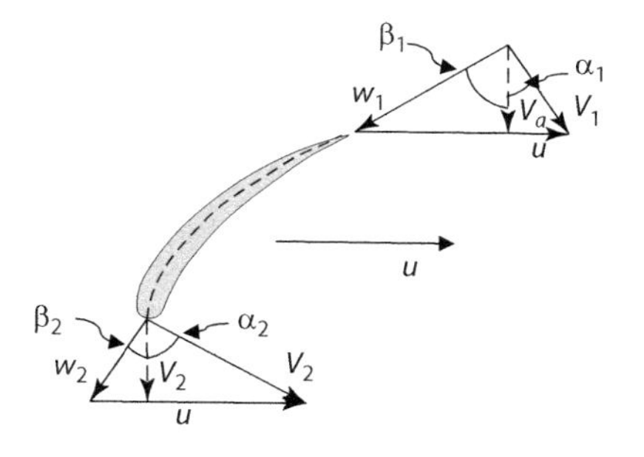

FIGURE 10.7
Velocity triangles at the inlet and exit of a typical rotor.

Thus, for fixed values of α_1 and β_2 the energy decreases with increasing axial velocity. Since the outlet angles of the stator and rotor determine α_1 and β_2, respectively, they can be regarded as fixed (as opposed to α_2, which depends on the axial velocity). This decreased energy is expressed through the work done factor λ. If Equation 10.12 represents the ideal energy involved, the actual energy would be given by

$$E = \lambda\left(u_2 V_{u2} - u_1 V_{u1}\right) = \lambda u\left(V_{u2} - V_{u1}\right) \tag{10.13}$$

Several comments need to be made regarding the preceding explanation. First, an increase in the axial velocity near the central part of the blade height is partially compensated for by the decrease at the tip and the root. However, this increased work at the root and hub cannot compensate for the central part, and the net effect is an overall decrease in E. Secondly, the flow is highly three-dimensional, rotating, and turbulent, so such an explanation is tenuous at best. Thirdly, a similar effect should be expected in turbines also. However, it is an accelerating flow, the boundary layers are thin, and the effect is not that pronounced. Hence, there is no work done factor needed for turbines as a correction in performance calculations. Finally, although the axial velocity changes from stage to stage initially, it has been found experimentally by several researchers that beyond the fourth stage, the axial velocity profile remains practically the same for all subsequent stages. Howell (1945) and Smith (1970) have worked extensively on this, and recommend a value between 0.98 and 0.85 for the first four stages and 0.85 for the remainder. In this connection, it is worthwhile to mention Howell's formula that relates the number of stages and work done factor, which can be written as

$$\lambda = 0.85 + 0.15\exp\left(-\left(n_s - 1\right)/2.73\right) \tag{10.13a}$$

where n_s is the stage number.

Example 10.4

The design of an axial flow compressor stage is being evaluated with the following parameters:

Degree of reaction	0.65
Flow coefficient	0.55
Blade velocity	1000 ft./s

The axial velocity is constant throughout the stage. What would the expected range of blade loading coefficients, blade angles, and flow angles be if the stagnation temperature rise per stage is altered from 40°F to 50°F?

Solution: The velocity triangles are given. First, the case when temperature rise is 50°F is considered.

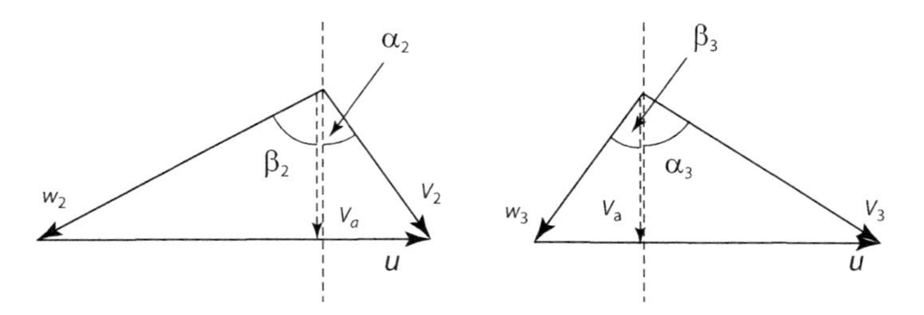

$$u = 1000\,\text{ft./s};\ \Delta T_0 = 50°\text{R}$$

$$\varphi \equiv \frac{V_a}{u} = 0.55 \Rightarrow V_a = (0.55)(1000) = 550\,\text{ft./s}$$

$$E = C_p \Delta T_0 = (6006)(50) = 300{,}300\,\text{ft} \cdot \text{lbf/slug}$$

$$\psi = \frac{E}{u^2} = \frac{300{,}300}{(1000)^2} = 0.3003$$

From the geometry of the triangles,

$$u = w_{u2} + V_{u2} = w_{u3} + V_{u3} \Rightarrow V_{u3} - V_{u2} = w_{u2} - w_{u3}$$

$$E = u(V_{u3} - V_{u2}) = 300{,}300 \Rightarrow (V_{u3} - V_{u2}) = \frac{300{,}300}{1000} = 300.3 = w_{u2} - w_{u3}$$

$$R = \frac{w_2^2 - w_3^2}{2E} = \frac{(V_a^2 + w_{u2}^2) - (V_a^2 + w_{u3}^2)}{2u(V_{u3} - V_{u2})} = \frac{(w_{u2}^2 - w_{u3}^2)}{2u(w_{u2} - w_{u3})}$$

$$= \frac{(w_{u2}^2 - w_{u3}^2)}{2u(w_{u2} - w_{u3})} = \frac{(w_{u2} + w_{u3})}{2u}$$

$$w_{u2} + w_{u3} = 2uR = 1300$$

Thus,

$$w_{u2} = \frac{1300 + 300.3}{2} = 800.15\,\text{ft./s}$$

$$w_{u3} = \frac{1300 - 300.3}{2} = 499.85\,\text{ft./s}$$

The angles can now be calculated:

$$\tan(\beta_2) = \frac{w_{u2}}{V_a} = \frac{800.15}{550} = 1.456 \Rightarrow \beta_2 = 55.5°$$

$$\tan(\beta_3) = \frac{w_{u3}}{V_a} = \frac{499.85}{550} = 0.909 \Rightarrow \beta_3 = 42.3°$$

$$\tan(\alpha_2) = \frac{V_{u2}}{V_a} = \frac{u - w_{u2}}{V_a} = \frac{1000 - 800.15}{550} = 0.363 \Rightarrow \alpha_2 = 20.0°$$

$$\tan(\alpha_3) = \frac{V_{u2}}{V_a} = \frac{u - w_{u3}}{V_a} = \frac{1000 - 499.85}{550} = 0.909 \Rightarrow \alpha_3 = 42.3°$$

These calculations can be repeated for a temperature rise of 40°F, and the following results are obtained:

$$\psi = 0.24;\ \beta_2 = 54.7°;\ \ \beta_3 = 43.9°;\ \ \alpha_2 = 22.7°;\ \ \alpha_3 = 40.5°$$

Comment: For the range of values considered for stage temperature rise, the differences in the angles are quite minimal.

10.5 Design of Axial Compressors

Modern axial compressors, especially those used in aircraft engines, use multiple spool configurations and are quite complex to design. Also, not all the stages are identical, making the process more involved. What will be presented in the following are the rudimentary steps used in the design. Typical *input specifications* are as follows:

1. Inlet pressure and temperature—these could be either stagnation or static. It is common practice to specify the stagnation conditions.
2. The overall stagnation pressure ratio between the inlet and exit of the compressor or the exit stagnation pressure.
3. Mass flow rate $\dot{m}$, which would be constant for all the stages.

To avoid confusion between conditions for a particular stage and the overall compressor, the following notation will be used in this section. The overall inlet and exit states for the compressor are denoted by the subscripts i and e, respectively; subscripts 1, 2 and 2, 3 refer to the inlet and exit states for the rotor and stator, respectively.

The designer has a wide range of options to select from since there are usually fewer equations than the number of variables to choose. However, since axial compressors have been in use for more than 70 years, several criteria are based on experience. Some of the *design guidelines* are as follows:

1. Blade tip velocity: There is no equation that enables the designer to pick the blade tip velocity. The major restriction on this is the blade stresses. Based on experience, this is around 350 m/s (~1150 ft./s) for acceptable stresses.
2. The axial velocity should be between 150 and 200 m/s (500–650 ft./s). The common practice is to assume no inlet guide vanes, which makes V_1 purely axial at the inlet.
3. Hub to tip ratio at the inlet is in the range 0.4–0.6.
4. Stagnation temperature rise per stage: This varies between 10 and 30 K (~20–55°R) for subsonic stages. However, for higher-performance stages it could be as high as 45 K (~80°R).
5. Work done factor changes from 97% to 85% during the first four stages and levels off at 85% for subsequent stages.
6. Relative Mach number at the inlet rotor tip should be transonic, with a value between 0.8 and 1.2.
7. The deHaller number, which is defined as the ratio w_2/w_1, should be greater than 0.72 if possible.

The *design procedure* can be summarized in the following steps. The first step is to estimate the number of stages required to achieve the required pressure rise.

1. A reasonable set of values for the tip speed u_t, axial velocity V_a, and hub to tip ratio r_h/r_t are assumed. For preliminary calculations, it is assumed that there are no inlet guide vanes. Hence, the inlet velocity $V_1 = V_a$

2. The inlet static temperature is calculated using $T_{01} = T_1 + \frac{V_1^2}{2C_p}$
3. Assuming isentropic conditions, the inlet static pressure is calculated using $\frac{p_1}{p_{01}} = \left(\frac{T_1}{T_{01}}\right)^{k/(k-1)}$
4. The density at the inlet is calculated using $\rho_1 = \frac{p_1}{RT_1}$
5. Using the mass flow rate equation, the tip radius is calculated as shown:

$$\dot{m} = \rho_1 V_a A_1 = \rho_1 V_a \pi\left(r_t^2 - r_h^2\right) = \rho_1 V_a \pi r_t^2 \left\{1 - \left(\frac{r_h}{r_t}\right)^2\right\}$$

$$r_t = \sqrt{\frac{\dot{m}}{\rho_1 V_a \pi \left\{1 - \left(\frac{r_h}{r_t}\right)^2\right\}}}$$

6. From the assumed value of blade speed, the speed of rotation N is calculated using $u_t = \frac{2\pi N}{60} r_t$
7. The inlet velocity triangle is completed and the relative Mach number at the tip is calculated:

$$w_t = \sqrt{u_t^2 + V_{1t}^2}; \quad a_1 = \sqrt{kRT_1}; \quad M_{1t} = \frac{w_t}{a_1}$$

M_{1t} should be in the transonic range, that is, between 0.8 and 1.2.
8. The temperature rise across the compressor can be estimated by assuming a compressor efficiency, typically 0.87–0.91. Thus,

$$T_{0e} = T_{0i}\left(\frac{p_{0e}}{p_{0i}}\right)^{(k-1)/k\eta_c} \quad \text{and} \quad \left(\Delta T_o\right|_{overall} = T_{0e} - T_{0i}$$

9. Stage temperature rise is needed to estimate the number of stages. Since the inlet velocity triangle is completely known, the exit velocity conditions for the first stage are needed. One procedure is to assume a suitable of degree of reaction and complete the exit triangle. However, an alternative procedure is to use the smallest limiting value of the deHaller number. The following steps can be used to calculate the stage temperature rise:

$$\frac{w_2}{w_1} = 0.72 \Rightarrow w_2 = 0.72 w_1$$

$$\frac{V_a}{w_2} = \cos\beta_2 \Rightarrow \beta_2 = \arccos\left(\frac{V_a}{w_2}\right)$$

$$w_{u2} = w_2 \sin\beta_2; \qquad V_{u2} = u - w_{u2}$$

$$E = \lambda |u_1 V_{u1} - u_2 V_{u2}| = \lambda u V_{u2} = C_p \langle \Delta T_0 |_{\text{stage}}$$

10. The number of stages is $\frac{\langle \Delta T_o |_{\text{overall}}}{\langle \Delta T_o |_{\text{stage}}}$. It is rounded to the nearest higher integer.

    This completes the first part of the design. The next part is to perform detailed analysis of each stage. The design of the first stage will be slightly different than the rest of the stages since it was used to estimate the number of stages earlier. These steps are demonstrated in the following points.

11. The stage temperature rise $\langle \Delta T_o |_{\text{stage}}$ is somewhat uniformly divided across each stage, with the first and last stages having slightly lower than average temperature rises. Similarly, the work done factor λ is also set for approximately the first four stages, with the value decreasing from 0.98 to 0.88. All stages numbered five and above will have a work done factor of 0.85.
12. From the assumed value of stage temperature rise $\langle \Delta T_o |_{\text{stage}}$ and work done factor λ, the remaining parameters are calculated as shown:

$$C_p \Delta T_0 = E = \lambda u (V_{u2} - V_{u1}) = \lambda u \Delta V_u$$

$$\Delta V_u = \frac{C_p \Delta T_0}{\lambda u}; \qquad \Delta V_u = V_{u2} - V_{u1}$$

$$\alpha_1 = \arctan\left(\frac{V_{u1}}{V_a}\right); \quad \beta_1 = \arctan\left(\frac{u - V_{u1}}{V_a}\right)$$

$$\beta_2 = \arctan\left(\frac{u - V_{u2}}{V_a}\right); \quad \alpha_2 = \arctan\left(\frac{V_{u2}}{V_a}\right)$$

    For the first stage, V_{u1} will be zero. Complete the inlet and exit velocity triangles.

13. The deHaller number for the stage is calculated next using the following expression. Its value should be greater than 0.7.

$$DH = \frac{w_2}{w_1} = \frac{V_a / \cos\beta_2}{V_a / \cos\beta_1} = \frac{\cos\beta_1}{\cos\beta_2}$$

14. Temperature and pressure at the exit of the stage are calculated next:

$$T_{02} = T_{01} + \Delta T_0; \quad \eta_{st} = \frac{T_{02'} - T_{01}}{T_{02} - T_{01}} \Rightarrow T_{02'} = \eta_{st}(T_{02} - T_{01}) + T_{01}$$

$$\frac{p_{02}}{p_{01}} = \frac{p_{02'}}{p_{01}} = \left(\frac{T_{02'}}{T_{01}}\right)^{\eta_{st} k / k-1} \Rightarrow p_{02} = p_{01} \left(\frac{T_{02'}}{T_{01}}\right)^{\eta_{st} k / k-1}$$

15. Degree of reaction is calculated using any of the several formulas available. One such formula is $R = 1 - \frac{V_{u2} + V_{u1}}{2u}$. Usually, the degree of reaction for the first stage is quite high.
16. The remaining stages are designed with an assumed value for the degree of reaction. Suggested values are 0.7 for the second stage and 0.5 subsequently. The calculations proceed as shown:

$$\tan\beta_1 = \frac{1}{2}\left(\frac{C_p\Delta T_0}{\lambda u V_a} + \frac{2Ru}{V_a}\right)$$

$$\tan\beta_2 = \frac{1}{2}\left(\frac{C_p\Delta T_0}{\lambda u V_a} - \frac{2Ru}{V_a}\right)$$

$$V_{u1} = u - V_a \tan\beta_1; \quad V_{u2} = u - V_a \tan\beta_2$$

$$\alpha_1 = \arctan\left(\frac{V_{u1}}{V_a}\right); \quad \alpha_2 = \arctan\left(\frac{V_{u2}}{V_a}\right)$$

17. The rest of the design procedure is identical to that for stage one.
18. As a final check, the stage pressure ratios are used to see if the overall pressure ratio is achieved.

There are several choices available to the designer to meet the specific requirements. The number of stages, the temperature rise per stage, and the work done factor can all be adjusted to achieve the required pressure rise. It should be realized that the procedure is not unique. The procedure outlined is demonstrated in Example 10.5.

Example 10.5

An axial flow compressor is to be designed to produce a mass flow of 25 kg/s and a pressure ratio of three. It takes inlet air at stagnation conditions of 100 kPa and 20°C. Perform the preliminary design of the compressor including the number of stages and also the specifications of each stage.

Solution: According to design guidelines, the blade speed should be restricted to about 350 m/s, axial velocity to 150–200 m/s, and the hub to tip ratio for the first stage should be between 0.4 and 0.6. Using these guidelines, and inlet conditions, the first step is to design the first and last stages to estimate the number of stages required. Let the axial velocity (assumed to be constant throughout the compressor) and the tip velocity at the inlet be 150 and 350 m/s, respectively. Without inlet guide vanes, the inlet absolute velocity would be equal to the axial velocity. Thus,

$$p_{01} = 100\,\text{kPa}; \; T_{01} = 20°\text{C} = 20 + 273 = 293\,\text{K}$$

$$V_a = V_1 = 150\,\text{m/s} \text{ and } u_t = 350\,\text{m/s}$$

$$T_1 = T_{01} - \frac{V_1^2}{2C_p} = 293 - \frac{(150)^2}{2(1004.5)} = 281.8\,\text{K}$$

Since the process 1 to 01 is isentropic,

$$\frac{p_1}{p_{01}} = \left(\frac{T_1}{T_{01}}\right)^{k/(k-1)} = \left(\frac{281.8}{293}\right)^{3.5} = 0.872 \Rightarrow$$

$$p_1 = (0.872)(100) = 87.2\,\text{kPa}$$

$$\rho_1 = \frac{p_1}{RT_1} = \frac{(87.2)(1000)}{(287)(281.8)} = 1.078\,\text{kg/m}^3$$

From the mass flow rate equation,

$$\dot{m} = \rho_1 V_1 A_1 = \rho_1 V_a \pi (r_t^2 - r_h^2) = \rho_1 V_a \pi r_t^2 \left(1 - \frac{r_h^2}{r_t^2}\right)$$

The radius ratio r_h/r_t can take a value between 0.4 and 0.6 for the first stage. Assume $r_h/r_t = 0.5$. Hence,

$$\dot{m} = 25 = (1.078)(150)\pi r_t^2 (1 - 0.5^2) \Rightarrow r_t = 0.256\,\text{m}$$

$$r_h = (0.5)(0.256) = 0.128\,\text{m};\; r_m = (0.5)(0.256 + 0.128) = 0.192\,\text{m}$$

Based on the assumed tip speed of 350 m/s and the just-calculated tip radius for stage one, the speed of rotation can be calculated. Thus,

$$N = \frac{350}{2\pi(0.256)} = 217.6\,\text{rev/s}$$

The speed can be rounded off to 220 rev/s. The tip blade speed can now be recalculated to ensure that it is within acceptable limits:

$$N = 220\,\text{rev/s} = 13{,}200\,\text{rpm} = 1382.3\,\text{rad/s}$$

$$u_t = (0.256)(1382.3) = 354\,\text{m/s which is acceptable.}$$

The next step is to estimate the number of stages required. To do this, the flow conditions at the exit or the last stage of the compressor are needed. Assume that all stages have the same efficiency and that the small stage efficiency $\eta_{st} = 0.92$. Thus, from Equation 10.10, and denoting the inlet and exit of the compressor by the subscripts i and e, respectively,

$$\frac{T_{0e}}{T_{0i}} = \left(\frac{p_{0e}}{p_{0i}}\right)^{(k-1)/k\eta_{st}} = (3)^{(1.4-1)/(1.4)(0.92)} = 1.41$$

$$T_{0e} = (1.41)(293) = 412.1\,\text{K}$$

$$\Delta T_{0,\text{overall}} = 412.1 - 293 = 119.1\,\text{K}$$

The temperature rise in the first stage is now calculated. The velocity triangles are constructed at the mean radius.

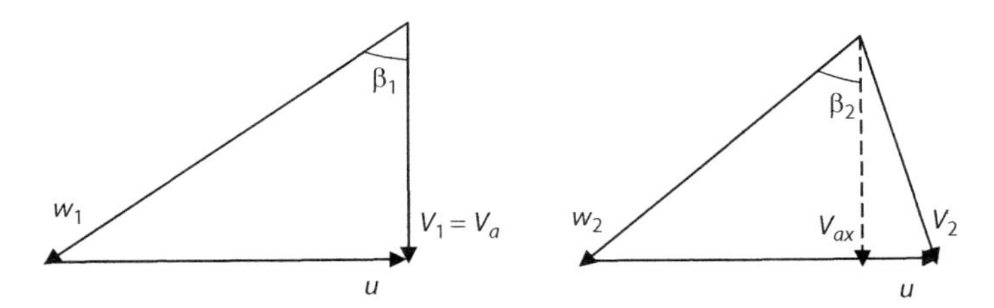

To get an estimate of the number of stages, all stages will be assumed to have the same temperature rise as the first. The calculations will be done at the mean radius. Thus, for the first stage,

$$u = (0.192)(1382.3) = 265.5\,\text{m/s}$$

$$\beta_1 = \arctan\left(\frac{u}{V_a}\right) = \arctan\left(\frac{265.5}{150}\right) = 60.5°;\ \alpha_1 = 0°$$

$$w_1 = \sqrt{u^2 + V_a^2} = \sqrt{(265.5)^2 + (150)^2} = 304.91\,\text{m/s};\ V_1 = 150\,\text{m/s}$$

To calculate the exit velocity triangle, either the degree of reaction or the deHaller number must be known. Assuming that the deHaller number is 0.72,

$$\frac{w_2}{w_1} = 0.72 \Rightarrow w_2 = (0.72)(304.91) = 219.5\,\text{m/s}$$

$$\frac{V_a}{w_2} = \cos\beta_2 \Rightarrow \beta_2 = \arccos\left(\frac{V_a}{w_2}\right) = 46.9°$$

$$w_{u2} = w_2 \sin\beta_2 = 160.3\,\text{m/s};\ V_{u2} = u - w_{u2} = 265.5 - 160.3 = 105.2\,\text{m/s}$$

$$E = |u_1 V_{u1} - u_2 V_{u2}| = uV_{u2} = (265.5)(105.2) = 27918\,\text{m}^2/\text{s}^2$$

The energy for the first stage can be related to the stagnation temperature rise for the stage. Ignoring the work done factor for now (since only the number of stages are being estimated), the temperature rise can be calculated as

$$C_p \Delta T_{0,\text{stage}} = E = 27{,}918\,\text{m}^2/\text{s}^2 \Rightarrow \Delta T_{0,\text{stage}} = \frac{E}{C_p} = \frac{27{,}918}{1004.5} = 27.8\,\text{K}$$

Thus, the number of stages is

$$\frac{\Delta T_{0,\text{overall}}}{\Delta T_{0,\text{stage}}} = \frac{119.1}{27.8} = 4.28\,\text{stages}$$

This is rounded up to five stages. Hence, the compressor would be required to have five stages, with the approximate temperature rise per stage being 119.1/5 = 23.8 K. This temperature rise is well within the limits.

The details of each stage are now presented. Since the total temperature rise is 119 K, let the rise in each successive stage be 23.5, 24, 24, 24, and 23.5 K, respectively. Assume that the work done factors are 0.98, 0.93, 0.88, 0.83, and 0.83, respectively.

Stage # 1:

Since the flow is assumed to be axial at the inlet, the velocity triangles are as shown earlier. Using the definition of E,

$$C_p\Delta T_0 = E = \lambda u(V_{u2} - V_{u1}) = \lambda u \Delta V_u$$

$$\Delta V_u = \frac{C_p\Delta T_0}{\lambda u} = \frac{(1004.5)(23.5)}{(0.98)(265.5)} = 90.7\,\text{m/s}$$

$$\Delta V_u = V_{u2} - V_{u1} = V_{u2} - 0 = V_{u2} = 90.7$$

$$\alpha_1 = \arctan\left(\frac{V_{u1}}{V_a}\right) = \arctan\left(\frac{0}{150}\right) = 0°$$

$$\beta_1 = \arctan\left(\frac{u - V_{u1}}{V_a}\right) = \arctan\left(\frac{265.5 - 0}{150}\right) = 60.5°$$

$$\beta_2 = \arctan\left(\frac{u - V_{u2}}{V_a}\right) = \arctan\left(\frac{265.5 - 90.7}{150}\right) = 49.4°$$

$$\alpha_2 = \arctan\left(\frac{V_{u2}}{V_a}\right) = \arctan\left(\frac{90.7}{150}\right) = 31.2°$$

The deHaller number DH ($=w_2/w_1$) for the rotor can now be calculated. Thus,

$$DH = \frac{w_2}{w_1} = \frac{V_a/\cos\beta_2}{V_a/\cos\beta_1} = \frac{\cos\beta_1}{\cos\beta_2} = \frac{\cos(60.5)}{\cos(49.4)} = 0.76$$

The deHaller number should be greater than 0.72, and hence the number for the first stage is satisfactory. Temperature and pressure at the exit of the first stage are calculated next:

$$T_{01} = 293;\ T_{02} = 293 + \Delta T_0 = 293 + 23.5 = 316.5\,\text{K}$$

$$\eta_{st} = \frac{T_{02}' - T_{01}}{T_{02} - T_{01}} = 0.92 \Rightarrow T_{02}' = \eta_{st}(T_{02} - T_{01}) + T_{01} = 314.62\,\text{K}$$

$$\frac{p_{02}}{p_{01}} = \frac{p_{02'}}{p_{01}} = \left(\frac{T_{02}'}{T_{01}}\right)^{\eta_{st}k/k-1} = 1.283 \Rightarrow p_{02} = (100)(1.283) = 128.3\,\text{kPa}$$

This stage and the corresponding states are shown on the h–s diagram that follows. The degree of reaction can be calculated to be

$$R = 1 - \frac{V_{u2} + V_{u1}}{2u} = 1 - \frac{90.7 + 0}{2(265.5)} = 0.83$$

This degree of reaction is quite high when compared with the desired value of 0.5; however, it will be reduced in the subsequent stages, and all the later stages will have values closer to 0.5.

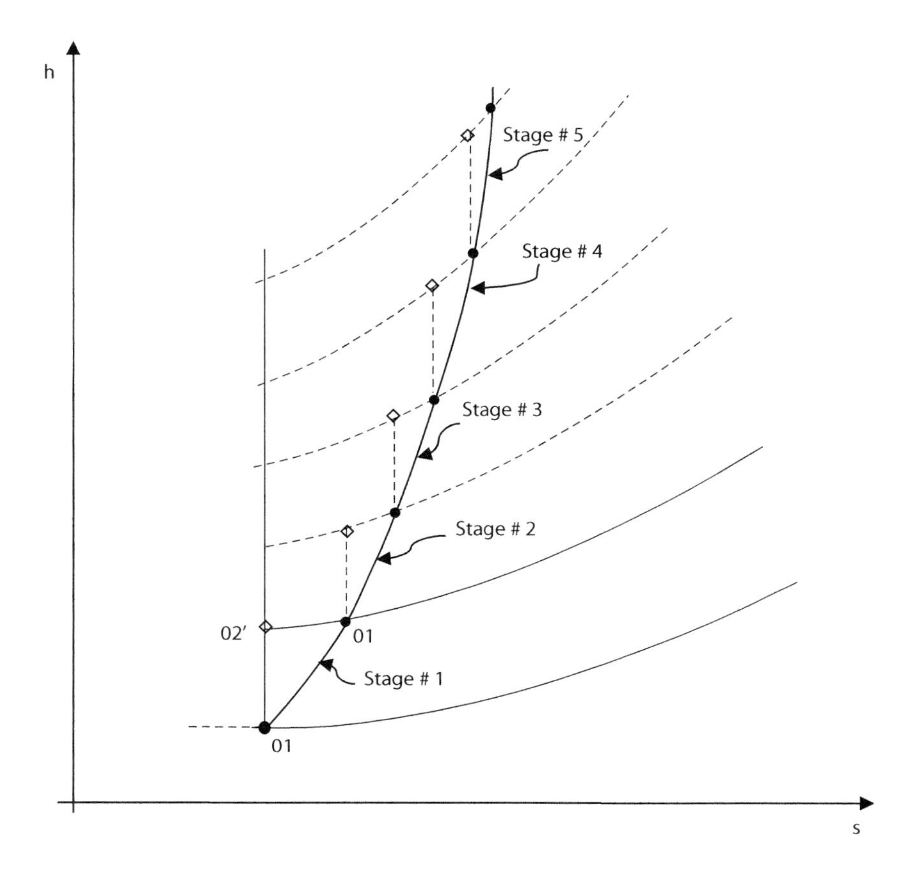

Stage # 2:

A value for degree of reaction can be assumed here. It will be taken to be 0.7, making it 0.5 for the rest of the stages. Also, the stagnation temperature rise ΔT_0 and work done factor λ are assumed to be 24 K and 0.93, respectively. Using the definition of E,

$$\Delta V_u = \frac{C_p \Delta T_0}{\lambda u} = \frac{(1004.5)(24)}{(0.93)(265.5)} = 97.3\,\text{m/s}$$

$$\tan\beta_1 = \frac{1}{2}\left(\frac{C_p \Delta T_0}{\lambda u V_a} + \frac{2Ru}{V_a}\right)$$

$$= \frac{1}{2}\left(\frac{(1004.5)(24)}{(0.93)(265.5)(150)} + \frac{2(0.7)(265.5)}{150}\right) = 1.564 \Rightarrow \beta_1 = 57.4^\circ$$

$$\tan\beta_2 = \frac{1}{2}\left(\frac{C_p \Delta T_0}{\lambda u V_a} - \frac{2Ru}{V_a}\right)$$

$$= \frac{1}{2}\left(\frac{(1004.5)(24)}{(0.93)(265.5)(150)} - \frac{2(0.7)(265.5)}{150}\right) = 0.913 \Rightarrow \beta_2 = 42.4^\circ$$

$$V_{u1} = u - V_a \tan\beta_1 = 265.5 - 150\tan 57.4 = 30.81\,\text{m/s}$$

$$V_{u2} = u - V_a \tan\beta_2 = 265.5 - 150\tan 42.4 = 129.4\,\mathrm{m/s}$$

$$\alpha_1 = \arctan\left(\frac{V_{u1}}{V_a}\right) = \arctan\left(\frac{30.81}{150}\right) = 11.6^\circ$$

$$\alpha_2 = \arctan\left(\frac{V_{u2}}{V_a}\right) = \arctan\left(\frac{129.4}{150}\right) = 40.8^\circ$$

The deHaller number DH ($=w_2/w_1$) for the rotor can now be calculated. Thus,

$$DH = \frac{w_2}{w_1} = \frac{V_a/\cos\beta_2}{V_a/\cos\beta_1} = \frac{\cos\beta_1}{\cos\beta_2} = \frac{\cos(57.4)}{\cos(42.4)} = 0.73$$

The deHaller number should be greater than 0.72, and hence the number for this stage is satisfactory. Temperature and pressure at the exit of the second stage are calculated next:

$$T_{01} = 316\,\mathrm{K};\ T_{02} = 316 + \Delta T_0 = 316 + 24 = 340\,\mathrm{K}$$

$$\eta_{st} = \frac{T_{02}' - T_{01}}{T_{02} - T_{01}} = 0.92 \Rightarrow T_{02}' = \eta_{st}(T_{02} - T_{01}) + T_{01} = 338.1\,\mathrm{K}$$

$$\frac{p_{02}}{p_{01}} = \frac{p_{02'}}{p_{01}} = \left(\frac{T_{02}'}{T_{01}}\right)^{\eta_{st}k/k-1} = 1.267 \Rightarrow p_{02} = (127.6)(1.267) = 161.7\,\mathrm{kPa}$$

Thus, the pressure at the exit of the second stage is 161.7 kPa.

Stage # 3, 4, 5

The calculations are repeated, and hence they are not shown here. However, the results for all the stages are tabulated as follows:

Stage #	ΔT_0 (K)	λ	R	DH	T_{01} (K)	T_{02} (K)	P_{01} (kPa)	P_{02}/P_{01}	P_{02}
1	23	0.98		0.76	293	316	100	1.276	127.6
2	24	0.93	0.7	0.73	316	340	127.6	1.267	161.7
3	24	0.88	0.5	0.72	340	364	161.7	1.246	201.5
4	24	0.83	0.5	0.70	364	388	201.5	1.228	247.6
5	23	0.83	0.5	0.70	388	412	247.6	1.213	300.6

A final check that needs to be made is whether the pressure ratio obtained by taking the products of each stage is equal to the overall pressure ratio as required. Thus,

$$\left(\frac{p_{02}}{p_{01}}\right)_{overall} = \left(\frac{p_{02}}{p_{01}}\right)_1\left(\frac{p_{02}}{p_{01}}\right)_2\left(\frac{p_{02}}{p_{01}}\right)_3\left(\frac{p_{02}}{p_{01}}\right)_4\left(\frac{p_{02}}{p_{01}}\right)_5$$
$$= (1.276)(1.267)(1.246)(1.228)(1.213) = 3.00$$

Comments: (1) From the mass flow rate, the height of the blades in each stage can be calculated. (2) This design is by no means unique. If some of the selected values such as inlet or tip speeds were different, a different design would be obtained that should be equally acceptable.

PROBLEMS

All pressures given are absolute unless specified otherwise. Use air properties unless specified otherwise. Take the work done factor to be equal to one unless specified otherwise.

10.1 Show that for axial flow compressors with constant axial velocity, the degree of reaction is given by the following expression (1 and 2 are inlet and exit states, respectively):

$$R = 1 - \frac{V_{u1} + V_{u2}}{2u}$$

10.2 Show that the degree of reaction R, flow coefficient ϕ, and blade angles are related by the following expression for axial flow compressors:

$$R = \frac{\varphi}{2}(\tan\beta_1 + \tan\beta_2)$$

10.3 Show that for axial flow compressors with constant axial velocity, the flow coefficient ϕ and the stage loading factor ψ are related by the following expression:

$$\psi = \varphi(\tan\beta_1 - \tan\beta_2)$$

10.4 It is instructive to compare compressor efficiencies based on static and total values. Consider a compressor that takes air at atmospheric pressure and 68°F at a speed of 1000 ft./s. The discharge temperature and pressure are 400°F and 60 psia, respectively. If the exit velocity is 880 ft./s, calculate the isentropic efficiency based on (a) static temperatures and (b) total temperatures.

Ans: 78.6%; 80.7%

10.5 Consider the flow in the rotor of a compressor stage. Air enters the compressor at a stagnation pressure of 101.4 kPa and 20°C while it rotates at 8000 rpm. The tip radius is 30 cm and the height of the blades is 5 cm. The axial velocity is a constant at 175 m/s. Absolute velocity at the rotor inlet is axial, that is, $V_1 = V_a$. The rotor blade turns the relative velocity at the inlet toward the axial direction by 30°, that is, $\beta_1 - \beta_2 = 30°$. If the compression is assumed to be isentropic, calculate (a) mass flow rate, (b) power, and (c) blade and fluid angles at the inlet and exit. Show the process on an h–s diagram. Assume $k = 1.35$.

Ans: mass flow = 15.88 kg/s; $P = 574$ kW; $\beta_1 = 52.8°$; $\beta_2 = 22.8°$

10.6 Solve Problem 10.5, but this time the flow is not isentropic and the rotor efficiency $\frac{h_2 - h_1}{h_{2'} - h_1}$ is 90%. Show the process on an h–s diagram.

10.7 Consider an axial flow compressor with a rotation of 6500 rpm. Inlet air is at 100 kPa and 25°C. The tip and hub diameters are 70 and 60 cm, respectively. As the air leaves the rotor, it turns through an angle of 30°. The axial velocity can be assumed to be a constant at 160 m/s. Calculate the power input, mass flow rate, and the degree of reaction. Assume there are no inlet guide vanes and that the air enters the rotor in the axial direction. If the expansion process is ideal, that is, isentropic, calculate the ideal pressure ratio for the stage. Show the processes (both the actual and ideal) on a *T–s* diagram.

Ans: 653 kW; 19.76 kg/s; 66.25%; 1.438

10.8 An axial flow compressor with 50% degree of reaction is to have an overall compression ratio of seven. The inlet and exit blade angles are 50° and 15°, both measured with respect to the axial direction. The inlet static temperature is 25°C and the axial velocity can be assumed to be a constant at 160 m/s. If the isentropic efficiency of the compressor is 0.88, estimate the stagnation temperature rise and the number of stages required to achieve the desired compression ratio.

Ans: $\Delta_{T0} = 34.4°$; Eight stages

10.9 (a) Consider a compressor consisting of two stages. The overall pressure ratio is four and the inlet conditions are 300 K and one atmosphere. Assume that each stage has the same stage efficiency of 85% and the pressure ratio is the same for each stage. Calculate the isentropic efficiency of the compressor and compare it with the stage efficiency.

Ans: $\eta_c = 83.5\%$; $\eta_c < \eta_{st}$

(b) Consider the same problem as (a), that is, a compressor consisting of two stages with an overall pressure ratio of four and inlet conditions of 300 K and one atmosphere. If the isentropic compressor efficiency is 85% and the stages are identical (pressure ratio is the same for each stage), calculate the stage efficiency.

Ans: $\eta_{st} = 86.3\%$; $\eta_c < \eta_{st}$

10.10 An axial flow compressor of 50% reaction is designed for a flow coefficient of 0.5. The inlet and exit stagnation pressures are 100 and 300 kPa, respectively. Inlet stagnation temperature is 25°C. The hub and tip radii are 0.2 and 0.4 m, respectively. If the stator exit angle is 30° and the rotor speed is 10,000 rpm, calculate the number of stages required and the mass flow rate. Assume compressor isentropic efficiency is 0.89.

Ans: Three stages; 60.11 kg/s

10.11 Using the expressions for the degree of reaction R, flow coefficient ϕ, and the stage loading coefficiet ψ, show that the blade angles can be expressed as

$$\tan\beta_1 = \frac{1}{\varphi}\left(R + \frac{\psi}{2}\right) \text{ and } \tan\beta_2 = \frac{1}{\varphi}\left(R - \frac{\psi}{2}\right)$$

10.12 This problem considers the effect of blade loading factor, Ψ, on number of stages. The inlet stagnation conditions are 17°C and 100 kPa. The overall pressure ratio is four. The blade velocity is 260 m/s. If the stage efficiency is 85% and all stages are identical, estimate the number of stages required if Ψ is (a) 0.2, (b) 0.3, and (c) 0.4.

Ans: 13, 9, and 7 stages

10.13 Assuming that the compression process takes place through infinitesimal stages, calculate the isentropic compressor efficiency under the following conditions. The compression takes place from an initial stage of 300 K and one atmosphere to a final stage of 500 K and four atmospheres. Compare this value with the small stage efficiency.

Ans: $\eta_c = 72.8\%$; $\eta_{st} = 77.5\%$

10.14 Consider an axial flow compressor stage with a degree of reaction of 50%. The mean diameter is 1 m and the rotational speed is 6000 rpm. The blade loading coefficient Ψ and the flow coefficient ϕ are 0.3 and 0.5, respectively. (a) Calculate the inlet and exit blade angles. If the inlet temperature is 25°C and the required pressure ratio is six, estimate the number of stages. Assume all the stages are identical and that each stage has an efficiency of 85%. (b) What would the number of stages be if the stage efficiency is 100%, i.e., stage losses are ignored?

Ans: 35°, 52.4°, and eight stages; seven stages

10.15 Obtain the various definitions of degree of reaction given by Equations 10.3 and 10.4.

10.16 Consider an axial flow compressor stage with the following data:

Inlet stagnation conditions	20°C and 1 atm
Blade velocity	210 m/s
Axial velocity	150 m/s
Exit rotor blade angle	−30°
Mass flow rate	10 kg/s
Total to total efficiency	0.85

Take the work done factor to be unity. Calculate the (a) power required to drive the compressor, (b) static pressure rise for the stage p_3/p_1, (c) stagnation pressure rise for the stage p_{03}/p_{01}, and (d) degree of reaction.

Ans: $P = 259$ kW; $p_3/p_1 = 1.300$; $p_{03}/p_{01} = 1.287$

10.17 An axial flow compressor has been designed for a degree of reaction of 0.5 and a mass flow rate of 10 kg/s. It is expected to produce a stagnation pressure ratio of 1.2 and the inlet stagnation conditions are one atmosphere and 20°C. The flow coefficient is 0.6 and the blade loading coefficient is expected to be 0.24. For a work done factor of one and a total to total efficiency of 0.85, calculate the (a) blade angles at the inlet and outlet of the rotor and (b) power needed to drive the compressor.

Ans: $P = 185$ kW

10.18 Consider the effect of changing the total to total efficiency on the power consumption of axial compressors. An axial compressor consisting of ten identical stages has a mass flow rate of 10 kg/s and produces a stagnation pressure ratio of four. The inlet stagnation temperature is 25°C. Assume the work done factor to be one.

Calculate the power required to drive the compressor when the total to total efficiency is (a) 75%, (b) 80%, (c) 85%, and (d) 90%.

10.19 An axial flow compressor is being designed to have a degree of reaction of 0.55 and the stage temperature rise limited to 20°C. If the blade velocity and flow coefficient are 230 m/s and 0.6, respectively, calculate the rotor and stator angles.

Ans: $\beta_2 = 54.3°$; $\beta_3 = 23.8°$; $\alpha_2 = 15.4°$; $\alpha_3 = 50.8°$

10.20 Design an axial flow compressor to meet the following specifications:

Mass flow rate	50 lbm/s
Inlet stagnation conditions	14.7 psi, 70°F
Pressure rise	2.5

Complete a preliminary design that includes the specifications of each stage of the compressor.

10.21 Derive the following equations in the context of axial compressors:

$$\tan\beta_1 = \frac{1}{2}\left(\frac{C_p \Delta T_0}{\lambda u V_a} + \frac{2Ru}{V_a}\right)$$

$$\tan\beta_2 = \frac{1}{2}\left(\frac{C_p \Delta T_0}{\lambda u V_a} - \frac{2Ru}{V_a}\right)$$

11

Steam Turbines

Steam turbines belong to the class of power-generating turbomachines, which are similar to gas turbines, but operate at lower speeds. Also, since the working fluid is superheated steam, there is a phase change of the fluid during power generation. In this sense, the design and operating features of steam turbines are somewhat different to those for gas turbines.

Upon completion of this chapter, the student will be able to

- Understand operating principles of steam turbines and their role in the Rankine cycle
- Understand the differences between velocity and pressure compounding
- Perform the preliminary design of steam turbines

11.1 Introduction

In steam power plants, the steam turbine is the prime mover that converts thermal energy into mechanical power. High-pressure steam from the boiler enters a set of stationary nozzles and expands to lower pressures, thus producing high-speed jets of steam that strike a set of blades to produce power. The low-pressure steam enters a condenser, and the condensed liquid is pumped back into the boiler. The plants operate on the Rankine cycle and may include modifications such as superheating, reheating, and feed water heating, all of which are aimed at increasing the thermal efficiency of the cycle. Another feature of steam plants is that they have a series of turbines through which the steam passes in series. They are the high-pressure (HP) turbine, which receives the steam from the boiler, intermediate-pressure (IP) turbine, which is usually used for reheating, and the low-pressure (LP) turbine, the exit from which leads into the condenser, in which the pressure is below atmospheric. Steam turbines can also be classified as noncondensing and condensing types. In the noncondensing (also called back-pressure) type of turbine, the steam leaves in a superheated state at pressures greater than atmospheric and the thermal energy is used for other industrial processes. In the condensing type, the steam leaves with a quality of less than one and exhausts into the condenser.

Steam turbines operate on principles that are similar to axial flow gas turbines, the major difference being the working fluid. They are usually significantly larger than gas turbines and also operate at higher pressures. Another feature that distinguishes steam turbines from gas turbines is that the latter stages operate at pressures below atmospheric. Typically, steam tables are used in the analysis and design of steam turbines. Since the Rankine cycle is usually a closed loop, a condenser is required as part of the cycle. This, along with the boiler, makes the steam turbine quite bulky and less attractive in comparison to gas turbines. The components and the temperature versus specific entropy (T–s) diagram for the Rankine cycle are shown in Figure 11.1.

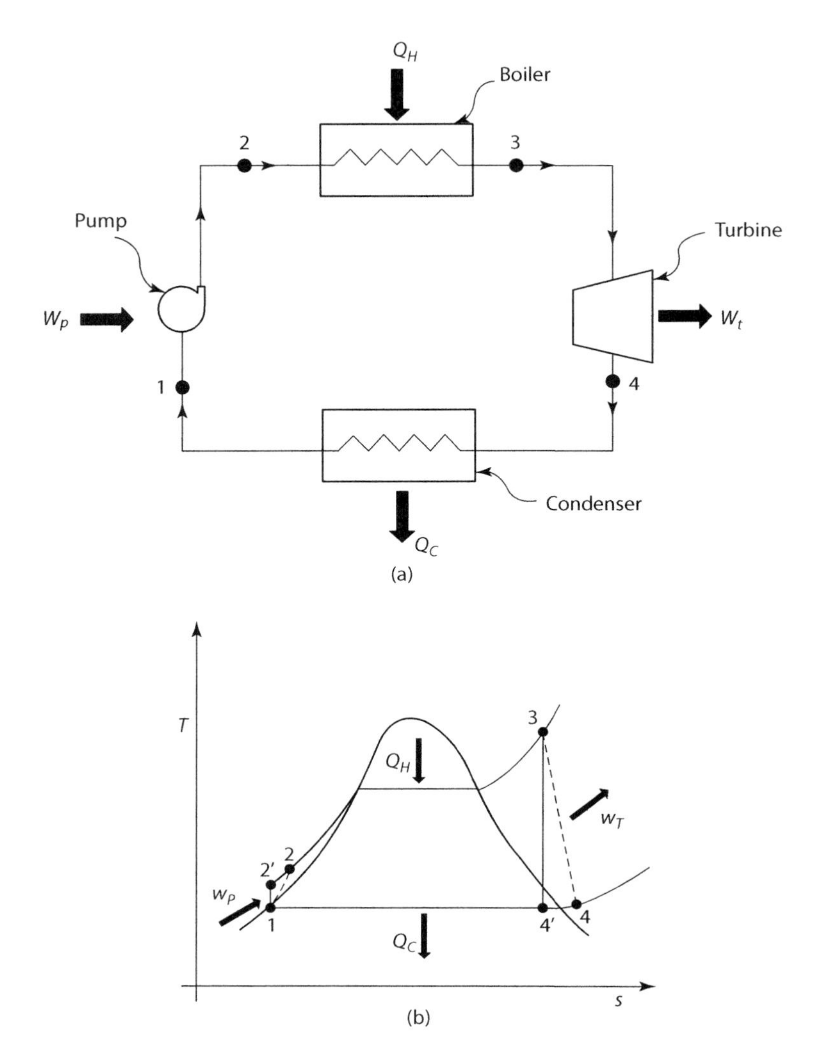

FIGURE 11.1
Components of the Rankine cycle: (a) components and (b) *T–s* diagram.

11.2 Multi-Staging of Steam Turbines

The pressure ratios used in steam turbines are much larger than in gas turbines, especially since the exit pressure from the turbine is below atmospheric. If the entire pressure drop takes place in a single stage, the natural choice would be an impulse stage, since impulse turbines produce twice the amount of power when compared with a reaction stage for the same blade speed. Such a turbine was designed and built by de Laval in 1889. The nozzle used in the turbine produced very high velocities, and the single-stage rotor ran at a speed in excess of 30,000 rpm. The high speed resulted in high blade stresses and large losses due to disk friction. Hence, single-stage impulse steam turbines resulted in low efficiencies. Since these turbines are used to run generators, which run at considerably lower speeds (a few thousand rpm), highly efficient gear trains with large gear ratios are required. Thus, single-stage impulse turbines are rarely used in practice. This necessitates the need for multiple impulse stages.

A method of obtaining high enthalpy drop and reasonable tip speeds in impulse turbines is through compounding of stages, which results in reduced blade speed for a given pressure drop. Two types of compounding are used, namely *velocity compounding* and *pressure compounding*. These are described next.

11.2.1 Velocity Compounding

Steam turbine stages using velocity compounding are called *Curtis stages*, named after Charles Gordon Curtis (1860–1953). Here, the steam expands through a single set of stationary nozzles and reaches high velocities, undergoing large pressure drops. The high-velocity steam then passes through several impulse stages consisting of stators and rotors. The total energy of the steam is absorbed by all the rows in succession until the kinetic energy becomes negligible in the last stage. Thus, the entire pressure drop takes place in the first nozzle, and no further reduction takes place as the steam passes through the succeeding stages. A two-row Curtis stage is shown in Figure 11.2a. It

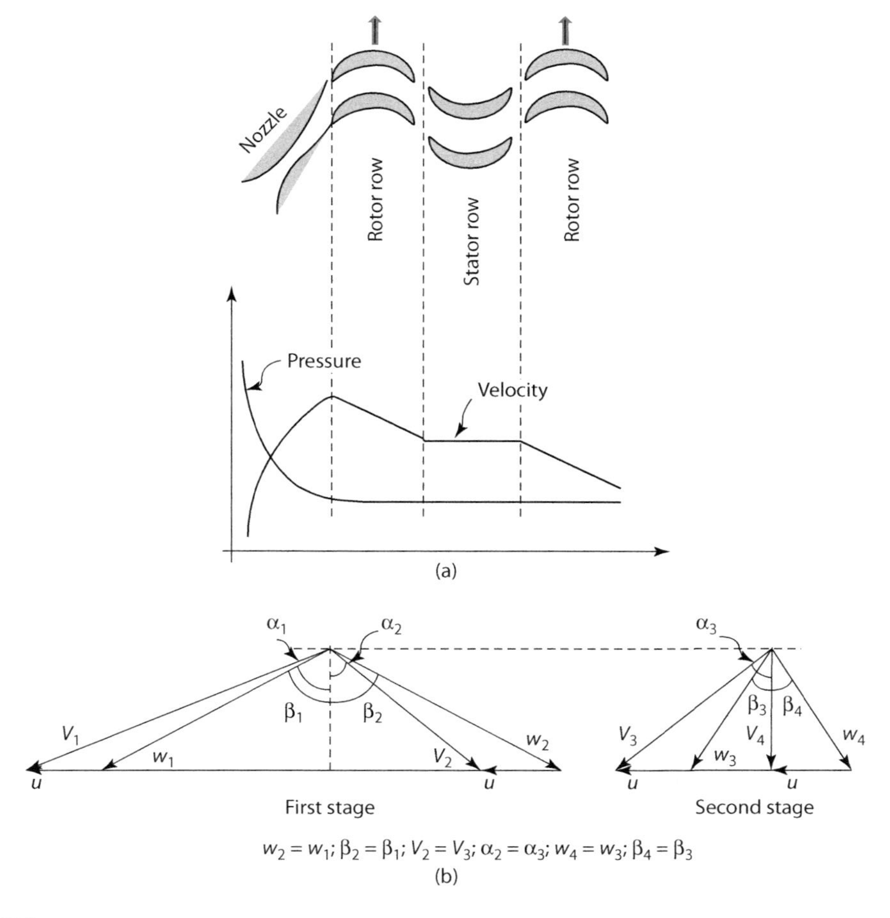

FIGURE 11.2
(a) Curtis stage and (b) velocity triangles.

consists of a convergent–divergent nozzle, a rotor, a stator whose purpose is to guide the fluid (but not to change pressure), and another rotor. The velocity diagrams are shown in Figure 11.2b. Note that the nozzle angle is measured with respect to the axial direction as opposed to the convention used in Chapter 3 wherein it was measured with respect to the tangential direction.

From the velocity triangles, the energy transfer for each stage can be calculated. Thus, for the first stage,

$$\begin{aligned} E_1 &= u(V_{u1}+V_{u2}) = u(V_1\sin\alpha_1 + V_2\sin\alpha_2) \\ &= u(V_1\sin\alpha_1 + w_2\sin\beta_2 - u) = u(V_1\sin\alpha_1 + w_1\sin\beta_1 - u) \\ &= u(V_1\sin\alpha_1 + V_1\sin\alpha_1 - u - u) = 2u(V_1\sin\alpha_1 - u) \end{aligned} \tag{11.1}$$

For the second stage,

$$\begin{aligned} E_2 &= u(V_{u3}+V_{u4}) = u(V_3\sin\alpha_3 + 0) = 2u(w_3\sin\beta_3) \\ &= 2u(V_3\sin\alpha_3 - u) = 2u(V_2\sin\alpha_2 - u) \\ &= 2u(w_2\sin\beta_2 - u - u) = 2u(w_1\sin\beta_1 - u - u) \\ &= 2u(V_1\sin\alpha_1 - u - u - u) = 2u(V_1\sin\alpha_1 - 3u) \end{aligned} \tag{11.2}$$

Thus, the total energy from both stages can be written as

$$\begin{aligned} E = E_1 + E_2 &= 2u(V_1\sin\alpha_1 - u) + 2u(V_1\sin\alpha_1 - 3u) \\ &= 4u(V_1\sin\alpha_1 - 2u) \end{aligned} \tag{11.3}$$

The total available energy for the turbine is $\frac{V_1^2}{2}$. Using the definition of utilization factor, the energy E can be combined with the total available energy as

$$\varepsilon = \frac{E}{E+\frac{1}{2}V_{exit}^2} = \frac{E}{E+\frac{1}{2}V_4^2} = \frac{E}{\frac{1}{2}V_1^2} \tag{11.3a}$$

This equation can be obtained by writing E in terms of the velocities for each stage and adding them (see Problem 11.11):

$$\varepsilon = \frac{E}{\frac{1}{2}V_1^2} = \frac{4u(V_1\sin\alpha_1 - 2u)}{\frac{1}{2}V_1^2} = 8\left(\frac{u}{V_1}\right)\left[\sin\alpha_1 - 2\left(\frac{u}{V_1}\right)\right] \tag{11.4}$$

The ratio u/V_1 is the called the *speed ratio* or *peripheral speed factor* and has been discussed previously. To obtain the maximum utilization factor, the derivative of ε with respect u/V_1 is set to zero. This gives

$$\frac{u}{V_1} = \frac{\sin\alpha_1}{4} \tag{11.4a}$$

which is half the value for a single impulse stage. The maximum utilization would be

$$\varepsilon_{\max} = 8\left(\frac{\sin\alpha_1}{4}\right)\left[\sin\alpha_1 - 2\left(\frac{\sin\alpha_1}{4}\right)\right]$$
$$= 2\sin\alpha_1\left[\sin\alpha_1 - \left(\frac{\sin\alpha_1}{2}\right)\right] = \sin^2\alpha_1 \qquad (11.5)$$

The maximum energy can be obtained from Equation 11.3 as

$$E_{\max} = 4u(4u - 2u) = 8u^2 \qquad (11.6)$$

The corresponding values for a single impulse stage and single reaction stage are $E_{max}=2u^2$ and $E_{max}=u^2$, respectively. The advantage of velocity compounding is obvious. However, the main disadvantage is the poor aerodynamic efficiency of the blades and the ensuing losses due to excessive fluid speeds. This is true even if the blade loss coefficients are low. Hence, velocity compounding is used in the initial stages for high-pressure turbines where the advantages of rapid pressure reduction (without which there will be high leakage losses and complex seal designs) and decreased number of stages outweigh its poor efficiency. The overall efficiency can still be maintained at a high value by using higher efficiency reaction stages later on.

This analysis can be extended to more rows (see Problems 11.1 and 11.2). The energy transfer for the *ith* stage in a turbine consisting of n impulse stages can be written as

$$E_i = 2\left[2(n-i)+1\right]u^2 \qquad (11.7)$$

The total energy for the n stages can be written as

$$E_n = \sum_{i=1}^{n} E_i = 2n^2u^2; \quad \frac{u}{V_1} = \frac{\sin\alpha_1}{2n} \qquad (11.8)$$

From these equations, it can be observed that the total turbine energy can be drastically increased by velocity compounding. Note that E_n takes on values of $2u^2$, $8u^2$, $18u^2$, $32u^2$, and $50u^2$ as the number of stages is increased from two to five. This has the advantage that the pressure of the steam decreases quickly, resulting in a smaller number of stages to produce a given amount of total energy. It will also be noticed that the work in the last stage is always $2u^2$, substantially smaller than in the first stage; one-fifth of the first stage for three stages. Thus, velocity compounding, although used quite frequently in steam turbines, seldom involves more than three stages. However, velocity compounding is not used for gas turbines. This is because the pressure ratios are relatively smaller in gas turbines, which leads to simpler seal designs. This, coupled with the fact that higher velocities result in larger losses, makes the overall efficiency too low to be acceptable.

11.2.2 *Pressure Compounding*

The major drawback of velocity compounding is the need for convergent–divergent nozzles to achieve high steam velocity. These are more expensive and difficult to design. Also, higher velocities mean higher cascade losses. This problem can be avoided by having successive impulse stages so that a pressure drop occurs in each stage. This arrangement is called *pressure-compounded staging* or *Rateau staging*, named after the French engineer Autuste Rateau. Due to the comparatively lower pressure drop, the nozzles are subsonic and hence have lower losses. The arrangement of the stages and accompanying pressure and velocity are shown in Figure 11.3. A common feature of Rateau staging is relatively higher efficiencies since lower velocities are used. In large turbines, it is usual to have pressure-compounded stages following velocity-compounded stages.

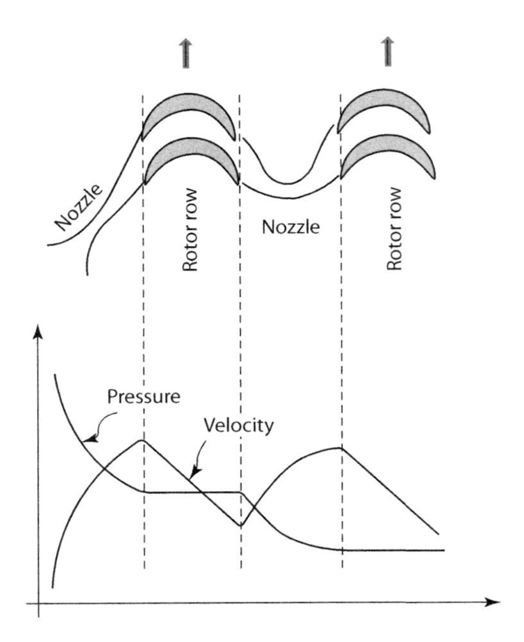

FIGURE 11.3
Pressure compounding of impulse stages.

Example 11.1

A two-stage Curtis turbine operates under the following conditions:

Mean rotor speed	800 ft./s
Nozzle angle	73°
Blade angle at inlet of the first stage	68°
Mass flow rate	2 lbm/s

The axial velocity decreases by 5% in each stage from inlet to outlet. Calculate the (a) ratio of the energy produced by the two stages, (b) total power, and (c) axial thrust. Assume that the impulse stages have the same blade angles at the inlet and outlet for each stage.

Solution: The velocity diagrams for each stage are shown. From trigonometry,

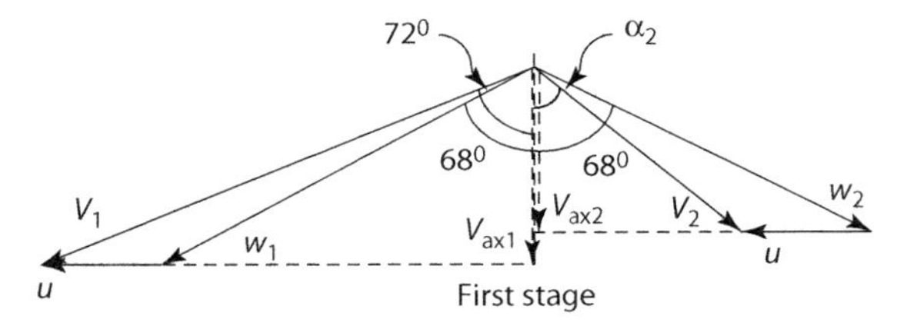

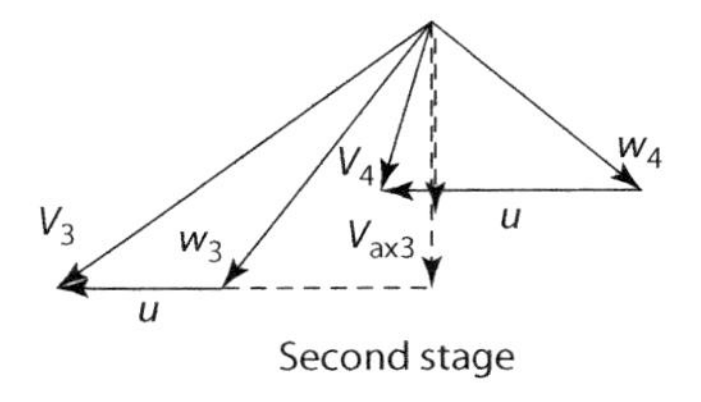

$$V_{u1} - w_{u1} = u = 800; \; V_{u1} = V_{ax1} \tan \alpha_1; \; w_{u1} = V_{ax1} \tan \beta_1$$

$$V_{ax1} \tan \alpha_1 - V_{ax1} \tan \beta_1 = u = 800 \tag{a}$$

$$V_{ax1} = \frac{800}{\tan \alpha_1 - \tan \beta_1} = 1005 \text{ ft./s}$$

Since the axial velocity drops by 5% from inlet to exit,

$$V_{ax2} = (0.95)(1005) = 955 \text{ ft./s} \tag{b}$$

From the velocity triangles, the following can be calculated:

$$V_1 = \frac{V_{ax1}}{\cos \alpha_1} = 3438.5 \text{ ft./s}; \; w_1 = \frac{V_{ax1}}{\cos \beta_1} = 2683.7 \text{ ft./s}$$

$$V_{u1} = V_1 \sin \alpha_1 = 3288 \text{ ft./s} \tag{c}$$

Since the blade angles are the same at the inlet and exit for the first stage, the exit velocity triangle can be completed. Hence,

$$w_{u2} = V_{ax2} \tan \beta_2 = 2363 \text{ ft./s}; \; w_2 = \left(V_{ax2}^2 + w_{u2}^2\right)^{0.5} = 2549 \text{ ft./s}$$

$$V_{u2} = -w_{u2} + u = -1563 \text{ ft./s}; \; V_2 = \left(V_{ax2}^2 + V_{u2}^2\right)^{0.5} = 1832 \text{ ft./s} \tag{d}$$

$$\alpha_2 = \arctan \frac{V_{u2}}{V_{ax2}} = -58.6°$$

Next, from the velocity triangles for the second stage, the following can be calculated:

$$|\alpha_3| = |\alpha_2| = 58.6°; \; V_3 = V_2 = 1832 \text{ ft./s}; \; V_{ax3} = V_{ax2} = 955 \text{ft./s}$$

$$V_{u3} = V_{u2} = 1563 \text{ ft./s}; \; w_{u3} = V_{u3} - u = 763 \text{ ft./s}$$

$$w_3 = \left(V_{ax3}^2 + w_{u3}^2\right)^{0.5} = 1223 \text{ ft./s} \tag{e}$$

$$\beta_3 = \arctan \frac{w_{u3}}{V_{ax3}} = 38.65° = \beta_4$$

Again, the blade angles at the inlet and exit are the same, and the axial velocity at the exit is reduced by five percent. Thus,

$$V_{ax4} = (0.95)(V_{ax3}) = 907 \text{ft./s}$$

$$w_4 = \frac{V_{ax4}}{\cos \beta_4} = 1162 \text{ ft./s}; \; w_{u4} = w_4 \sin \beta_4 = 726 \text{ ft./s} \tag{f}$$

$$V_{u4} = -w_{u4} + u = 74 \text{ ft./s}; \; V_4 = \left(V_{ax4}^2 + V_{u4}^2\right)^{0.5} = 910 \text{ ft./s}$$

$$\alpha_4 = \arcsin \frac{V_{u4}}{V_4} = 4.68°$$

The energy associated with each stage can now be calculated:

$$E_1 = u(V_{u1} - V_{u2}) = (3.88)10^6 \text{ ft.}^2/\text{s}^2;\ E_2 = u(V_{u3} - V_{u4}) = (1.19)10^6 \text{ ft.}^2/\text{s}^2$$

$$\frac{E_1}{E_2} = 3.26;\ E = E_1 + E_2 = (5.07)10^6 \text{ ft.}^2/\text{s}^2$$

$$P = \dot{m}E = (3.15)10^5 \text{ ft} \cdot \text{lbf/s} = 573 \text{hp}$$

$$\frac{u}{V_1} = 0.232$$

Comment: From the velocity triangle for the second stage, the utilization factor is not a maximum, since the exit velocity is not purely axial, that is, α_4 is small but not zero. It can be verified that the velocity ratio is not optimum nor is the utilization factor maximum. See Problem 11.3.

Example 11.2

Superheated steam is supplied to a two-stage Curtis turbine from a boiler at a stagnation pressure of 20 MPa and 500°C. It enters a nozzle and expands isentropically in the nozzle to a pressure of 10.13 MPa as it enters the first stage, where the mean diameter is 0.4 m. The turbine is designed to produce 320 kW when the mass flow rate is 2 kg/s. Assume that the axial velocity is a constant through both stages. Calculate the (a) nozzle angle at the inlet, (b) blade angles for both stages, (c) utilization factor, and (d) rotor speed.

Solution: The velocity diagrams are shown. From steam tables, the enthalpies at the inlet and the exit of the nozzle can be calculated. Thus, at the inlet of the nozzle,

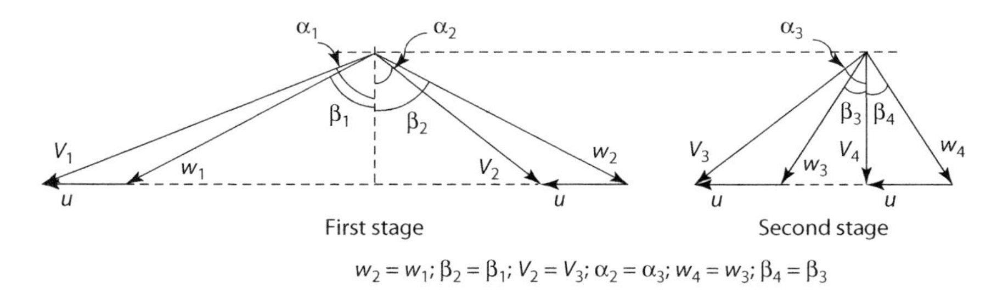

$$p_{0i} = 20 \text{MPa and } T_{0i} = 500\ C \Rightarrow h_{0i} = 3239 \text{ kJ/kg and } s_{0i} = 6.142 \text{ kJ/kg} \cdot \text{K}$$

Since the nozzle expansion is given to be isentropic, enthalpy at the exit (or inlet to the first stage) can be calculated as

$$p_1 = 10.13 \text{MPa and } s_1 = 6.142 \text{ kJ/kg} \cdot \text{K} \Rightarrow h_1 = 3053 \text{ kJ/kg}$$

The inlet velocity can now be calculated from the energy balance across the nozzle. Thus,

$$h_{0i} = h_1 + \frac{V_1^2}{2} \Rightarrow V_1 = \sqrt{2(h_{0i} - h_1)} = 610.3 \text{ m/s}$$

For ideal velocity compounding involving two stages, the maximum power output is $8u^2$. Hence,

$$E = \frac{P}{\dot{m}} = \frac{(320)(1000)}{2} = 160{,}000 = 8u^2 \Rightarrow u = 141.4 \text{ m/s}$$

$$N = \frac{u}{r} = \frac{141}{0.2} = 707 \text{ rad/s} = 6752 \text{ rpm}$$

Also under ideal conditions, for two stages,

$$\frac{u}{V_1} = \frac{\sin(\alpha_1)}{4} \Rightarrow \alpha_1 = 68°$$

The parameters for the inlet and exit velocity triangles for the first stage can be calculated. Thus,

$$V_{u1} = V_1 \sin(\alpha_1) = 610.3 \sin 68 = 565.7 \text{ m/s}$$

$$V_{ax1} = V_1 \cos(\alpha_1) = 610.3 \cos 68 = 228.6 \text{ m/s} = V_{ax2}$$

$$w_{u1} = V_{u1} - u = 424.3 \text{ m/s} = w_{u2}$$

$$w_1 = \left(w_{u1}^2 + V_{ax1}^2\right)^{0.5} = 481.9 \text{ m/s} = w_2$$

$$\beta_1 = \arctan\left(\frac{w_{u1}}{V_{ax1}}\right) = 61.7° = \beta_2$$

The parameters for the second stage can now be calculated:

$$V_{u2} = w_{u2} - u = 282.8 \text{ m/s} \Rightarrow V_2 = \left(V_{u2}^2 + V_{ax2}^2\right)^{0.5} = 363.6 \text{ m/s}$$

$$\alpha_2 = \arctan\left(\frac{V_{u2}}{V_{ax2}}\right) = -51.1°$$

Next, from the velocity triangles for the second stage, the following can be calculated:

$$|\alpha_3| = |\alpha_2| = 51.1°; \ V_3 = V_2 = 363.6 \text{ m/s}; \ V_{ax3} = V_{ax2} = 228.6 \text{ m/s}$$

$$w_{u3} = V_{u3} - u = 282.8 - 141.4 = 141.4 \text{ m/s} = w_{u4}$$

$$w_3 = \left(V_{ax3}^2 + w_{u3}^2\right)^{0.5} = 268.7 \text{ m/s} = w_4$$

$$\beta_3 = \arctan\frac{w_{u3}}{V_{ax3}} = 31.7°; \beta_4 = -31.7°$$

$$V_{u4} = w_{u4} - u = 141.4 - 141.4 = 0; \ V_4 = V_{ax3} = 228.6 \text{ m/s}$$

Since V_{u4} is zero, the absolute velocity V_4 at the exit of second stage is purely in the axial direction and is equal to V_{ax1} (which is incidentally constant through both stages). The utilization factor can now be calculated as

$$\varepsilon = \frac{E}{E + \frac{V_4^2}{2}} = 0.86$$

Comment: The utilization factor is maximum since the exit velocity is purely in the axial direction.

Example 11.3

An axial flow steam turbine consisting of six impulse stages has a mean diameter of 1.7 m and rotates at 3000 rpm. The mass flow rate of steam is 2 kg/s, and it enters the first stage at static pressure and temperature of 5000 kPa and 600°C, respectively. The exit pressure is 100 kPa. Calculate the (a) isentropic efficiency of the turbine and (b) power produced. Assume each of the impulse stages is operating at maximum utilization.

Solution: Since each stage is at maximum utilization, the absolute velocity at the exit of each stage will be axial, that is, there is no momentum. Denoting 1 and 2 to be the inlet and exit for each stage, respectively, the inlet and exit velocity triangles for maximum utilization are shown in the diagram, along with the *T*–*s* diagram for the entire turbine. Here, i and e are the inlet and exit of the entire turbine, respectively.

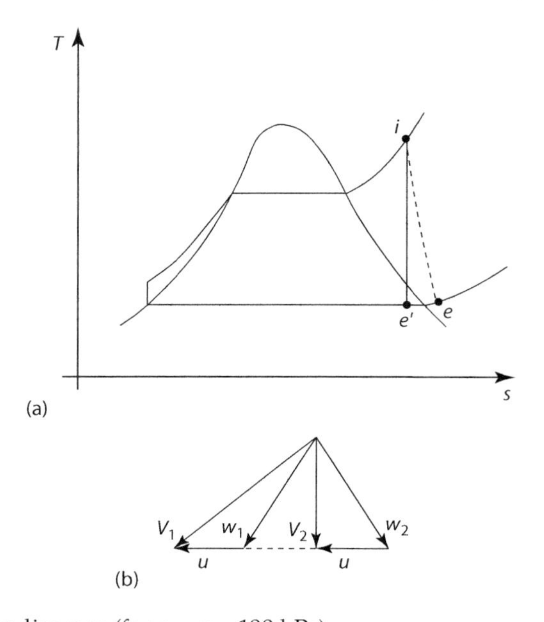

From the Mollier diagram (for $p_{e'} = p_e = 100$ kPa),

$$p_i = 5000\,\text{kPa and } T_{0i} = 600°\text{C} \Rightarrow h_i = 3666\,\text{kJ/kg and } s_i = 7.259\ \text{kJ/kg}\cdot\text{K}$$

$$p_{e'} = p_e = 100\,\text{kPa and } s_e = s_i = 7.259\,\text{kJ/kg}\cdot\text{K} \Rightarrow h_{e'} = 2638\,\text{kJ/kg}$$

From the velocity triangles, the energy per stage can be calculated. Thus,

$$N = 3000\,\text{rpm} = 314\,\text{rad/s};\ u = (0.85)(314) = 267\,\text{m/s}$$

$$E = 2u^2 = 2(267)^2 = 142.6\,\text{kJ/kg}$$

For six stages, the actual exit enthalpy can be calculated as

$$h_i - h_e = 6E \Rightarrow h_e = 3666 - (6)142.6 = 2810\,\text{kJ/kg}$$

The adiabatic or isentropic efficiency of the turbine can be calculated as

$$\eta = \frac{h_i - h_e}{h_i - h_{e'}} = \frac{3666 - 2810}{3666 - 2638} = 83.2\%$$

Power can be calculated as

$$P = \dot{m}E = 2(142.6) = 285.2\,\text{kW}$$

Comment: If the utilization factor for each stage were not the maximum, the velocity triangles would be different.

11.3 Performance Parameters for Steam Turbines

The parameters that affect the performance of steam turbines are the degree of reaction, utilization factor, reheat factor, and quality factor. Some of these such as degree of reaction and utilization factor have been discussed extensively in the earlier chapters. Hence, they will be mentioned only briefly here.

Although it is not possible to use ideal gas laws since steam does not behave like an ideal gas, the difference between small stage efficiency and overall efficiency can be seen in steam turbines also. In the case of gas turbines, this was seen as the effect of treating the overall expansion as a series of small expansions in successive stages, and their cumulative effect on the overall efficiency. However, the terminology used in steam turbines is slightly different. It is common practice to use the term *reheat factor* R_F as a measure of the inefficiency of the expansion. Consider a typical expansion process for a steam turbine shown in Figure 11.4, in which three stages have been used to achieve the overall expansion.

The total isentropic enthalpy drop through all the stages taking place stage-wise would be

$$\sum_{stages} \Delta h_i = (h_1 - h_a) + (h_c - h_d) + (h_f - h_g) \tag{11.8}$$

The reheat factor is defined as the ratio of cumulative enthalpy drop through all the stages to the overall isentropic enthalpy drop of the turbine. Thus,

$$R_F = \frac{(h_1 - h_a) + (h_c - h_d) + (h_f - h_g)}{(h_1 - h_{2'})} = \frac{\sum_{stages} \Delta h_i}{(h_1 - h_{2'})} \tag{11.9}$$

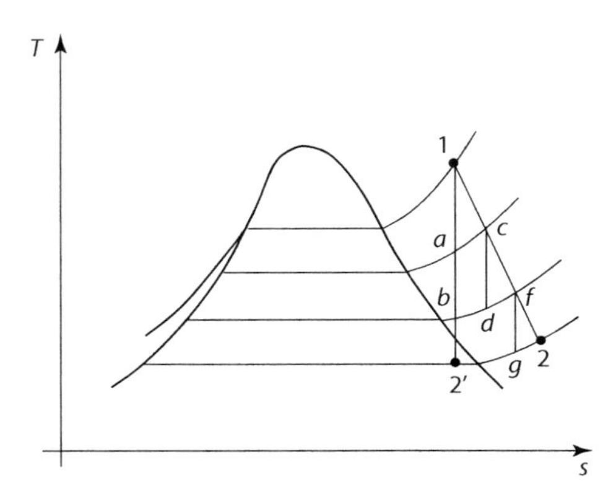

FIGURE 11.4
Expansion process through a turbine to show effects of reheat.

The adiabatic or isentropic efficiency of the turbine can be related to the reheat factor as follows:

$$\eta_t = \frac{(h_1 - h_2)}{(h_1 - h_{2'})} = \left(\frac{(h_1 - h_2)}{\sum_{stages} \Delta h_i} \right) \left(\frac{\sum_{stages} \Delta h_i}{(h_1 - h_{2'})} \right) = \eta_{st} R_F \tag{11.10}$$

where η_{st} is the stage efficiency or polytropic efficiency that has been previously discussed in Chapters 8 and 10 on axial flow gas turbines and axial flow compressors, respectively. Typical values of R_F range from 1.03 to 1.08. Similar to gas turbines, the turbine efficiency is greater than the stage efficiency for steam turbines also.

The degree of reaction has the same meaning as has been described in connection with gas turbines, both radial and axial. It is the ratio of enthalpy drop in the rotor to the enthalpy across the entire stage. It will not be discussed any further here except to note that unlike gas turbines where temperatures could be used, enthalpies need to be used here, since steam is not an ideal gas. The meaning of utilization factor is also the same as was defined and discussed extensively in earlier chapters. It is a measure of the amount of energy that has been used when compared with the total energy that has been supplied. Thus,

$$\varepsilon = \frac{E}{E + \frac{V_{exit}^2}{2}} \tag{11.11}$$

It may be recalled that the utilization factor for axial flow turbines is a maximum when the exit kinetic energy is a minimum, or when the exit velocity is purely in the axial direction. Two other variables that have relevance are the stator and rotor velocity coefficients. The stator coefficient is the ratio of actual and ideal absolute velocities. For the rotor, it is defined as the ratio of the actual and ideal relative velocities at the exit (see Equation 7.18).

Another variable that is used frequently in the discussion of multistage steam turbines is the so-called carry over ratio or the carry over efficiency. The kinetic energy leaving each stage (except the last) of the turbine is not completely lost, as might be the case for a single stage. The velocity leaving one stage becomes the velocity entering nozzles for the following stage. Carry over ratio is defined as the ratio of kinetic energy actually utilized as entering energy of a stage to the kinetic energy of the steam leaving the previous stage. Depending on the physical distance between successive stages, this ratio can take a value between zero and unity. When the wheels and diaphragms are packed closely together, there is likely to be a low carry over loss and the ratio is likely closer to unity.

It will be recalled that the efficiency of pumps and their general shape is strongly influenced by the specific speed. As the specific speed increases, it was found that the design changes from purely radial to mixed to purely axial flow machines. It was also found that for a given value of specific speed, one particular type of pump was most efficient. A similar situation exists for steam turbines also. However, the parameter that is used is called the *quality factor* Q_F, which is a ratio of the total blade kinetic energy of all stages to the enthalpy rise across the turbine. Thus,

$$Q_F = \frac{\sum u^2}{\Delta h_s} \tag{11.11a}$$

It has been shown by Peng (2008) that the quality factor can be expressed in terms of other variables as

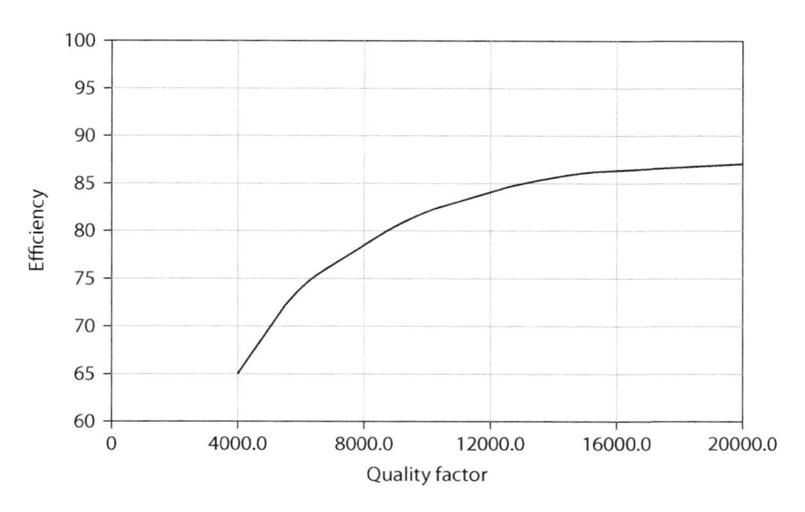

FIGURE 11.5
Variation of efficiency with quality factor.

$$Q_F = \frac{\sum u^2}{\Delta h_s} = 2\varphi R_H \tag{11.11b}$$

where:
φ is the speed ratio as defined earlier
R_H is the reheat factor

Extensive work has been done in the past on multistage steam turbines correlating the quality factor and the efficiency. Gartman (1970) has quantified the relationship between the quality factor and efficiency which is shown in Figure 11.5.

11.4 Impulse and Reaction Staging

The basic concepts underlying impulse and reaction turbines have been discussed in detail in Chapter 3 and will be mentioned only briefly here. In impulse staging, the entire pressure drop occurs in the nozzles, and the purpose of the rotors is mainly to decrease the absolute velocity of the fluid. In the ideal case, that is, when the exiting kinetic energy is minimum, the energy produced is $2u^2$. In reaction stages, pressure drops occur in both the stator and rotor. The situation is somewhat similar to pressure compounding, wherein pressure drops occur in each stage. The preferred design value for degree of reaction is 50%, since this leads to symmetrical blades. Such stages are called *Parsons stages,* named after the designer Sir Charles Parsons. The energy produced in ideal conditions is u^2, half that of a similar impulse stage when operating at the same blade speed and utilization factor. However, the efficiency of a 50% reaction stage is higher than for both Curtis and Rateau stages. Thus, the designer has the choice of using fewer impulse stages that exhibit lower efficiency or more reaction stages that exhibit higher efficiency. The compromise between excessive number of stages and unacceptably low efficiency would be to use one or two impulse stages followed by reaction stages. This would lead to reasonable overall turbine efficiency without having to use too many stages.

Example 11.4

Consider a 50% reaction low-pressure steam turbine stage rotating at a speed of 6800 rpm. The mean diameter of the blades is 2 ft. and the height of the blades is 3 in. The inlet static pressure and temperature of the stage are 30 psia and 300°F, respectively. The exit pressure of the stage is 18 psia. The blade angle at the inlet is 35° measured with respect to the axial direction. The speed ratio (blade speed with respect to rotor inlet velocity) is 0.72. Calculate the stage efficiency, utilization factor, and power output.

Solution: The blade speed, inlet speed, and area can be calculated from the given data. Also, the stage is 50% reaction and hence the velocity triangles are symmetric. Thus,

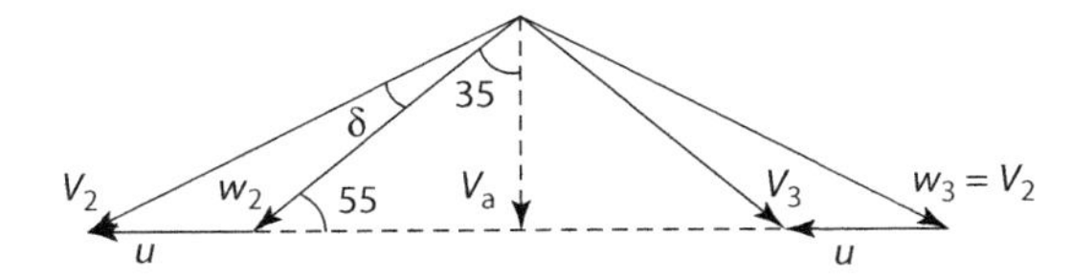

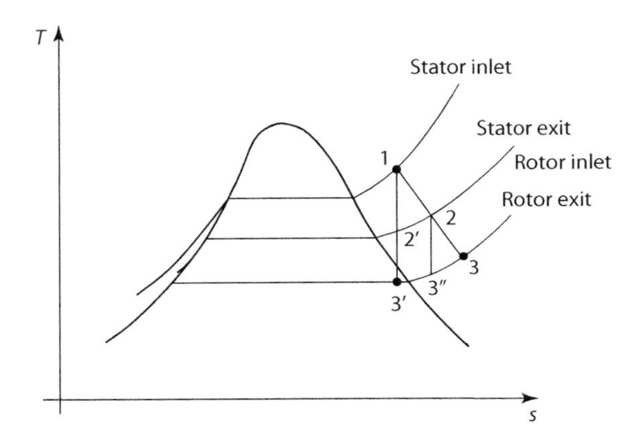

$$N = 6800\,\text{rpm} = 712\,\text{rad/s};\quad u = Nr = (712)(1) = 712\,\text{ft./s}$$

$$\text{Area} = \pi Dh = \pi(2)(3/12) = 1.57\,\text{ft.}^2$$

$$\frac{u}{V_2} = 0.72 \Rightarrow V_2 = \frac{712}{0.72} = 989\,\text{ft./s} = w_3$$

From the Mollier diagram,

$$\text{For } p_1 = 30\,\text{psia} \text{ and } T_1 = 300°\text{F}$$

$$h_1 = 1189 \text{ Btu/lbm}$$

$$s_1 = 1.733\,\text{Btu/lbm}\cdot°\text{R};\ \rho_1 = 0.0678\,\text{lbm/ft.}^3 = 0.002097\,\text{slugs/ft.}^3$$

The exit enthalpy can be calculated assuming the process is isentropic. Thus, from the Mollier diagram,

$$\text{For } p_3 = 18\,\text{psia} \text{ and } s_{3'} = s_1 = 1.733 \text{ Btu/lbm}\cdot°\text{R}$$

$$h_{3'} = 1150 \text{ Btu/lbm}$$

From the velocity triangles, the following values can be calculated. Thus,

$$\frac{V_2}{\sin 125} = \frac{u}{\sin \delta} \Rightarrow \frac{989}{\sin 125} = \frac{712}{\sin \delta} \Rightarrow \delta = 36.14^\circ$$

$$\alpha_2 = 36.14 + 35 = 71.14^\circ = \beta_3$$

$$V_{ax2} = V_2 \cos 71.14 = 319.7\,\text{ft./s}; \; w_2 = \frac{V_{ax2}}{\cos 35} = 390\,\text{ft./s} = V_3$$

$$V_{u2} = V_2 \sin 71.14 = 935\,\text{ft./s}; \; V_{u3} = -w_{u2} = -w_2 \sin 35 = -223\,\text{ft./s}$$

$$E = u\left(V_{u2} - V_{u3}\right) = 712(935 + 223) = 825{,}865\,\text{ft.}^2/\text{s}^2$$

Since the density is known, the mass flow rate and power can be calculated as

$$\dot{m} = \rho_2 A_2 V_{ax2} = (0.002097)(1.57)(319) = 1.052\,\text{slugs/s}$$

$$P = \dot{m}E = (1.052)(825{,}865)/550 = 1581\,\text{hp}$$

The adiabatic or isentropic efficiency and utilization factor of the turbine can be calculated as

$$\eta_t = \frac{E}{h_1 - h_{3'}} = \frac{825{,}865}{(1189 - 1150)(25{,}037)} = 84.6\%$$

$$\varepsilon = \frac{E}{E + \frac{V_3^2}{2}} = \frac{825{,}865}{825{,}865 + \frac{390^2}{2}} = 91.5\%$$

Comment: The utilization factor is not the maximum, since the exit velocity triangle is not a right triangle.

Example 11.5

A steam turbine consists of two stages, each with an efficiency of 82%. The inlet conditions are 2000 psia and 900°F, and it expands to a pressure of 200 psia. If the inlet pressure of the intermediate stage is 500 psia, calculate the reheat factor.

Solution: From the Mollier diagram shown in the figure, the following can be calculated for the first stage:

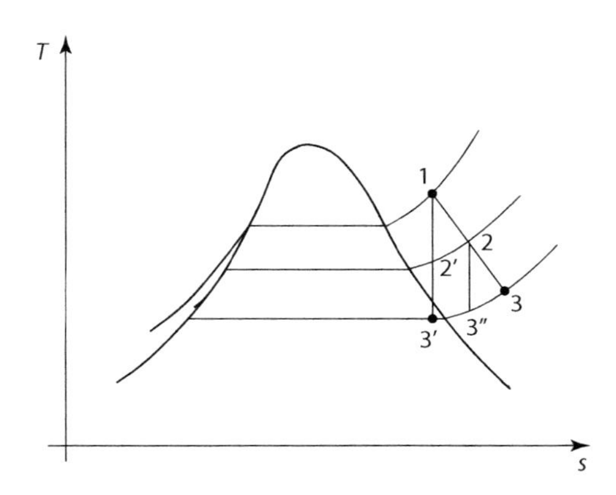

For $p_1 = 2000$ psia and $T_1 = 900°F$

$h_1 = 1408$ Btu/lbm; $s_1 = 1.513$ Btu/lbm · °R $= s_{2'} = s_{3'}$

For $p_{2'} = 500$ psia and $s_{2'} = 1.513$ Btu/lbm · °R

$h_{2'} = 1252$ Btu/lbm; $h_1 - h_{2'} = 156$ Btu/lbm

For $p_{3'} = 200$ psia and $s_{3'} = 1.513$ Btu/lbm · °R

$h_{3'} = 1171$ Btu/lbm; $h_1 - h_{3'} = 237$ Btu/lbm

From the stage efficiency, enthalpy at the second stage can be calculated. Hence,

$$\eta_{st} = \frac{h_1 - h_2}{h_1 - h_{2'}} = 0.82 \Rightarrow h_1 - h_2 = (0.82)(h_1 - h_{2'}) = 127.9 \text{ Btu/lbm}$$

Hence, $h_2 = h_1 - 127.9 = 1280$ Btu/lbm

Since process 2′-3″ is isentropic, the enthalpies can be calculated:

For $p_2 = 500$ psia and $h_2 = 1280$ Btu/lbm

$s_2 = 1.541$ Btu/lbm · °R $= s_{3''}$

For $s_{3''} = 1.541$ Btu/lbm-R and $p_{3''} = 200$ psia, $h_{3''} = 1194$ Btu/lbm

$h_2 - h_{3''} = 86$ Btu/lbm

Again, from the definition of stage efficiency,

$$\eta_{st} = \frac{h_2 - h_3}{h_2 - h_{3''}} = 0.82 \Rightarrow h_2 - h_3 = (0.82)(h_2 - h_{3''}) = 70.5 \text{ Btu/lbm}$$

Hence, $h_3 = h_2 - 70.5 = 1209.5$ Btu/lbm

The reheat factor can now be calculated:

$$R_F = \frac{(h_1 - h_{2'}) + (h_2 - h_{3''})}{(h_1 - h_{3'})} = \frac{156 + 86.08}{237} = 1.021$$

$$\eta_t = \frac{(h_1 - h_3)}{(h_1 - h_{3'})} = \frac{198.5}{237} = 0.8376$$

As a check, the relationship between reheat factor, turbine efficiency, and stage efficiency can be verified:

$$\eta_t = R_F \eta_{st} = (1.021)(0.82) = 0.8372$$

The quality at 3′ is 0.967, which is reasonably high. However, the state of the fluid at 3 is superheated vapor.

Comment: As expected, the turbine efficiency is larger than stage efficiency. This is also true for gas turbines, as was shown in Chapter 8.

Example 11.6

A stage of a steam turbine is described by the following data:

Inlet static conditions at the stator	400°C, 1200 kPa
Inlet velocity at the stator	300 m/s
Blade speed	200 m/s
Stator inlet angle	–20°
Pressure at rotor inlet	800 kPa
Stator efficiency	85%
Rotor efficiency	88%
Mass flow rate	3 kg/s

Show the process on a Mollier diagram and calculate the degree of reaction, utilization factor, power produced, and the inlet and exit temperatures of the rotor. Assume axial velocity to be a constant. Assume that the stage is a repeating stage.

Solution: From the Mollier diagram shown in the figure, the following can be calculated for the first stage:

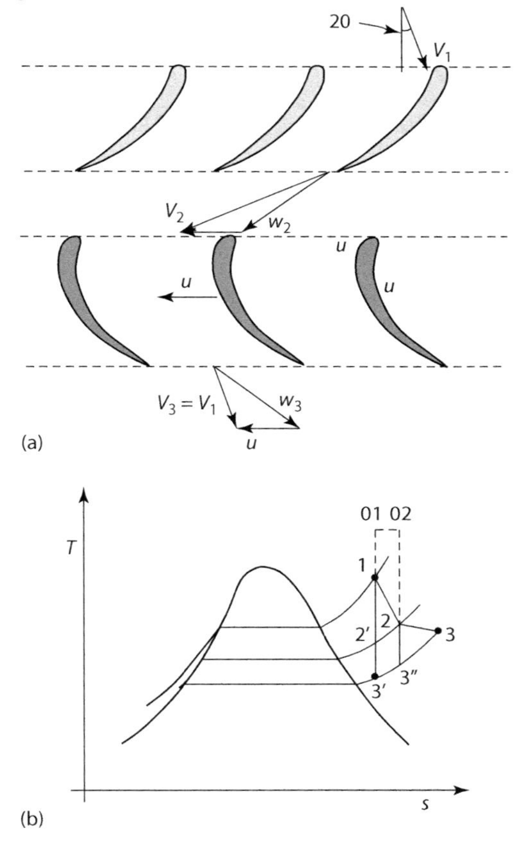

From the velocity at the inlet,

$$V_{ax} = V_1 \cos 20 = 300 \cos 20 = 281.9\,\text{m/s}$$

From the steam tables,

$$p_1 = 1200\,\text{kPa} \quad \text{and} \quad T_1 = 400°\text{C}$$

$$h_1 = 3260.6\,\text{kJ/kg}; \; s_1 = 7.377\,\text{kJ/kg-K} = s_{2'} = s_{3'}$$

$$h_{01} = h_1 + \frac{V_1^2}{2} = 3260.6 + \frac{300^2}{2} = 3305.6\,\text{kJ/kg} = h_{02}$$

The process 1-2′ is isentropic, and pressure at 2′ is 800 kPa. Hence, from the steam tables,

$$h_{2'} = 3142\,\text{kJ/kg}; \; T_{2'} = 340.8\,\text{K}$$

$$h_1 - h_{2'} = 118.6\,\text{kJ/kg}$$

From the definition of stator efficiency,

$$\eta_{stator} = 0.85 = \frac{h_1 - h_2}{h_1 - h_{2'}} = \frac{h_1 - h_2}{118.6} \Rightarrow h_1 - h_2 = (0.85)(118.6) = 100.81\,\text{kJ/kg}$$

$$h_2 = 3260.6 - 100.8 = 3159.8\,\text{kJ/kg}; \; T_{2'} = 340.8\,\text{K}$$

Since the stagnation enthalpies at 1 and 2 are the same, the exit velocity from the stator can be calculated:

$$\frac{V_2^2}{2} = h_{02} - h_2 \Rightarrow V_2 = \sqrt{2(h_{02} - h_2)} = \sqrt{2(1000)(3305.6 - 3159.8)} = 540\,\text{m/s}$$

Since axial velocity is a constant, the inlet velocity triangle can be completed:

$$V_{u2} = \sqrt{(V_2^2 - V_{ax}^2)} = 460.6\,\text{m/s};$$

$$w_{u2} = V_{u2} - u = 460.6 - 200 = 260.6\,\text{m/s}$$

$$w_2 = \sqrt{(w_{u2}^2 + V_{ax}^2)} = 383.9\,\text{m/s}$$

Since the stage is a repeating stage, the exit velocity triangle can be completed:

$$w_{u3} = V_{u3} + u = 102.6.6 + 200 = 302.6\,\text{m/s}$$

$$w_3 = \sqrt{(w_{u3}^2 + V_{ax}^2)} = 413.6\,\text{m/s}$$

$$V_3 = V_1 = 300\,\text{m/s}$$

Since the velocities are known, energy transferred, degree of reaction, and utilization factor can be calculated. Thus,

$$E = u(V_{u2} - V_{u3}) = 200(460.6 + 102.6) = 112{,}640\,\text{J/kg}$$

$$P = \dot{m}.E = \frac{3(112{,}640)}{1000} = 337.9\,\text{kW}$$

$$R = \frac{w_3^2 - w_2^2}{2E} = \frac{413.6^2 - 383.9^2}{2(112{,}640)} = 0.105$$

$$\varepsilon = \frac{E}{E + \frac{V_3^2}{2}} = \frac{112{,}640}{112{,}640 + \frac{300^2}{2}} = 0.714$$

$$E = h_{02} - h_{03} \quad \Rightarrow h_{03} = h_{02} - E = 3305.6 - 112{,}640 / 1000 = 3192.96\,\text{kJ/kg}$$

$$h_{03} = h_3 + \frac{V_3^2}{2} \quad \Rightarrow h_3 = h_{03} - \frac{V_3^2}{2} = 3192.96 - \frac{300^2}{2(1000)} = 3147.96\,\text{kJ/kg}$$

From the rotor efficiency, the enthalpy at 3″ can be calculated. Thus,

$$\eta_{rotor} = 0.88 = \frac{h_2 - h_3}{h_2 - h_{3''}} \Rightarrow h_2 - h_{3''} = \frac{3159.8 - 3147.9}{0.88} = 13.44\,\text{kJ/kg}$$

$$h_{3''} = 3159.8 - 13.44 = 3146.36\,\text{kJ/kg};$$

From the Mollier diagram, the following can be noted:

$$p_2 = p_{2'}; p_3 = p_{3'} = p_{3''}$$

The processes 1-2′-3′ and 2-3″ are isentropic. Hence, from the steam tables,

$$p_1 = 1200\,\text{kPa},\ T_1 = 400\,°\text{C} \Rightarrow h_1 = 3260.6\,\text{kJ/kg};\ s_1 = 7.377\,\text{kJ/kg}\cdot\text{K}$$

$$p_{2'} = 800\,\text{kPa},\ s_{2'} = 7.377\,\text{kJ/kg-K} \Rightarrow h_{2'} = 3142\,\text{kJ/kg},\ T_{2'} = 340.8\ \text{K}$$

$$p_2 = 800\,\text{kPa},\ h_2 = 3159.79\,\text{kJ/kg} \Rightarrow T_2 = 349.3\ \text{K},\ s_2 = 7.406\,\text{kJ/kg}\cdot\text{K}$$

$$h_{3''} = 3146.3\,\text{kJ/kg-K},\ s_{3''} = 7.406\,\text{kJ/kg-K} \Rightarrow p_{3''} = 761.4\,\text{kPa},\ T_{3''} = 342.3°\text{C}$$

$$h_3 = 3147.96\,\text{kJ/kg-K},\ p_3 = 761.4\,\text{kPa},\ \Rightarrow T_3 = 343.3\text{K},\ s_3 = 7.409\,\text{kJ/kg}\cdot\text{K}$$

$$p_{3'} = 761.4\,\text{kPa},\ s_{3'} = 7.377\,\text{kJ/kg-K} \Rightarrow h_{3'} = 3128.3\,\text{kJ/kg},\ T_{3'} = 333.9\text{K}$$

The overall efficiency of the stage can now be calculated as

$$\eta = \frac{h_1 - h_3}{h_1 - h_{3'}} = \frac{3260.6 - 3147.9}{3260.6 - 3128.3} = 85.1\%$$

Comment: Some of the other definitions for utilization factor and degree of reaction can be used and the same answers would be obtained.

11.5 Design Aspects of Steam Turbines

The design of steam turbines proceeds along similar lines to gas turbines, with the major difference being that the Mollier diagram needs to be used since steam is not an ideal gas. The typical inputs would be the inlet conditions to the first stage of the turbine (or outlet conditions from the boiler), exit pressure of the turbine, rotational speed, mass flow rate of steam, power produced, or some combination of these variables. Some of these variables are related and may be redundant. For example, if the power is specified along with the inlet and exit conditions, then mass flow rate would be redundant. Some of the guidelines for design are given here. For simple impulse turbines, the speed ratio φ $(=u/V)$ is the in the range 0.45–0.48. For pressure-compounded impulse stages, it is taken to be 0.47. For velocity-compounded stages, it is 0.23 for two-stage turbines and 0.17 for three-stage turbines. It should be recalled that for Pelton wheels that are impulse turbines, the optimum value of φ was 0.5 and the practical range was similar to what is required in steam turbines. For reaction turbines, the speed ratio φ is in the range 0.7–1.3. The nozzle angle is between 65° and 75°. The stator and rotor velocity coefficients are usually quite high; in the range 0.93–0.99. The blade velocity is limited to 1250 ft./s to avoid excessive stresses. The blade angle at the exit is 15°–30° for high- and intermediate-pressure stages, and can be up to 50° for low-pressure stages. From metallurgical considerations, the inlet conditions to the turbine should be restricted to 620°C (1150°F). There is also the possibility of water molecules dissociating at high temperatures into hydrogen and oxygen ions. Although this process takes place at very high temperatures, it is found that even at lower temperatures such as 600°C–620°C, steam begins to dissociate within a period of a little over a year. Hence, the highest recommended temperature for inlet steam is around 620°C. Also, to prevent excessive erosion in the blades for low-pressure stages, the quality of steam should be higher than 0.92–0.94.

PROBLEMS

All pressures given are absolute unless specified otherwise.

11.1 By formally drawing the velocity triangles for each stage obtain the analogous expressions for Equations 11.4 and 11.6 for three impulse stages, that is, show that

$$\frac{u}{V_1} = \frac{\sin\alpha_1}{6} \quad \text{and} \quad E_{max} = 18u^2$$

11.2 Generalize the results for Problem 11.1 for n stages, that is, prove Equations 11.7 and 11.8.

11.3 Verify that in Example 11.1, the velocity ratio and total energy are not maximum.

11.4 A velocity-compounded two-stage Curtis turbine is rotating at 7000 rpm when the mass flow rate is 2.5 kg/s. The mean rotor diameter is 0.5 m and the nozzle angle at the inlet to the first stage is 72°. Under ideal conditions, that is, the exit velocity from second stage is purely axial, calculate the (a) power produced, (b) utilization factor, and (c) blade angles for the first and second stages.

Ans: a) 671 kW; b) 73.7%; c) 66.6° and 37.6°

11.5 Repeat Example 11.1 with the axial velocity being a constant through both stages. Also calculate the utilization factor.

11.6 The preliminary design of an axial flow steam turbine to produce 1500 kW is to be performed. At the inlet to the turbine, the static pressure and temperature are 6000 kPa and 500°C, respectively. The turbine isentropic efficiency is 88%, while the stage efficiency can be assumed to be 100%. The blade angle is 55° and the expected mass flow rate is 2 kg/s. The mean radius of the turbine is 0.5 m and it rotates at 3600 rpm. Estimate the required number of impulse stages. Assume maximum utilization for each stage. Also calculate the following: (a) exit pressure of the last stage, (b) the boiler exit temperature and pressure if the inlet steam to the turbine is supplied from a boiler.

Ans: 11 stages; a) 157 kPa; b) 7537 kPa and 540°C

11.7 A boiler supplies steam to an axial flow steam turbine of several impulse stages. The mass flow rate is 3 kg/s. Steam from the boiler has stagnation conditions of 7200 kPa and 580°C. It enters a nozzle, where it expands to a pressure of 6000 kPa and enters the first stage of the turbine. Each stage is an impulse stage and operates at 3600 rpm, and the mean radius of the rotor is 0.4 m. If the nozzle angle is 68°, estimate the number of stages required to produce 1200 kW. Also, calculate the utilization factor of each stage and the exit pressure from the last stage of the turbine.

Ans: 7 stages; 84.8%; 139.9 kPa

11.8 Consider a 50% reaction axial flow steam turbine in which the inlet conditions are 3000 kPa and 400°C. The steam exits at 200 kPa. The axial velocity can be taken to be a constant and the blade speed can be taken to be 160 m/s. If the blade angle and nozzle angle at the inlet are 55° and 72°, respectively (measured with respect to the axial direction) calculate the number of stages when the stage efficiency is (a) 80%, (b) 90%, and (c) 100%.

Ans: a) 11; b) 9; c) 8

11.9 A reaction turbine is used to produce power when rotating at 5800 rpm. Steam is supplied at the stage inlet at 30 psia and 270°F, and exits at 18 psia. The mean diameter of the wheel is 2.25 ft. and the height of the blades is 0.2 ft. The speed ratio u/V_1 is 0.68. If the degree of reaction is 0.5, calculate the power developed, efficiency, and utilization factor.

Ans: 1684 hp; 83.7%

11.10 The inlet conditions for a steam turbine are 600 kPa and 200°C. It expands in three stages, and the pressures are 400, 200, and 100 kPa (which is the exit pressure). Each stage has an efficiency of 86%. Calculate the reheat factor and the overall turbine efficiency.

Ans: 1.01; 86.7%

11.11 Utilization factor can be written in terms of the inlet kinetic energy as given by Equation 11.3a. Derive it by writing the energy for each stage in terms of the velocities.

11.12 Calculate the blade and fluid angles for the stator and rotor in Example 11.6.

11.13 Consider a repeating steam turbine stage with the following data:

Pressure at rotor inlet	115 psi
Stator efficiency	86%
Rotor efficiency	87%
Mass flow rate	7 lbm/s
Inlet static conditions at the stator	750°F, 175 psi
Inlet velocity at the stator	980 ft./s
Blade speed	650 ft./s
Stator inlet angle	−20°

Assuming the axial velocity to be a constant, calculate the degree of reaction, utilization factor, power produced, stator and rotor angles, and the inlet and exit temperatures of the rotor. Assume axial velocity to be a constant. Show the process on a Mollier diagram

11.14 If the rotor efficiency is 86%, calculate the stator exit (rotor inlet) enthalpy, pressure, and temperature in Example 11.4.

12

Wind Turbines

Windmills are called *extended gas turbines*. However, unlike gas turbines, they operate at considerably lower speeds. Their design features and operating principles are considerably different to those for gas turbines. Since the speeds are small, incompressible flow principles are used. Thus, temperature is not a variable and the analysis is considerably simpler.

Upon completion of this chapter, the student will be able to

- Understand propeller theory
- Perform analysis of simple wind turbines

12.1 Introduction

Wind is a naturally available source of energy that is clean, renewable, quite abundant, and accessible at almost no cost. It can be argued that wind power is a form of solar energy, since portions of the Earth are heated by the Sun, and the convective currents in the atmosphere produce wind of reasonable speeds. Large wind forces in nature and their devastating effects on property have led to efforts by humans to harness this energy since early times. Wind power has been used in sails for propulsion, and modern windmills have been used to grind grain and pump water. Such uses can be traced back almost 2000 years to the windmills used in the Middle East. With the development of electric power, windmills are now used mainly for the production of electricity through generators which make their use more versatile. Today, wind-powered generators are available in various sizes ranging from small units for battery charging at isolated residences to gigawatt-sized offshore wind farms providing power to national electric grids.

In the United States, wind power is abundant. It is estimated that with today's technology there is enough wind power to meet all the nation's electricity needs. However, according to the American Wind Energy Association, the total installed wind capacity at the end of 2015 in the United States was only 73,992 MW. Although there has been a rapid increase in wind capacity from less than 100 MW in 1999 to almost 74,000 MW in 2015, electricity generated from wind still represents only a small fraction (about 1%) of the total electrical power generated from all sources. The primary reason is that electricity from wind is less competitive than electricity from coal and gas. However, with increasing emphasis on clean energy and a continual decrease in the cost of wind power, it is expected that use of wind power will increase in the future. Despite representing a small percentage of the total installed capacity, the United States produces 26.2% of the world's total wind-generated electric power. A comparison of the percentages is shown in Figure 12.1.

To get a true picture of the use of wind power as compared to other sources of power, the term *wind energy penetration* is often used. Wind energy penetration is defined as the fraction

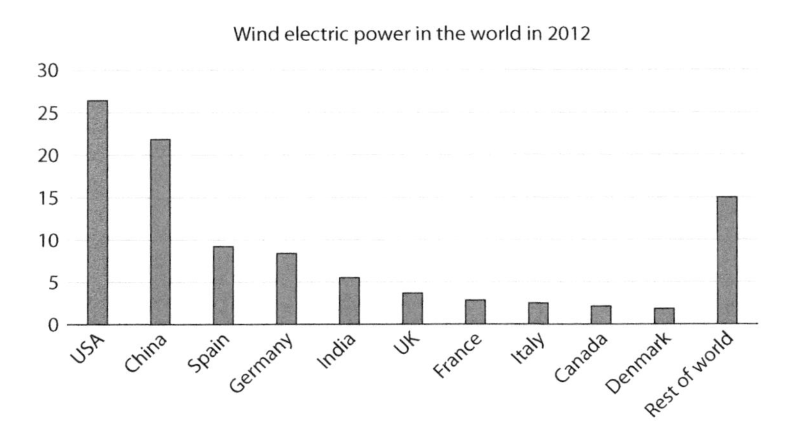

FIGURE 12.1
Percentage distribution of wind electric power in the world in 2012.

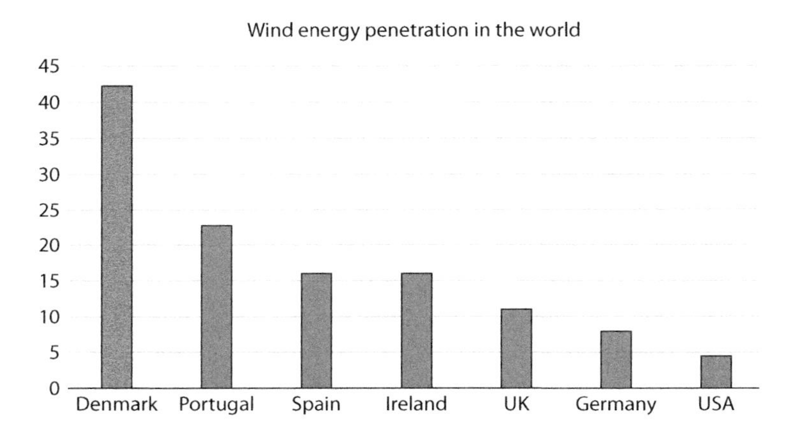

FIGURE 12.2
Percentage of wind electric energy compared with total electricity produced from other sources.

of the energy produced by wind compared with the total generation. A comparison of wind energy penetration across the world is shown in Figure 12.2. Compared with the United States, Europe has a significantly higher percentage of power produced through wind.

12.2 Classification of Wind Turbines

There are several ways to classify wind turbines. The most common classification is based on the axis of rotation: horizontal axis machines and vertical axis machines.

12.2.1 Horizontal Axis Wind Turbines (HAWT)

These wind turbines have their axis of rotation horizontal and almost parallel to the wind stream. They have a relatively high power coefficient, and most commercial wind turbines fall under this category. Because of their orientation, the generator and gear box have to

FIGURE 12.3
Horizontal axis wind turbine with three blades. (From US Department of Agriculture [USDA], https://www.usda.gov/energy/maps/resources/brochure/$file/renewable_energy_brochure.pdf.)

be placed above the tower which makes their design complex and expensive. Depending on the number of blades, these can be further classified as single-bladed, two-bladed, and three-/multi-bladed turbines. As the number of blades increases, so too does the material cost and drag. Single-bladed turbines are not popular due to the problems associated with rotor balance. Also, along with two-bladed rotors, they face the problem of aesthetic appeal. Thus, the most common HAWT used for power generation have three-bladed rotors. However, HAWT with more blades (six, eight, ten, or higher) are also available. Although aerodynamic losses increase with increased number of blades, some applications such as water pumping need higher starting torque. For such applications, HAWT with more than three blades are needed. A picture of a three-bladed rotor is shown in Figure 12.3.

12.2.2 Vertical Axis Wind Turbines (VAWT)

The axis of rotation for these turbines is vertical and almost perpendicular to the wind direction. These turbines can receive wind from any direction, and as such the generator and gear box can be housed at the ground level. This makes the design as well as their maintenance simple and economical. Some types of VAWT are described next.

12.2.2.1 Darrieus Rotor

In 1931, Georges Jeans Darrieus, a French aeronautical engineer, patented a turbine consisting of a number of curved airfoil blades mounted on a vertical rotating shaft. The original design consisted of blades that were shaped like egg beaters or turning rope so that they were under pure tension while spinning, thereby minimizing bending stresses. There are several variations to the original Darrieus design, many including straight vertical blades. Such designs are collectively called *giromills*. Darrieus rotors operate at high tip speed ratios and are most suitable for electric power generation. A picture of a Darrieus rotor is shown in Figure 12.4. The principle of operation of such a turbine is shown in Figure 12.5. When the rotor is spinning, the airfoils move through the air in a circular path. When the relative motion due to this is added vectorially to the wind speed, the resultant

FIGURE 12.4
Darrieus rotor. (From Illustrated History of Wind Power Development, USDA Agricultural Research Station in Amarillo, Texas, http://telosnet.com/wind/govprog.html.)

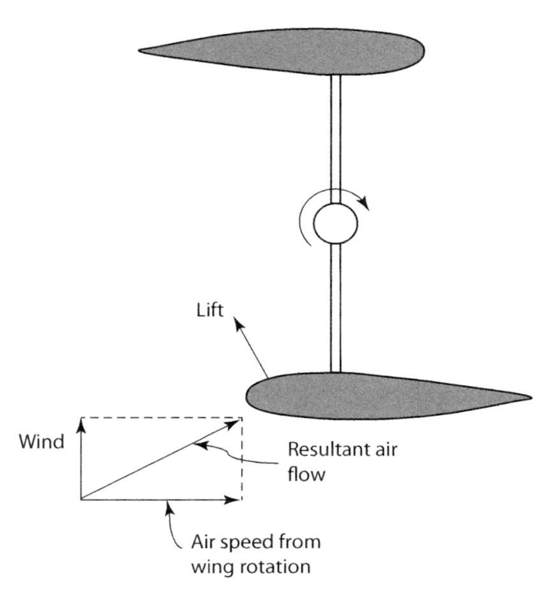

FIGURE 12.5
Principle of operation of the Darrieus rotor.

air speed creates a small positive angle of attack with the blade that gives rise to lift force, causing the rotor to spin.

12.2.2.2 Savonius Rotor

The Savonius wind turbine, invented by the Finnish engineer Sigurd Johannes Savonius in 1922, is a vertical axis turbine that consists of two half-cylindrical drums arranged in the shape of an "S" (see Figures 12.6 and 12.7), such that the convex and concave sides of the half cylinders are facing the wind at one time. Since the drag on the concave side is higher than on the convex side, the rotor will experience a net torque and rotate. The

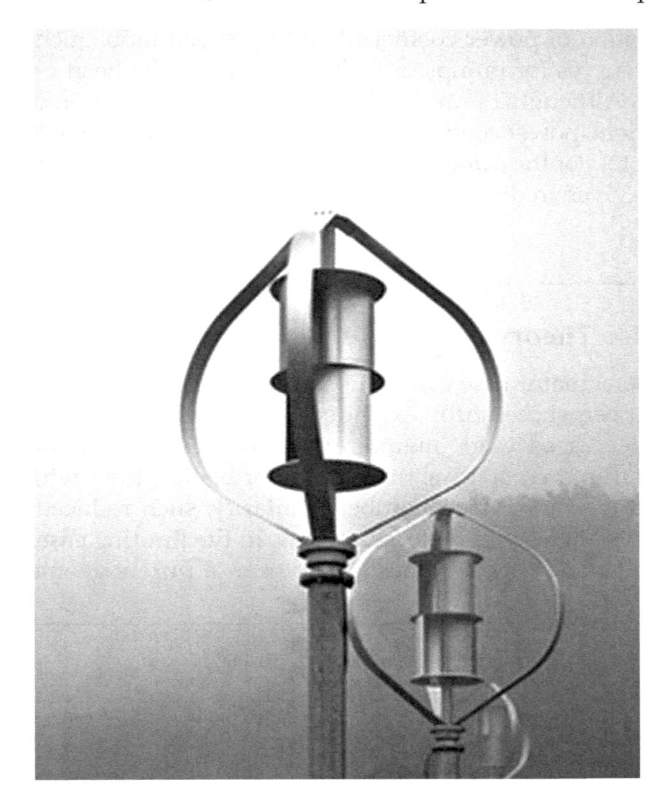

FIGURE 12.6
Savonius wind turbine enclosed in Darrieus rotor. (From Fred Hsu, Wikimedia Commons. File: Taiwan 2009 JinGuaShi Historic Gold Mine Combined Darrieus Savonius Wind Turbines FRD 8638.jpg)

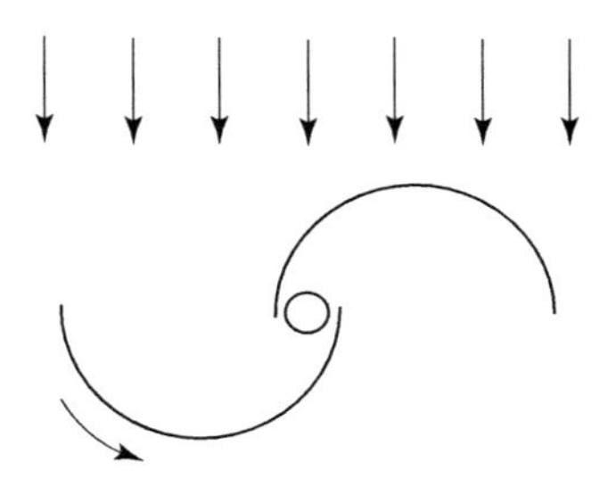

FIGURE 12.7
Principle of Savonius rotor.

power coefficient for such rotors is relatively small and they have high starting torque. They also work with lower tip speed to blade speed ratios. In spite of these less attractive factors, Savonius rotors are quite popular due to their simplicity of construction.

12.3 Performance Characteristics of Wind Turbines

The efficiency with which a wind turbine rotor extracts power from wind depends on the dynamics between the rotor and the fluid. The performance of the wind turbine is expressed in terms of the variation of power coefficient and tip speed ratio. Such curves are similar to the performance curves for pumps or turbines, wherein the head or power are plotted against the flow rate. Although it is desirable to express the variables in dimensionless form, namely, head coefficient, power coefficient, and flow coefficient, it is customary in the pump and power industry to plot the dimensional variables. However, for wind turbines, the performance curves are given in dimensionless terms. These are shown in Figure 12.8.

12.4 Actuator Disc Theory

The concept of the actuator disc can be used to explain the three-dimensional flow in propellers and has been successfully extended to turbomachinery. If the axial width of each blade row is decreased while maintaining consistency of the other variables such as the blade angles, number, spacing, and shape of blades, then within the limitations of the Reynolds number and Mach number similarity, such reduced-width blade rows affect the flow in the same way as the original row. In the limiting case, as the axial width approaches zero, the blade row becomes a disc from a purely geometrical perspective.

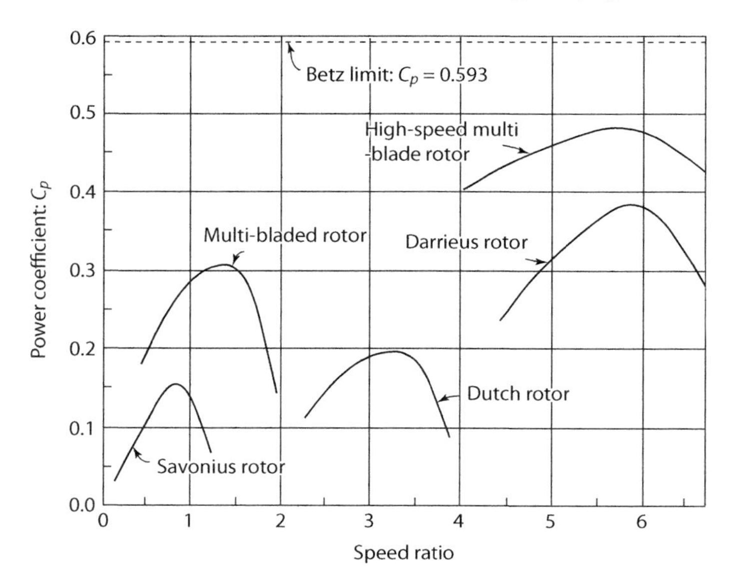

FIGURE 12.8
Variation of power coefficient with speed ratio for various wind turbine rotors.

Such a disc is called the *actuator disc*. It is assumed that tangential velocity undergoes a discontinuity while the axial and radial velocities are continuous across the disc. Shown in Figure 12.9 is the flow through a windmill that is replaced by an actuator disc, and the flow is contained in the boundary stream tube. The following analysis uses simplifying assumptions, but gives useful approximate results.

The flow is considered to be steady and incompressible, with no flow rotation produced by the disc. It is also assumed that the flow is uniform upstream and downstream of the disc. The wind velocities upstream and downstream of the disc are denoted by V_u and V_d, respectively. As the wind stream approaches the disc, the axial velocity continuously decreases with a corresponding increase in pressure, a consequence of the Bernoulli principle. Letting V be the velocity at any section of the tube, conservation of mass and linear momentum can be written as (with the control volume being the entire stream tube)

$$\dot{m} = \rho A V = \rho A_u V_u = \rho A_d V_d \tag{12.1}$$

$$F = \dot{m}(V_d - V_u) = \rho A V (V_d - V_u) \tag{12.2}$$

The Bernoulli equation can be applied from the inlet to the actuator and from the actuator to the exit, to get

$$\begin{aligned} p_a + \frac{1}{2}\rho V_u^2 &= p_1 + \frac{1}{2}\rho V_1^2 = p_1 + \frac{1}{2}\rho V^2 \\ p_a + \frac{1}{2}\rho V_d^2 &= p_2 + \frac{1}{2}\rho V_2^2 = p_2 + \frac{1}{2}\rho V^2 \end{aligned} \tag{12.3}$$

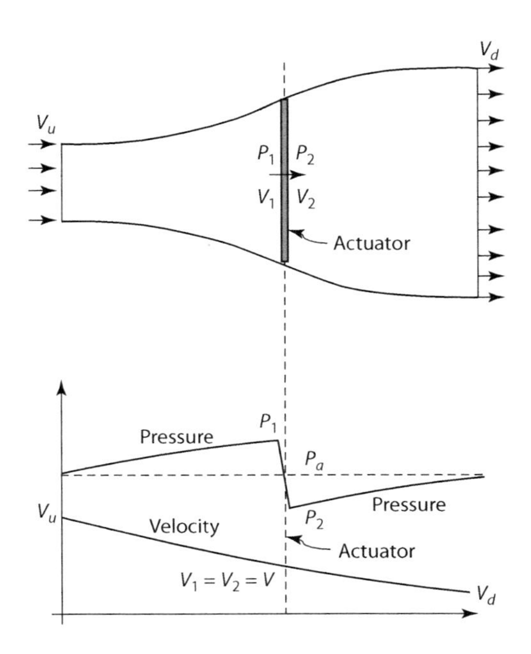

FIGURE 12.9
Flow through wind turbine actuator disc.

Subtracting,

$$p_1 - p_2 = \frac{1}{2}\rho\left(V_u^2 - V_d^2\right) \tag{12.4}$$

The force on the actuator due to pressure difference is

$$F = A\left(p_1 - p_2\right) = A\frac{1}{2}\rho\left(V_u^2 - V_d^2\right) \tag{12.5}$$

Combining Equations 12.2 and 12.5, and noticing the force on the fluid is the reverse of the force on the actuator, the equations can be simplified:

$$F = A\frac{1}{2}\rho\left(V_u^2 - V_d^2\right) = \rho AV\left(V_u - V_d\right) \Rightarrow V = \frac{1}{2}\left(V_u + V_d\right) \tag{12.6}$$

Thus, the speed at the actuator is the average of the upstream and downstream speeds. The power can now be calculated as

$$\begin{aligned} P = FV = A\frac{1}{2}\rho\left(V_u^2 - V_d^2\right)V = \frac{1}{4}\rho A\left(V_u^2 - V_d^2\right)\left(V_u + V_d\right) \\ = \frac{1}{4}\rho AV_u^3\left[1 - \left(\frac{V_d}{V_u}\right)^2\right]\left[1 + \left(\frac{V_d}{V_u}\right)\right] \end{aligned} \tag{12.7}$$

Introducing a new variable, the velocity ratio $z\,(=V_d/V_u)$, Equation 12.7 can be rewritten as

$$P = \frac{1}{4}\rho AV_u^3\left[1 - z^2\right]\left[1 + z\right] = \frac{1}{4}\rho AV_u^3\left[1 - z\right]\left[1 + z\right]^2 \tag{12.8}$$

For maximum power, it can be shown that (see Problem 12.1)

$$z = \frac{V_d}{V_u} = \frac{1}{3} \tag{12.9}$$

Combining Equations 12.6 and 12.9, we get

$$V = \frac{1}{2}\left(V_u + V_d\right) = \frac{1}{2}V_u\left(1 + \frac{V_d}{V_u}\right) = \frac{2}{3}V_u \tag{12.10}$$

That is, maximum power occurs when the speed at the actuator is two-thirds the upstream speed. The maximum power under these conditions will be

$$P_{max} = \frac{8}{27}\rho AV_u^3 = \frac{16}{27}\left(\frac{1}{2}\rho AV_u^3\right) \tag{12.11}$$

The pressure coefficient can be defined by considering unperturbed upstream speed V_u and an equivalent actuator area so that the available energy would be

$$P_{available} = \dot{m}\frac{1}{2}V_u^2 = \rho AV_u\frac{1}{2}V_u^2 = \frac{1}{2}\rho AV_u^3 \tag{12.12}$$

Now, the dimensionless variables, power coefficient, and thrust coefficient are introduced. The power coefficient can be defined as the ratio of power obtained at the rotor to the available power, that is,

$$C_P = \frac{P}{P_{available}} = \frac{\frac{1}{4}\rho A V_u^3 [1-z][1+z]^2}{\frac{1}{2}\rho A V_u^3} = \frac{1}{2}[1-z][1+z]^2 \tag{12.13}$$

Combining Equations 12.11 and 12.12, the maximum power coefficient can be written as

$$C_{P,\max} = \frac{P_{\max}}{P_{available}} = \frac{16}{27} = 0.593 \tag{12.14}$$

This is the theoretical maximum value in the absence of aerodynamic and mechanical losses. It is seldom achieved and the actual value is considerably lower.

An accurate estimate of the axial thrust is quite important in the design and installation of HAWT in view of their massive size. From Equation 12.6, the expression for axial thrust on the disc is

$$\begin{aligned} F &= \frac{1}{2}A\rho\left(V_u^2 - V_d^2\right) = \frac{1}{2}A\rho V_u^2\left[1 - \frac{V_d^2}{V_u^2}\right] \\ &= \frac{1}{2}A\rho V_u^2(1-z)(1+z) \end{aligned} \tag{12.15}$$

with z being defined as in Equation 12.8. Similarly, thrust ratio is defined as the ratio of actual thrust at the rotor to the ideal thrust that would be available if the inlet area were equal to the rotor area. Thus,

$$C_T = \frac{F}{F_{available}} = \frac{\frac{1}{2}\rho A V_u^2 [1-z][1+z]}{\frac{1}{2}\rho A V_u^2} = [1-z][1+z] \tag{12.16}$$

For the condition of maximum power, that is, $z=1/3$, the axial thrust and thrust coefficient would be

$$F = \frac{4}{9}A\rho V_u^3; \quad C_T = \frac{8}{9} \tag{12.16}$$

Although this is the value for conditions of maximum power, during gusts, the force would be considerably higher, and the actual design would be based on gust conditions.

A new variable called the *axial induction factor* q can be defined as

$$q = \frac{1}{2}\left(1 - \frac{V_d}{V_u}\right) = \frac{1-z}{2} \tag{12.17}$$

An alternative set of formulas corresponding the power and thrust coefficients can be written in terms of the axial induction factor. This is left as an exercise (see Problem 12.2).

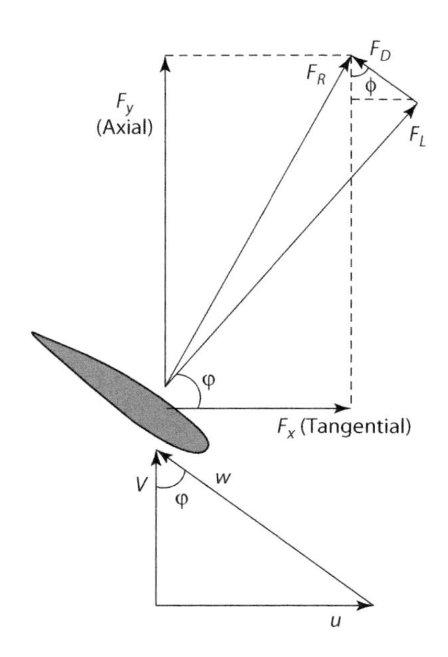

FIGURE 12.10
Geometry and forces on a blade element of HAWT.

A simple expression for the efficiency of HAWT can be obtained by considering the lift and drag on the section of the blade. Since the blades of HAWT are quite long, it is convenient to consider a small section of the wind and the forces acting on it, as shown in Figure 12.10.

The approach wind V makes an angle ϕ with the blade. The relative velocity w is the approach velocity that needs to be used to calculate lift and drag forces, F_L and F_D respectively. The resultant force of F_L and F_D is F_R and it is split along the axial and tangential directions as F_x and F_y respectively. From the velocity triangles, the blade to wind ratio can be written as

$$\lambda = \frac{u}{V} = \tan\varphi \tag{12.18}$$

The axial and tangential forces can be written in terms of the lift and drag as

$$\begin{aligned} F_x &= F_L \sin\varphi + F_D \cos\varphi \\ F_y &= F_L \cos\varphi - F_D \sin\varphi \end{aligned} \tag{12.19}$$

Using drag and lift coefficients, Equation 12.19 can be rewritten as

$$\begin{aligned} F_x &= \frac{1}{2}\rho A w^2 \left(C_L \sin\varphi + C_D \cos\varphi\right) \\ F_y &= \frac{1}{2}\rho A w^2 \left(C_L \cos\varphi - C_D \sin\varphi\right) \end{aligned} \tag{12.20}$$

The efficiency of the blade element can be defined as the ratio of work done on the blade to the energy input rate. Thus,

$$\eta = \frac{F_y u}{F_x V} = \frac{F_y}{F_x}\tan\varphi = \frac{(C_L\cos\varphi - C_D\sin\varphi)}{(C_L\sin\varphi + C_D\cos\varphi)}\tan\varphi$$

$$= \frac{\left(1-\frac{C_D}{C_L}\tan\varphi\right)}{\left(\tan\varphi+\frac{C_D}{C_L}\right)}\tan\varphi = \frac{\left(1-\frac{C_D}{C_L}\tan\varphi\right)}{\left(1+\frac{C_D}{C_L}\cot\varphi\right)} \tag{12.21}$$

Combining Equations 12.18 and 12.21 gives

$$\eta = \frac{\left(1-\frac{C_D}{C_L}\frac{u}{V}\right)}{\left(1+\frac{C_D}{C_L}\frac{V}{u}\right)} = \frac{\left(1-\frac{C_D}{C_L}\lambda\right)}{\left(1+\frac{C_D}{C_L}\frac{1}{\lambda}\right)} \tag{12.22}$$

For a given value of the ratio C_D/C_L, a plot of efficiency versus velocity ratio λ can be plotted. A typical plot is shown in Figure 12.11.

For low-drag and high-lift sections, the efficiency approaches 100%. Also, the efficiency drops off rapidly for lower and higher values of λ. Although there is an optimal value of λ for greatest efficiency, it is not possible to achieve it consistently, since the windmills operate over a wide range of wind speeds. Another feature of propeller turbines is the large swept area, as result of which they capture more wind energy and thus produce more power. This feature, in addition to their operation at relatively higher speeds, results in such turbines having relatively higher power coefficients and efficiency than VAWT which are discussed next.

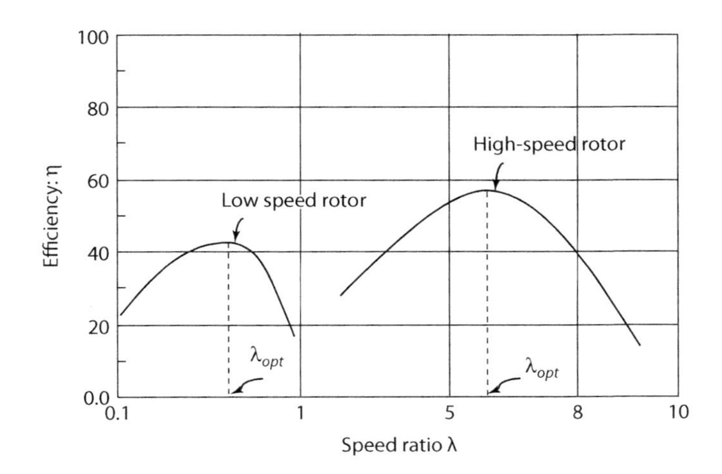

FIGURE 12.11
Variation of wind turbine efficiency with velocity ratio.

12.5 Performance of Vertical Axis Wind Turbines

As described earlier, VAWT are wind turbines that generate power from rotors that are vertical. In such machines, since the torque-generating surfaces move in the wind direction, the blade speeds are always less than wind speed. Consequently, the speeds of VAWT are lower than those of the horizontal axis-type turbines. Also, during part of the revolution, the motion of the blades is against the wind which results in lower power output. This can be corrected partially by the use of a blanking arc. All these features make VAWT suitable for lower-power applications when compared with the horizontal type of turbines.

As mentioned earlier, there are several types of vertical axis turbines, and each operates on slightly different principles. Hence, it is not possible to show the analysis of each of them. Only the panemone rotor will be discussed here. Details of others can be found in Wortman (1983). A typical rotor of the panemone type is shown in Figure 12.12.

The relative velocity of the wind is given by $w = V - u$, and the force on the blades is given by

$$F_y = C_F \frac{1}{2}\rho A w^2 = C_F \frac{1}{2}\rho A (V-u)^2 \tag{12.23}$$

The power developed is then

$$\begin{aligned} P = F_y u &= C_F \frac{1}{2}\rho A (V-u)^2 u \\ &= C_F \frac{1}{2}\rho A V^3 \left(1 - \frac{u}{V}\right)^2 \frac{u}{V} = C_F \frac{1}{2}\rho A V^3 (1-\lambda)^2 \lambda \end{aligned} \tag{12.24}$$

By differentiation, the optimum value with respect to λ can be found as $\lambda = 1/3$. The maximum power can be calculated from Equation 12.24 as

$$P_{\max} = \frac{4}{27} C_F \frac{1}{2}\rho A V^3 \tag{12.25}$$

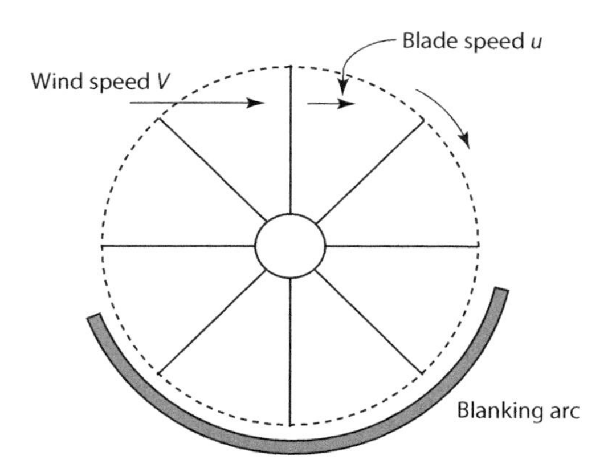

FIGURE 12.12
Wind forces on the blades of a vertical axis turbine (panemone).

The energy available to the rotor due to the wind would be

$$
\begin{aligned}
P_{avail} &= F_y V = C_F \frac{1}{2}\rho A (V-u)^2 V \\
&= C_F \frac{1}{2}\rho A V^3 \left(1-\frac{u}{V}\right)^2 = C_F \frac{1}{2}\rho A V^3 (1-\lambda)^2
\end{aligned}
\tag{12.26}
$$

At the optimum value of the blade speed ratio, the available power is obtained by setting $\lambda = 1/3$ in Equation 12.26. Thus,

$$
P_{avail} = \frac{4}{9} C_F \frac{1}{2}\rho A V^3 \tag{12.27}
$$

The maximum power coefficient is then

$$
C_{P,\max} = \frac{1}{3} = 0.333 \tag{12.28}
$$

which is considerably lower than the value of 0.593 that was attainable for horizontal axis machines. It should be pointed out that the analysis presented here is very elementary, but gives an indication of the lower power coefficients that are obtainable for vertical axis machines.

12.6 Wind Power Advantages and Disadvantages

During the early days, wind power was used mostly for agricultural applications such as corn grinding, water pumping during irrigation, and the driving of sawmills. Currently, wind energy is used for several other applications where electric power is not available. One of the main drawbacks of wind power is the variable velocities due to fluctuating wind speeds. Thus, windmills are used to drive DC generators that generate electric power at varying voltages which in turn can be used for heating and battery charging, among other applications.

As with any technology, wind turbines have their advantages and disadvantages, which are described below:

Advantages

1. Wind power is environmentally friendly and has no fuel cost.
2. Since there is no fuel to be transported, stored, or handled, wind power plants are very simple to operate.
3. Being easy to operate, and with the abundance of wind energy, wind turbines are ideal for isolated places where other sources are unavailable.
4. Unlike hydraulic turbines, for which the materials must be cavitation and corrosion resistant, wind turbines can be manufactured from a wide variety of materials that may be easily available.

Disadvantages

1. Since wind energy is intermittent, wind turbines cannot be used in places that need a continuous supply of power.
2. Wind power plants tend to have low load factors.
3. Due to their low power density, large numbers of turbines are required for bulk generation. This in turn requires large areas of land.
4. Capital costs for wind turbines are high.

PROBLEMS

All pressures given are absolute unless specified otherwise. Use air properties unless specified otherwise.

12.1 Complete the details of derivation of Equation 12.9 from Equation 12.8

12.2 Derive Equations 12.13 through 12.16 using the axial induction factor. Also, obtain a plot similar to Figure 12.2 which shows the variation of C_P and C_T in terms of q.

12.3 The power P from a wind turbine depends on density ρ, viscosity μ, speed of rotation N, size of rotor D, and wind speed V. Using dimensional analysis, show that the relevant dimensionless variables are power coefficient, tip speed to wind speed, and Reynolds number given by

$$C_P = \frac{P}{\rho N^3 D^5};\ \lambda = \frac{ND}{V};\ \mathrm{Re} = \frac{\rho N D^2}{\mu}$$

Similarly, obtain an expression for the thrust coefficient as

$$C_T = \frac{T}{\rho N^2 D^4}$$

References

Ainley, D. G. and Mathieson, G. C. R., (1957) A method of performance estimation for axial-flow turbines, Aeronautical Research Council, Reports and Memoranda, 2974.

Balje, O. E., (1960) A study of design criteria and matching of turbomachines, pt. B.—compressor and pump performance and matching of turbo components, ASME Paper No. 60–W A-231.

Balje, O. E., (1981) *Turbomachines: A Guide to Design, Selection, and Theory*, John Wiley, New York.

Bathie, W. W., (1984) *Fundamentals of Gas Turbines*, John Wiley, New York.

Biederman, B. P., Mulugeta, J., Zhang, L., and Brasz, J. J., (2004) Cycle analysis and turbo-compressor sizing with ketone C6F as working fluid for water-cooled chiller applications, *International Compressor Engineering Conference*, Paper No. 1626.

Brown, L. P. and Whippen, W. E., (1976) *Hydraulic Turbines, Parts I–III*, International Correspondence Schools, Scranton, PA 18515.

Craig, H. R. M. and Cox, H. J. A, (1970) Performance estimation of axial flow turbines, *Proceedings of the Institute of Mechanical Engineers*, 185: 407–424.

Csanady, G. T., (1964) *Theory of Turbomachines*, McGraw-Hill, New York.

Dixon, S. L. and Hall, C. A., (2010) *Fluid Mechanics and Thermodynamics of Turbomachinery*, 6th edition, Elsevier, Amsterdam.

Dougherty, R. L., Franzini, J. B., and Finnemore, E. J., (1985) *Fluid Mechanics with Engineering Applications*, 8th edition, McGraw-Hill, New York.

Dunham, J. and Came, P., (1970) Improvements to the Ainley–Mathieson method of turbine performance prediction, *Transactions of ASME*, Series A, 92.

Eck, B., (1973) *Fans: Design and Operation of Centrifugal, Axial-Flow, and Cross-Flow Fans*, Pergamon Press, London.

Elger, D. F., Williams, B. C., Crowe, C. C., and Roberson, J. A., (2012) *Engineering Fluid Mechanics*, 9th Edition, John Wiley, New York.

El-Sayed A. F., (2008) *Aircraft Propulsion, and Gas Turbine Engines*, CRC Press, Boca Raton.

Ferguson, T. B., (1963) *The Centrifugal Compressor Stage*, Butterworth, London.

Gartman, H., (1970) *De Laval Engineering Handbook*, McGraw-Hill, New York.

Glassman, A. J., (1976) *Computer Program for Design and Analysis of Radial Infolw Turbines*, TN-8164, NASA.

Hassan, A. Y., (1976) Tawi al-Din and Arabic Mechanical Engineering, *Institute for the History of Arabic Science*, Aleppo University, Syria.

Howell, A. R., (1945) Fluid dynamics of axial compressors, *Proceeding of Institution of Mechanical Engineers*, London, No. 153.

Hydraulic Institute, (1994) *Centrifugal Pumps*, Hydraulic Institute, Parsippany, New Jersey.

Jamieson, A. W. H., (1955) The Radial Turbine. Chapter 9, *Gas Turbine Principles and Practice*, Ed. Sir H. Roxbe-Cox, Newnes.

Kadambi, V. and Prasad, M., (1977) *An Introduction to Energy Conversion Vol III, Turbomachinery*, Wiley Eastern, New Delhi.

Lal, J., (1986) *Hydraulic Turbines*, 6th Edition, Metropolitan Book, New Delhi.

Logan, E. Jr., (1993) *Turbomachinery: Basic Theory and Applications*, 2nd Edition, Marcel Dekker, New York.

Lyman, F. A., (2004) A practical hero: or how an obscure New York mechanic got a steam-powered toy to drive sawmills, *Mechanical Engineering – CIME, ASME*.

Mattingly, J. D., Heiser, W. H., and Daley, D. H., (1987) *Air Craft Engine Design*, AIAA, New York.

Moody, L. F., (1926) The propeller type turbine, *Transactions of ASCE*, 89: 628.

Nechleba, M., (1957) *Hydraulic Turbines: Their Design and Equipment*, Artia, Prague.

Peng, W. H., (2008) *Fundamentals of Turbomachinery*, John Wiley, New York.

Rohlik, H. E., (1968) Analytical determination of radial inflow turbine design geometry for maximum efficiency, NASA TN D-4384.
Sarvanamuttoo, H. I. H., Rogers, G. F. C., Cohen, H., and Straznicky, P. V., (2009) *Gas Turbine Theory*, 6th Edition, Prentice Hall, New Jersey.
Shepherd, D. G., (1956) *Principles of Turbomachinery*, Macmillan, New York.
Smith, L. H. Jr., (1970) Casing boundary layers in multistage compressors, *Proceedings of Symposium on Flow Research in Blading*, Editor: Dzung, L. S., Elsevier, Amsterdam.
Soderberg, C. R., (1949) Unpublished report, Gas Turbine Laboratory, MIT, Cambridge, MA.
Stanitz, J. D., (1952) Some theoretical aerodynamic investigations of impeller in radial and mixed flow centrifugal compressors, *Transactions of ASME*, 74(4).
Stodola, A., (1927) *Steam and Gas Turbines, Vol I and II*, McGraw-Hill, New York.
Streeter, V. L., and Wylie, E. B., (1985) *Fluid Mechanics*, 8th Edition, McGraw-Hill, New York.
Van Wylen, G. J., and Sonntag, G. J., (1985) *Fundamentals of Classical Thermodynamics*, 4th Edition, John Wiley, New York.
Vavra, M. H., (1960) *Aerothermodynamics and Flow in Turbomachines*, John Wiley & Sons, New York.
Walsh, P. P., and Fletcher, P., (1998) *Gas Turbine Performance*, Blackwell Science, Hoboken, NJ.
Walters, R. N. and Bates, C. G., (1976) *Selecting Hydraulic Reaction Turbines*, Engineering Monograph No. 20, Bureau of Reclamation, U.S. Department of the Interior, Washington, D.C.
Warnick, C. C., in collaboration with Mayo, H. A., Carson, J. L., and Sheldon, L. H., (1984) *Hydropower Engineering*, Prentice Hall, New Jersey.
Whitfield, A. and Baines, N. C., (1990) *Design of Radial Turbomachines*, Longman, Essex.
Wilson, D. G., (1984) *The Design of High Efficiency Turbomachinery and Gas Turbines*, MIT Press, Cambridge, MA.
Wislicenus, G. F., (1965) *Fluid Mechanics of Turbomachiery*, 2nd Edition, McGraw-Hill, New York.
Wortman, A., (1983) *Introduction to Wind Turbine Engineering*, Butterworth Publishers, Woburn, MA.
Yahya, S. M., (2011) *Turbines, Compressors, and Fans*, Tata McGraw-Hill, New Delhi.

Appendix I

TABLE AI.1

Physical Properties of Water at Standard Sea-Level Atmospheric Pressure

Temperature, T	Specific Weight, γ	Density, ρ	Viscosity, μ	Kinematic Viscosity, ν	Surface Tension, σ	Saturation Vapor Pressure, p_v	Satur'n Vapor Pressure Head, p_v/γ	Bulk Modulus of Elasticity, E_v
°F	lb/ft³	Slugs/ft³	10^{-6} lb·s/ft²	10^{-6} ft²/s	lb/ft	psia	ft abs	psi
32	62.42	1.940	37.46	19.31	0.005 18	0.09	0.20	293,000
40	62.43	1.940	32.29	16.64	0.005 14	0.12	0.28	294,000
50	62.41	1.940	27.35	14.10	0.005 09	0.18	0.41	305,000
60	62.37	1.938	23.59	12.17	0.005 04	0.26	0.59	311,000
70	62.30	1.936	20.50	10.59	0.004 98	0.36	0.84	320,000
80	62.22	1.934	17.99	9.30	0.004 92	0.51	1.17	322,000
90	62.11	1.931	15.95	8.26	0.004 86	0.70	1.61	323,000
100	62.00	1.927	14.24	7.39	0.004 80	0.95	2.19	327,000
110	61.86	1.923	12.84	6.67	0.004 73	1.27	2.95	331,000
120	61.71	1.918	11.68	6.09	0.004 67	1.69	3.91	333,000
130	61.55	1.913	10.69	5.58	0.004 60	2.22	5.13	334,000
140	61.38	1.908	9.81	5.14	0.004 54	2.89	6.67	330,000
150	61.20	1.902	9.05	4.76	0.004 47	3.72	8.58	328,000
160	61.00	1.896	8.38	4.42	0.004 41	4.74	10.95	326,000
170	60.80	1.890	7.80	4.13	0.004 34	5.99	13.83	322,000
180	60.58	1.883	7.26	3.85	0.004 27	7.51	17.33	318,000
190	60.36	1.876	6.78	3.62	0.004 20	9.34	21.55	313,000
200	60.12	1.868	6.37	3.41	0.004 13	11.52	26.59	308,000
212	59.83	1.860	5.93	3.19	0.004 04	14.70	33.90	300,000
°C	kN/m³	kg/m³	N·s/m²	10^{-6} m²/s	N/m	kN/m² abs	m abs	10^6 kN/m²
0	9.805	999.8	0.001 781	1.785	0.0756	0.61	0.06	2.02
5	9.807	1000.0	0.001 518	1.519	0.0749	0.87	0.09	2.06
10	9.804	999.7	0.001 307	1.306	0.0742	1.23	0.12	2.10
15	9.798	999.1	0.001 139	1.139	0.0735	1.70	0.17	2.14
20	9.789	998.2	0.001 002	1.003	0.0728	2.34	0.25	2.18
25	9.777	997.0	0.000 890	0.893	0.0720	3.17	0.33	2.22
30	9.764	995.7	0.000 798	0.800	0.0712	4.24	0.44	2.25
40	9.730	992.2	0.000 653	0.658	0.0696	7.38	0.76	2.28
50	9.689	988.0	0.000 547	0.553	0.0679	12.33	1.26	2.29
60	9.642	983.2	0000 466	0.474	0.0662	19.92	2.03	2.28
70	9.589	977.8	0.000 404	0.413	0.0644	31.16	3.20	2.25
80	9.530	971.8	0.000 354	0.364	0.0626	47.34	4.96	2.20
90	9.466	965.3	0.000 315	0.326	0.0608	70.10	7.18	2.14
100	9.399	958.4	0.000 282	0.294	0.0589	101.33	10.33	2.07

TABLE AI.2

The ICAO[a] Standard Atmosphere

Elevation above Sea Level	Temperature, T	Absolute Pressure, P	Specific Weight, γ	Density, ρ	Viscosity, μ	Kinematic Viscosity, ν	Speed of Sound, c	Gravitational Acceleration, g
ft	°F	psia	lb/ft³	Slug/rt³	10⁻⁶ lb s/ft²	10⁻³ ft²/s	ft/s	ft/s²
0	59.000	14.695 9	0.076 472	0.002 376 8	0.373 72	0.157 24	1116.45	32.1740
5,000	41.173	12.228 3	0.065 864	0.002 048 1	0.363 66	0.177 56	1097.08	32.158
10,000	23.355	10.108 3	0.056 424	0.001 755 5	0.353 43	0.201 33	1077.40	32.142
15,000	5.545	8.297 0	0.048 068	0.001 496 1	0.343 02	0.229 28	1057.35	32.129
20,000	−12.255	6.758 8	0.040 694	0.001 267 2	0.332 44	0.262 34	1036.94	32.113
25,000	−30.048	5.460 7	0.034 224	0.001 066 3	0.321 66	0.301 67	1016.11	32.097
30,000	−47.832	4.372 6	0.028 573	0.000 890 65	0.310 69	0.348 84	994.85	32.081
35,000	−65.607	3.467 6	0.023 672	0.000 738 19	0.299 52	0.405 75	973.13	32.068
40,000	−69.700	2.730 0	0.018 823	0.000 587 26	0.296 91	0.505 59	968.08	32.052
45,000	−69.700	2.148 9	0.014 809	0.000 462 27	0.296 91	0.642 30	968.08	32.036
50,000	−69.700	1.691 7	0.011 652	0.000 363 91	0.296 91	0.815 89	968.08	32.020
60,000	−69.700	1.048 8	0.007 217 5	0.000225 61	0.296 91	1.316 0	968.08	31.991
70,000	−67.425	0.650 87	0.004 448 5	0.000 139 20	0.298 36	2.1434	970.90	31.958
80,000	−61.976	0.406 32	0.002 736 6	0.000 085 707	0.301 82	3.521 5	977.62	31.930
90,000	−56.535	0.255 40	0.001 695 2	0.000 053 145	0.305 25	5.743 6	984.28	31.897
100,000	−51.099	0.161 60	0.001 057 5	0.000 033 182	0.308 65	9.301 8	990.91	31.868
km	°C	kPa abs	N/m³	kg/m³	10⁻⁶ N · s/m	10⁻⁶ m²/s	m/s	m/s²
0	15.000	101.325	12.0131	1.225 0	17.894	14.607	340.294	9.80665
1	8.501	89.876	10.8987	1.111 7	17.579	15.813	336.43	9.8036
2	2.004	79.501	9.8652	1.006 6	17.260	17.147	332.53	9.8005
3	−4.500	70.121	8.9083	0.909 25	16.938	18.628	328.58	9.7974
4	−10.984	61.660	8.0250	0.819 35	16.612	20.275	324.59	9.7943
5	−17.474	54.048	7.2105	0.736 43	16.282	22.110	320.55	9.7912
6	−23.963	47.217	6.4613	0.660 11	15.949	24.161	316.45	9.7882
8	−36.935	35.651	5.1433	0.525 79	15.271	29.044	308.11	9.7820
10	−49.898	26.499	4.0424	0.413 51	14.577	35.251	299.53	9.7759
12	−56.500	19.399	3.0476	0.311 94	14.216	45.574	295.07	9.7697
14	−56.500	14.170	2.2247	0.227 86	14.216	62.391	295.07	9.7636
16	−56.500	10.352	1.6243	0.166 47	14.216	85.397	295.07	9.7575
18	−56.500	7.565	1.1862	0.12165	14.216	116.86	295.07	9.7513
20	−56.500	5.529	0.8664	0.088 91	14.216	159.89	295.07	9.7452
25	−51.598	2.549	0.3900	0.040 08	14.484	361.35	298.39	9.7300
30	−46.641	1.197	0.1788	0.018 41	14.753	801.34	301.71	9.7147

[a] International Civil Aviation Organization.

TABLE AI.3

Properties of Air at Standard Atmospheric Pressure

Temperature	Density (kg/m³)	Specific Weight (N/m³)	Dynamic Viscosity (N · s/m²)	Kinematic Viscosity m²/s
−20°C	1.40	13.70	1.61×10^{-5}	1.16×10^{-5}
−10°C	1.34	13.20	1.67×10^{-5}	1.24×10^{-5}
0°C	1.29	12.70	1.72×10^{-5}	1.33×10^{-5}
10°C	1.25	12.20	1.76×10^{-5}	1.41×10^{-5}
20°C	1.20	11.80	1.81×10^{-5}	1.51×10^{-5}
30°C	1.17	11.40	1.86×10^{-5}	1.60×10^{-5}
40°C	1.13	11.10	1.91×10^{-5}	1.69×10^{-5}
50°C	1.09	10.70	1.95×10^{-5}	1.79×10^{-5}
60°C	1.06	10.40	2.00×10^{-5}	1.89×10^{-5}
70°C	1.03	10.10	2.04×10^{-5}	1.99×10^{-5}
80°C	1.00	9.81	2.09×10^{-5}	2.09×10^{-5}
90°C	0.97	9.54	2.13×10^{-5}	2.19×10^{-5}
100°C	0.95	9.28	2.17×10^{-5}	2.29×10^{-5}
120°C	0.90	8.82	2.26×10^{-5}	2.51×10^{-5}
140°C	0.85	8.38	2.34×10^{-5}	2.74×10^{-5}
160°C	0.81	7.99	2.42×10^{-5}	2.97×10^{-5}
180°C	0.78	7.65	2.50×10^{-5}	3.20×10^{-5}
200°C	0.75	7.32	2.57×10^{-5}	3.44×10^{-5}
	slugs/ft³	**lbf/ft³**	**lbf-s/ft²**	**ft²/s**
0°F	0.00269	0.0866	3.39×10^{-7}	1.26×10^{-4}
20°F	0.00257	0.0828	3.51×10^{-7}	1.37×10^{-4}
40°F	0.00247	0.0794	3.63×10^{-7}	1.47×10^{-4}
60°F	0.00237	0.0764	3.74×10^{-7}	1.58×10^{-4}
80°F	0.00228	0.0735	3.85×10^{-7}	1.69×10^{-4}
100°F	0.00220	0.0709	3.96×10^{-7}	1.80×10^{-4}
120°F	0.00213	0.0685	4.07×10^{-7}	1.91×10^{-4}
150°F	0.00202	0.0651	4.23×10^{-7}	2.09×10^{-4}
200°F	0.00187	0.0601	4.48×10^{-7}	2.40×10^{-4}
300°F	0.00162	0.0522	4.96×10^{-7}	3.05×10^{-4}
400°F	0.00143	0.0462	5.40×10^{-7}	3.77×10^{-4}

TABLE AI.4

Physical Properties of Gases ($T = 15°C$ [59°F], $p = 1$ atm)

Gas	Density kg/m³ (slugs/ft³)	Kinematic Viscosity m²/s (ft²/s)	R Gas Constant J/kg K (ft-lbf/slug-°R)	$c_p \frac{J}{kg\,K}$ $\left(\frac{Btu}{lbm-°R}\right)$	$k = \frac{c_p}{c_v}$	S Sutherland's Constant K (°R)
Air	1.22 (0.00237)	1.46×10^{-5} (1.58×10^{-4})	287 (1716)	1004 (0.240)	1.40	111 (199)
Carbon dioxide	1.85 (0.0036)	7.84×10^{-6} (8.48×10^{-5})	189 (1130)	841 (0.201)	1.30	222 (400)
Helium	0.169 (0.00033)	1.14×10^{-4} (1.22×10^{-3})	2077 (12,419)	5187 (1.24)	1.66	79.4 (143)
Hydrogen	0.0851 (0.00017)	1.01×10^{-4} (1.09×10^{-5})	4127 (24,677)	14,223 (3.40)	1.41	96.7 (174)
Methane (natural gas)	0.678 (0.0013)	1.59×10^{-5} (1.72×10^{-4})	518 (3098)	2208 (0.528)	1.31	198 (356)
Nitrogen	1.18 (0.0023)	1.45×10^{-5} (1.56×10^{-4})	297 (1776)	1041 (0.249)	1.40	107 (192)
Oxygen	1.35 (0.0026)	1.50×10^{-5} (1.61×10^{-4})	260 (1555)	916 (0.219)	1.40	

TABLE AI.5

Approximate Physical Properties of Common Liquids at Atmospheric Pressure

Liquid and Temperature	Density kg/m³ (slugs/ft³)	Specific Gravity	Specific Weight N/m³ (lbf/ft³)	Dynamic Viscosity N·s/m² (lbf-s/ft²)	Kinematic Viscosity m²/s (ft²/s)	Surface Tension N/m* (lbf/ft)
Ethyl alcohol[(1) (3)] 20°C (68°F)	799 (135)	0.79	7,850 (50.0)	1.2×10^{-3} (2.5×10^{-5})	1.5×10^{-6} (1.6×10^{-5})	2.2×10^{-2} (1.5×10^{-3})
Carbon tetrachloride[(3)] 20°C (68°F)	1,590 (3,09)	1.59	15,600 (99.5)	9.6×10^{-4} (2.0×10^{-5})	6.0×10^{-7} (6.5×10^{-6})	2.6×10^{-2} (1.8×10^{-3})
Glycerine[(3)] 20°C (68°F)	1,260 (2.45)	1.26	12,300 (78.5)	1.41 (2.95×10^{-2})	1.12×10^{-3} (1.22×10^{-2})	6.3×10^{-2} (4.3×10^{-3})
Kerosene[(1)(2)] 20°C (68°F)	814 (1.58)	0.81	8,010 (51)	1.9×10^{-3} (4.0×10^{-5})	2.37×10^{-6} (2.55×10^{-5})	2.9×10^{-2} (2.0×10^{-3})
Mercury[(1)(3)] 20°C (68°F)	13,550 (26.3)	13.55	133,000 (847)	1.5×10^{-3} (3.1×10^{-5})	1.2×10^{-7} (1.3×10^{-6})	4.8×10^{-1} (3.3×10^{-2})
Sea water 10°C at 3.3% salinity	1,026 (1.99)	1.03	10,070 (64.1)	1.4×10^{-3} (2.9×10^{-5})	1.4×10^{-6} (1.5×10^{-5})	
Oils—38°C (100°F) SAE 10W[(4)]	870 (1.69)	0.87	8,530 (54.4)	3.6×10^{-2} (7.5×10^{-4})	4.1×10^{-5} (4.4×10^{-4})	
SAE 10W-30[(4)]	880 (1.71)	0.88	8,630 (55.1)	6.7×10^{-2} (1.4×10^{-3})	7.6×10^{-5} (8.2×10^{-4})	
SAE 30[(4)]	880 (1.71)	0.88	8,630 (55.1)	1.0×10^{-1} (2.1×10^{-3})	1.1×10^{-4} (1.2×10^{-3})	

TABLE AI.6A

Properties of Saturated Water (Liquid–Vapour): Temperature Table (SI units)

		Specific Volume (m³/kg)		Internal Energy kJ/kg		Enthalpy kJ/kg			Entropy kJ/kg · K		
Temperature (°C)	Press Bar	Sat. Liquid $v_f \times 10^3$	Sat. Vapor v_g	Sat. Liquid u_f	Sat. Vapor u_g	Sat. Liquid h_f	Evap. h_{fg}	Sat. Vapor h_g	Sat. Liquid s_f	Sal. Vapor s_g	Temperature (°C)
0.01	0.00611	1.0002	206.136	0.00	2375.3	0.01	2501.3	2501.4	0.0000	9.1562	.01
4	0.00813	1.0001	157.232	16.77	2380.9	16.78	2491.9	2508.7	0.0610	9.0514	4
5	0.00872	1.0001	147.120	20.97	2382.3	20.98	2489.6	2510.6	0.0761	9.0257	5
6	0.00935	1.0001	137.734	25.19	2383.6	25.20	2487.2	2512.4	0.0912	9.0003	6
8	0.01072	1.0002	120.917	33.59	2386.4	33.60	2482.5	2516.1	0.1212	8.9501	8
10	0.01228	1.0004	106.379	42.00	2389.2	42.01	2477.7	2519.8	0.1510	8.9008	10
11	0.01312	1.0004	99.857	46.20	2390.5	46.20	2475.4	2521.6	0.1658	8.8765	11
12	0.01402	1.0005	93.784	50.41	2391.9	50.41	2473.0	2523.4	0.1806	8.8524	12
13	0.01497	1.0007	88.124	54.60	2393.3	54.60	2470.7	2525.3	0.1953	8.8285	13
14	0.01598	1.0008	82.848	58.79	2394.7	58.80	2468.3	2527.1	0.2099	8.8048	14
15	0.01705	1.0009	77.926	62.99	2396.1	62.99	2465.9	2528.9	0.2245	8.7814	15
16	0.01818	1.0011	73.333	67.18	2397.4	67.19	2463.6	2530.8	0.2390	8.7582	16
17	0.01938	1.0012	69.044	71.38	2398.8	71.38	2461.2	2532.6	0.2535	8.7351	17
18	0.02064	1.0014	65.038	75.57	2400.2	75.58	2458.8	2534.4	0.2679	8.7123	18
19	0.02198	1.0016	61.293	79.76	2401.6	79.77	2456.5	2536.2	0.2823	8.6897	19
20	0.02339	1.0018	57.791	83.95	2402.9	83.96	2454.1	2538.1	0.2966	8.6672	20
21	0.02487	1.0020	54.514	88.14	2404.3	88.14	2451.8	2539.9	0.3109	8.6450	21
22	0.02645	1.0022	51.447	92.32	2405.7	92.33	2449.4	2541.7	0.3251	8.6229	22
23	0.02810	1.0024	48.574	96.51	2407.0	96.52	2447.0	2543.5	0.3393	8.6011	23
24	0.02985	1.0027	45.883	100.70	2408.4	100.70	2444.7	2545.4	0.3534	8.5794	24
25	0.03169	1.0029	43.360	104.88	2409.8	104.89	2442.3	2547.2	0.3674	8.5580	25
26	0.03363	1.0032	40.994	109.06	2411.1	109.07	2439.9	2549.0	0.3814	8.5367	26
27	0.03567	1.0035	38.774	113.25	2412.5	113.25	2437.6	2550.8	0.3954	8.5156	27
28	0.03782	1.0037	36.690	117.42	2413.9	117.43	2435.2	2552.6	0.4093	8.4946	28
29	0.04008	1.0040	34.733	121.60	2415.2	121.61	2432.8	2554.5	<0.4231	8.4739	29

(Continued)

TABLE AI.6A (CONTINUED)

Properties of Saturated Water (Liquid–Vapour): Temperature Table (SI units)

Temperature (°C)	Press Bar	Specific Volume (m^3/kg)		Internal Energy kJ/kg		Enthalpy kJ/kg			Entropy kJ/kg · K		Temperature (°C)
		Sat. Liquid $v_f \times 10^3$	Sat. Vapor v_g	Sat. Liquid u_f	Sat. Vapor u_g	Sat. Liquid h_f	Evap. h_{fg}	Sat. Vapor h_g	Sat. Liquid s_f	Sal. Vapor s_g	
30	0.04246	1.0043	32.894	125.78	2416.6	125.79	2430.5	2556.3	0.4369	8.4533	30
31	0.04496	1.0046	31.165	129.96	2418.0	129.97	2428.1	2558.1	0.4507	8.4329	31
32	0.04759	1.0050	29.540	134.14	2419.3	134.15	2425.7	2559.9	0.4644	8.4127	32
33	0.05034	1.0053	28.011	138.32	2420.7	138.33	2423.4	2561.7	0.4781	8.3927	33
34	0.05324	1.0056	26.571	142.50	2422.0	142.50	2421.0	2563.5	0.4917	8.3728	34
35	0.05628	1.0060	25.216	146.67	2423.4	146.68	2418.6	2565.3	0.5053	8.3531	35
36	0.05947	1.0063	23.940	150.85	2424.7	150.86	2416.2	2567.1	0.5188	8.3336	36
38	0.06632	1.0071	21.602	159.20	2427.4	159.21	2411.5	2570.7	0.5458	8.2950	38
40	0.07384	1.0078	19.523	167.56	2430.1	167.57	2406.7	2574.3	0.5725	8.2570	40
45	0.09593	1.0099	15.258	188.44	2436.8	188.45	2394.8	2583.2	0.6387	8.1648	45
50	0.1235	1.0121	12.032	209.32	2443.5	209.33	2382.7	2592.1	.7038	8.0763	50
55	0.1576	1.0146	9.568	230.21	2450.1	230.23	2370.7	2600.9	.7679	7.9913	55
60	0.1994	1.0172	7.671	251.11	2456.6	251.13	2358.5	2609.6	.8312	7.9096	60
65	0.2503	1.0199	6.197	272.02	2463.1	272.06	2346.2	2618.3	.8935	7.8310	65
70	0.3119	1.0228	5.042	292.95	2469.6	292.98	2333.8	2626.8	.9549	7.7553	70
75	0.3858	1.0259	4.131	313.90	2475.9	313.93	2321.4	2635.3	1.0155	7.6824	75
80	0.4739	1.0291	3.407	334.86	2482.2	334.91	2308.8	2643.7	1.0753	7.6122	80
85	0.5783	1.0325	2.828	355.84	2488.4	355.90	2296.0	2651.9	1.1343	7.5445	85
90	0.7014	1.0360	2.361	376.85	2494.5	376.92	2283.2	2660.1	1.1925	7.4791	90
95	0.8455	1.0397	1.982	397.88	2500.6	397.96	2270.2	2668.1	1.2500	7.4159	95
100	1.014	1.0435	1.673	418.94	2506.5	419.04	2257.0	2676.1	1.3069	7.3549	100
110	1.433	1.0516	1.210	461.14	2518.1	461.30	2230.2	2691.5	1.4185	7.2387	110
120	1.985	1.0603	0.8919	503.50	2529.3	503.71	2202.6	2706.3	1.5276	7.1296	120
130	2.701	1.0697	0.6685	546.02	2539.9	546.31	2174.2	2720.5	1.6344	7.0269	130
140	3.613	1.0797	0.5089	588.74	2550.0	589.13	2144.7	2733.9	1.7391	6.9299	140

(Continued)

TABLE AI.6A (CONTINUED)

Properties of Saturated Water (Liquid–Vapour): Temperature Table (SI units)

		Specific Volume (m^3/kg)		Internal Energy kJ/kg		Enthalpy kJ/kg			Entropy kJ/kg · K		
Temperature (°C)	Press Bar	Sat. Liquid $v_f \times 10^3$	Sat. Vapor v_g	Sat. Liquid u_f	Sat. Vapor u_g	Sat. Liquid h_f	Evap. h_{fg}	Sat. Vapor h_g	Sat. Liquid s_f	Sal. Vapor s_g	Temperature (°C)
150	4.758	1.0905	0.3928	631.68	2559.5	632.20	2114.3	2746.5	1.8418	6.8379	150
160	6.178	1.1020	0.3071	674.86	2568.4	675.55	2082.6	2758.1	1.9427	6.7502	160
170	7.917	1.1143	0.2428	718.33	2576.5	719.21	2049.5	2768.7	2.0419	6.6663	170
180	10.02	1.1274	0.1941	762.09	2583.7	763.22	2015.0	2778.2	2.1396	6.5857	180
190	12.54	1.1414	0.1565	806.19	2590.0	807.62	1978.8	2786.4	2.2359	6.5079	190
200	15.54	1.1565	0.1274	850.65	2595.3	852.45	1940.7	2793.2	2.3309	6.4323	200
210	19.06	1.1726	0.1044	895.53	2599.5	897.76	1900.7	2798.5	2.4248	6.3585	210
220	23.18	1.1900	0.08619	940.87	2602.4	943.62	1858.5	2802.1	2.5178	6.2861	220
230	27.95	1.2088	0.07158	986.74	2603.9	990.12	1813.8	2804.0	2.6099	6.2146	230
240	33.44	1.2291	0.05976	1033.2	2604.0	1037.3	1766.5	2803.8	2.7015	6.1437	240
250	39.73	1.2512	0.05013	1080.4	2602.4	1085.4	1716.2	2801.5	2.7927	6.0730	250
260	46.88	1.2755	0.04221	1128.4	2599.0	1134.4	1662.5	2796.6	2.8838	6.0019	260
270	54.99	1.3023	0.03564	1177.4	2593.7	1184.5	1605.2	2789.7	2.9751	5.9301	270
280	64.12	1.3321	0.03017	1227.5	2586.1	1236.0	1543.6	2779.6	3.0668	5.8571	280
290	74.36	1.3656	0.02557	1278.9	2576.0	1289.1	1477.1	2766.2	3.1594	5.7821	290
300	85.81	1.4036	0.02167	1332.0	2563.0	1344.0	1404.9	2749.0	3.2534	5.7045	300
320	112.7	1.4988	0.01549	1444.6	2525.5	1461.5	1238.6	2700.1	3.4480	5.5362	320
340	145.9	1.6379	0.01080	1570.3	2464.6	1594.2	1027.9	2622.0	3.6594	5.3357	340
360	186.5	1.8925	0.006945	1725.2	2351.5	1760.5	720.5	2481.0	3.9147	5.0526	360
374.14	220.9	3.155	0.003155	2029.6	2029.6	2099.3	0	2099.3	4.4298	4.4298	374.14

Source: Tables A.2 through A.5 are extracted from Keunan, J. H. et al. *Swam Tables*, Wiley, New York, 1969.

TABLE AI.6B

Properties of Saturated Water (Liquid–Vapor): Temperature Table (English Units)

Temperature (°F)	Press lbf/in.²	Specific Volume ft³/lb		Internal Energy Blu/lb		Enthalpy Blu/lb			Entropy Blu/lb · °R		Temperature (°F)
		Sat. Liquid v_f	Sat. Vapor v_g	Sal. Liquid u_f	Sat. Vapor u_g	Sat. Liquid h_t	Evap. h_{tg}	Sat. Vapor h_g	Sat. Liquid s_f	Sal. Vapor s_g	
32	0.0886	0.01602	3305	−0.01	1021.2	−0.01	1075.4	1075.4	−0.00003	2.1870	32
35	0.0999	0.01602	2948	2.99	1022.2	3.00	1073.7	1076.7	0.00607	2.1764	35
40	0.1217	0.01602	2445	8.02	1023.9	8.02	1070.9	1078.9	0.01617	2.1592	40
45	0.1475	0.01602	2037	13.04	1025.5	13.04	1068.1	1081.1	0.02618	2.1423	45
50	0.1780	0.01602	1704	18.06	1027.2	18.06	1065.2	1083.3	0.03607	2.1259	50
52	0.1917	0.01603	1589	20.06	1027.8	20.07	1064.1	1084.2	0.04000	2.1195	52
54	0.2064	0.01603	1482	22.07	1028.5	22.07	1063.0	1085.1	0.04391	2.1131	54
56	0.2219	0.01603	1383	24.08	1029.1	24.08	1061.9	1085.9	0.04781	2.1068	56
58	0.2386	0.01603	1292	26.08	1029.8	26.08	1060.7	1086.8	0.05159	2.1005	58
60	0.2563	0.01604	1207	28.08	1030.4	28.08	1059.6	1087.7	0.05555	2.0943	60
62	0.2751	0.01604	1129	30.09	1031.1	30.09	1058.5	1088.6	0.05940	2.0882	62
64	0.2952	0.01604	1056	32.09	1031.8	32.09	1057.3	1089.4	0.06323	2.0821	64
66	0.3165	0.01604	988.4	34.09	1032.4	34.09	1056.2	1090.3	0.06704	2.0761	66
68	0.3391	0.01605	925.8	36.09	1033.1	36.09	1055.1	1091.2	0.07084	2.0701	68
70	0.3632	0.01605	867.7	38.09	1033.7	38.09	1054.0	1092.0	0.07463	2.0642	70
72	0.3887	0.01606	813.7	40.09	1034.4	40.09	1052.8	1092.9	0.07839	2.0584	72
74	0.4158	0.01606	763.5	42.09	1035.0	42.09	1051.7	1093.8	0.08215	2.0526	74
76	0.4446	0.01606	716.8	44.09	1035.7	44.09	1050.6	1094.7	0.08589	2.0469	76
78	0.4750	0.01607	673.3	46.09	1036.3	46.09	1049.4	1095.5	0.08961	2.0412	78
80	0.5073	0.01607	632.8	48.08	1037.0	48.09	1048.3	1096.4	0.09332	2.0356	80
82	0.5414	0.01608	595.0	50.08	1037.6	50.08	1047.2	1097.3	0.09701	2.0300	82
84	0.5776	0.01608	559.8	52.08	1038.3	52.08	1046.0	1098.1	0.1007	2.0245	84
86	0.6158	0.01609	527.0	54.08	1038.9	54.08	1044.9	1099.0	0.1044	2.0190	86
88	0.6562	0.01609	496.3	56.07	1039.6	56.07	1043.8	1099.9	0.1080	2.0136	88
90	0.6988	0.01610	467.7	58.07	1040.2	58.07	1042.7	1100.7	0.1117	2.0083	90

(Continued)

TABLE AI.6B (CONTINUED)

Properties of Saturated Water (Liquid–Vapor): Temperature Table (English Units)

		Specific Volume ft³/lb		Internal Energy Blu/lb		Enthalpy Blu/lb			Entropy Blu/lb · °R		
Temperature (°F)	Press lbf/in.²	Sat. Liquid v_f	Sat. Vapor v_g	Sal. Liquid u_f	Sat. Vapor u_g	Sat. Liquid h_t	Evap. h_{tg}	Sat. Vapor h_g	Sat. Liquid s_f	Sal. Vapor s_g	Temperature (°F)
92	0.7439	0.01611	440.9	60.06	1040.9	60.06	1041.5	1101.6	0.1153	2.0030	92
94	0.7914	0.01611	415.9	62.06	1041.5	62.06	1040.4	1102.4	0.1189	1.9977	94
96	0.8416	0.01612	392.4	64.05	1041.2	64.06	1039.2	1103.3	0.1225	1.9925	96
98	0.8945	0.01612	370.5	66.05	1042.8	66.05	1038.1	1104.2	0.1261	1.9874	98
100	0.9503	0.01613	350.0	68.04	1043.5	68.05	1037.0	1105.0	0.1296	1.9822	100
110	1.276	0.01617	265.1	78.02	1046.7	78.02	1031.3	1109.3	0.1473	1.9574	110
120	1.695	0.01621	203.0	87.99	1049.9	88.00	1025.5	1113.5	0.1647	1.9336	120
130	2.225	0.01625	157.2	97.97	1053.0	97.98	1019.8	1117.8	0.1817	1.9109	130
140	2.892	0.01629	122.9	107.95	1056.2	107.96	1014.0	1121.9	0.1985	1.8892	140
150	3.722	0.01634	97.0	117.95	1059.3	117.96	1008.1	1126.1	0.2150	1.8684	150
160	4.745	0.01640	77.2	127.94	1062.3	127.96	1002.2	1130.1	0.2313	1.8484	160
170	5.996	0.01645	62.0	137.95	1065.4	137.97	996.2	1134.2	0.2473	1.8293	170
180	7.515	0.01651	50.2	147.97	1068.3	147.99	990.2	1138.2	0.2631	1.8109	180
190	9.343	0.01657	41.0	158.00	1071.3	158.03	984.1	1142.1	0.2787	1.7932	190
200	11.529	0.01663	33.6	168.04	1074.2	168.07	977.9	1145.9	0.2940	1.7762	200
210	14.13	0.01670	27.82	178.1	1077.0	178.1	971.6	1149.7	0.3091	1.7599	210
212	14.70	0.01672	26.80	180.1	1077.6	180.2	9703	1150.5	0.3121	1.7567	212
220	17.19	0.01677	23.15	188.2	1079.8	188.2	965.3	1153.5	0.3241	1.7441	220
230	20.78	0.01685	19.39	198.3	1082.6	198.3	958.8	1157.1	0.3388	1.7289	230
240	24.97	0.01692	16.33	208.4	1085.3	208.4	952.3	1160.7	0.3534	1.7143	240
250	29.82	0.01700	13.83	218.5	1087.9	218.6	945.6	1164.2	0.3677	1.7001	250
260	35.42	0.01708	11.77	228.6	1090.5	228.8	938.8	1167.6	0.3819	1.6864	260
270	41.85	0.01717	10.07	238.8	1093.0	239.0	932.0	1170.9	0.3960	1.6731	270

(Continued)

TABLE AI.6B (CONTINUED)

Properties of Saturated Water (Liquid–Vapor): Temperature Table (English Units)

Temperature (°F)	Press lbf/in.^{2}	Specific Volume ft^3/lb Sat. Liquid v_f	Sat. Vapor v_g	Internal Energy Blu/lb Sal. Liquid u_f	Sat. Vapor u_g	Enthalpy Blu/lb Sat. Liquid h_t	Evap. h_{tg}	Sat. Vapor h_g	Entropy Blu/lb · °R Sat. Liquid s_f	Sal. Vapor s_g	Temperature (°F)
280	49.18	0.01726	8.65	249.0	1095.4	249.2	924.9	1174.1	0.4099	1.6602	280
290	57.53	0.01735	7.47	259.3	1097.7	259.4	917.8	1177.2	0.4236	1.6477	290
300	66.98	0.01745	6.472	269.5	1100.0	269.7	910.4	1180.2	0.4372	1.6356	300
310	77.64	0.01755	5.632	279.8	1102.1	280.1	903.0	1183.0	0.4507	1.6238	310
320	89.60	0.01765	4.919	290.1	1104.2	290.4	895.3	1185.8	0.4640	1.6123	320
330	103.00	0.01778	4.312	300.5	1106.2	300.8	887.5	1188.4	0.4772	1.6010	330
340	117.93	0.01787	3.792	310.9	1108.0	311.3	879.5	1190.8	0.4903	1.5901	340
350	134.53	0.01799	3.346	321.4	1109.8	321.8	871.3	1193.1	0.5033	1.5793	350
360	152.92	0.01811	2.961	331.8	1111.4	332.4	862.9	1195.2	0.5162	1.5688	360
370	173.23	0.01823	2.628	342.4	1112.9	343.0	854.2	1197.2	0.5289	1.5585	370
380	195.60	0.01836	2.339	353.0	1114.3	353.6	845.4	1199.0	0.5416	1.5483	380
390	220.2	0.01850	2.087	363.6	1115.6	364.3	836.2	1200.6	0.5542	1.5383	390
400	247.1	0.01864	1.866	374.3	1116.6	375.1	826.8	1202.0	0.5667	1.5284	400
410	276.5	0.01878	1.673	385.0	1117.6	386.0	817.2	1203.1	0.5792	1.5187	410
420	308.5	0.01894	1.502	395.8	1118.3	396.9	807.2	1204.1	0.5915	1.5091	420
430	343.3	0.01909	1.352	406.7	1118,9	407.9	796.9	1204.8	0.6038	1.4995	430
440	381.2	0.01926	1.219	417.6	1119.3	419.0	786.3	1205.3	0.6161	1.4900	440
450	422.1	0.01943	1.1011	428.6	1119.5	430.2	775.4	1205.6	0.6282	1.4806	450
460	466.3	0.01961	0.9961	439.7	1119.6	441.4	764.1	1205.5	0.6404	1.4712	460
470	514.1	0.01980	0.9025	450.9	1119.4	452.8	752.4	1205.2	0.6525	1.4618	470
480	565.5	0.02000	0.8187	462.2	1118.9	464.3	740.3	1204.6	0.6646	1.4524	480
490	620.7	0.02021	0.7436	473.6	1118.3	475.9	727.8	1203.7	0.6767	1.4430	490

(Continued)

TABLE AI.6B (CONTINUED)

Properties of Saturated Water (Liquid–Vapor): Temperature Table (English Units)

Temperature (°F)	Press lbf/in.²	Specific Volume ft³/lb		Internal Energy Blu/lb		Enthalpy Blu/lb			Entropy Blu/lb · °R		Temperature (°F)
		Sat. Liquid v_f	Sat. Vapor v_g	Sal. Liquid u_f	Sat. Vapor u_g	Sat. Liquid h_t	Evap. h_{tg}	Sat. Vapor h_g	Sat. Liquid s_f	Sal. Vapor s_g	
500	680.0	0.02043	0.6761	485.1	1117.4	487.7	714.8	1202.5	0.6888	1.4335	500
520	811.4	0.02091	0.5605	508.5	1114.8	511.7	687.3	1198.9	0.7130	1.4145	520
540	961.5	0.02145	0.4658	532.6	1111.0	536.4	657.5	1193.8	0.7374	1.3950	540
560	1131.8	0.02207	0.3877	548.4	1105.8	562.0	625.0	1187.0	0.7620	1.3749	560
580	1324.3	0.02278	0.3225	583.1	1098.9	588.6	589.3	1178.0	0.7872	1.3540	580
600	1541.0	0.02363	0.2677	609,9	1090.0	616.7	549.7	1166.4	0.8130	1.3317	600
620	1784.4	0.02465	0.2209	638.3	1078.5	646.4	505.0	1151.4	0.8398	1.3075	620
640	2057.1	0.02593	0.1805	668.7	1063.2	678.6	453.4	1131.9	0.8681	1.2803	640
660	2362	0.02767	0.1446	702.3	1042.3	714.4	391.1	1105.5	0.8990	1.2483	660
680	2705	0.03032	0.1113	741.7	1011.0	756.9	309.8	1066.7	0.9350	1.2068	680
700	3090	0.03666	0.0744	801.7	947.7	822.7	167.5	990.2	0.9902	1.1346	700
705.4	3204	0.05053	0.05053	872.6	872.6	902.5	0	902.5	1.0580	1.0580	705.4

Source: Tables A.2E through A.6E arc extracted from Keenan J.H. et al., *Steam Tables*, Wiley, New York. 1969.

TABLE AI.7A,

Properties of Saturated Water (Liquid–Vapor): Pressure Table (SI Units)

Press. Bar	Temperature (°C)	Specific Volume m³/kg		Internal Energy kJ/kg		Enthalpy kJ/kg			Entropy kJ/kg · K		Press. Bar
		Sat. Liquid $v_f \times 10^3$	Sat. Vapor v_g	Sal. Liquid u_f	Sat. Vapor u_g	Sat. Liquid h_t	Evap. h_{tg}	Sat. Vapor h_g	Sat. Liquid s_t	Sal. Vapor s_g	
0.04	28.96	1.0040	34.800	121.45	2415.2	121.46	2432.9	2554.4	0.4226	8.4746	0.04
0.06	36.16	1.0064	23.739	151.53	2425.0	151.53	2415.9	2567.4	0.5210	8.3304	0.06
0.08	41.51	1.0084	18.103	173.87	2432.2	173.88	2403.1	2577.0	0.5926	8.2287	0.08
0.10	45.81	1.0102	14.674	191.82	2437.9	191.83	2392.8	2584.7	0.6493	8.1502	0.10
0.20	60.06	1.0172	7.649	251.38	2456.7	251.40	2358.3	2609.7	0.8320	7.9085	0.20
0.30	69.10	1.0223	5.229	289.20	2468.4	289.23	2336.1	2625.3	0.9439	7.7686	0.30
0.40	75.87	1.0265	3.993	317.53	2477.0	317.58	2319.2	2636.8	1.0259	7.6700	0.40
0.50	81.33	1.0300	3.240	340.44	2483.9	340.49	2305.4	2645.9	1.0910	7.5939	0.50
0.60	85.94	1.0331	2.732	359.79	2489.6	359.86	2293.6	2653.5	1.1453	7.5320	0.60
0.70	89.95	1.0360	2.365	376.63	2494.5	376.70	2283.3	2660.0	1.1919	7.4797	0.70
0.80	93.50	1.0380	2.087	391.58	2498.8	391.66	2274.1	2665.8	1.2329	7.4346	0.80
0.90	96.71	1.0410	1.869	405.06	2502.6	405.15	2265.7	2670,9	1.2695	7.3949	0.90
1.00	99.63	1.0432	1.694	417.36	2506.1	417.46	2258.0	2675.5	1.3026	7.3594	1.00
1.50	111.4	1.0528	1.159	466.94	2519.7	467.11	2226.5	2693.6	1.4336	7.2233	1.50
2.00	120.2	1.0605	0.8857	504.49	2529.5	504.70	2201.9	2706.7	1.5301	7.1271	2.00
2.50	127.4	1.0672	0.7187	535.10	2537.2	535.37	2181.5	2716.9	1.6072	7.0527	2.50
3.00	133.6	1.0732	0.6058	561.15	2543.6	561.47	2163.8	2725.3	1.6718	6.9919	3.00
3.50	138.9	1.0786	0.5243	583.95	2546.9	584.33	2148.1	2732.4	1.7275	6.9405	3.50
4.00	143.6	1.0836	0.4625	604.31	2553.6	604.74	2133.8	2738.6	1.7766	6.8959	4.00
4.50	147.9	1.0882	0.4140	622.25	2557.6	62325	2120.7	2743.9	1.8207	6.8565	4.50
5.00	151.9	1.0926	0.3749	639.68	2561.2	640.23	2108.5	2748.7	1.8607	6.8212	5.00
6.00	158.9	1.1006	0.3157	669.90	2567.4	670.56	2086.3	2756.8	1.9312	6.7600	6.00
7.00	165.0	1.1080	0.2729	696.44	2572.5	697.22	2066.3	2763.5	1.9922	6.7080	7.00
8.00	170.4	1.1148	0.2404	720.22	2576.8	721.11	2048.0	2769.1	2.0462	6.6628	8.00
9.00	175.4	1.1212	0.2150	741.83	2580.5	742.83	2031.1	2773.9	2.0946	6.6226	9.00

(Continued)

TABLE AI.7A,

Properties of Saturated Water (Liquid–Vapor): Pressure Table (SI Units)

		Specific Volume m^3/kg		Internal Energy kJ/kg		Enthalpy kJ/kg			Entropy kJ/kg · K		
Press. Bar	Temperature (°C)	Sat. Liquid $v_f \times 10^3$	Sat. Vapor v_g	Sal. Liquid u_f	Sat. Vapor u_g	Sat. Liquid h_t	Evap. h_{tg}	Sat. Vapor h_g	Sat. Liquid s_t	Sal. Vapor s_g	Press. Bar
10.0	179.9	1.1273	0.1944	761.68	2583.6	762.81	2015.3	2778.1	2.1387	6.5863	10.0
15.0	198.3	1.1539	0.1318	843.16	2594.5	844.84	1947.3	2792.2	2.3150	6.4448	15.0
20.0	212.4	1.1767	0.09963	906.44	2600.3	908.79	1890.7	2799.5	2.4474	6.3409	20.0
25.0	224.0	1.1973	0.07998	959.11	2603.1	962.11	1841.0	2803.1	2.5547	6.2575	25.0
30.0	233.9	1.2165	0.06668	1004.8	2604.1	1008.4	1795.7	2804.2	2.6457	6.1869	30.0
35.0	242.6	1.2347	0.05707	1045.4	2603.7	1049.8	1753.7	2803.4	2.7253	6.1253	35.0
40.0	250.4	1.2522	0.04978	1082.3	2602.3	1087.3	1714.1	2801.4	2.7964	6.0701	40.0
45.0	257.5	1.2692	0.04406	1116.2	2600.1	1121.9	1676.4	2798.3	2.8610	6.0199	45.0
50.0	264.0	1.2859	0.03944	1147.8	2597.1	1154.2	1640.1	2794.3	2.9202	5.9734	50.0
60.0	275.6	1.3187	0.03244	1205.4	2589.7	1213.4	1571.0	2784.3	3.0267	5.8892	60.0
70.0	285.9	1.3513	0.02737	1257.6	2580.5	1267.0	1505.1	2772.1	3.1211	5.8133	70.0
80.0	295.1	1.3842	0.02352	1305.6	2569.8	1316.6	1441.3	2758.0	3.2068	5.7432	80.0
90.0	303.4	1.4178	0.02048	1350.5	2557.8	1363.3	1378.9	2742.1	3.2858	5.6772	90.0
100.0	311.1	1.4524	0.01803	1393.0	2544.4	1407.6	1317.1	2724.7	3.3596	5.6141	100.0
110.0	318.2	1.4886	0.01599	1433.7	2529.8	1450.1	1255.5	2705.6	3.4295	5.5527	110.0
120.0	324.8	1.5267	0.01426	1473.0	2513.7	1491.3	1193.6	2684.9	3.4962	5.4924	120.0
130.0	330.9	1.5671	0.01278	1511.1	2496.1	1531.5	1130.7	2662.2	3.5606	5.4323	130.0
140.0	336.8	1.6107	0.01149	1548.6	2476.8	1571.1	1066.5	2637.6	3.6232	5.3717	140.0
150.0	342.2	1.6581	0.01034	1585.6	2455.5	1610.5	1000.0	2610.5	3.6848	5.3098	150.0
160.0	347.4	1.7107	0.009306	1622.7	2431.7	1650.1	930.6	2580.6	3.7461	5.2455	160.0
170.0	352.4	1.7702	0.008364	1660.2	2405.0	1690.3	856.9	2547.2	3.8079	5.1777	170.0
180.0	357.1	1.8397	0.007489	1698.9	2374.3	1732.0	777.1	2509.1	3.8715	5.1044	180.0
190.0	361.5	1.9243	0.006657	1739.9	2338.1	1776.5	688.0	2464.5	3.9388	5.0228	190.0
200.0	365.8	2.036	0.005834	1785.6	2293.0	1826.3	583.4	2409.7	4.0139	4.9269	200.0
220.9	374.1	3.155	0.003155	2029.6	2029.6	2099.3	0	2099.3	4.4298	4.4298	220.9

TABLE AI.7B

Properties of Saturated Water (Liquid–Vapor): Pressure Table (English Units)

		Specific Volume ft³/lb		Internal Energy Blu/lb		Enthalpy Blu/lb			Entropy Blu/lb · °R			
Press lbf/in.²	Temperature (°F)	Sat. Liquid v_f	Sat. Vapor v_g	Sal. Liquid u_f	Sat. Vapor u_g	Sat. Liquid h_f	Evap. h_{fg}	Sat. Vapor h_g	Sat. Liquid s_f	Evap. s_{fg}	Sal. Vapor s_g	Press lbf/in.²
0.4	72.84	0.01606	792.0	40.94	1034.7	40.94	1052.3	1093.3	0.0800	1.9760	2.0559	0.4
0.6	85.19	0.01609	540.0	53.26	1038.7	53.27	1045.4	1098.6	0.1029	1.9184	2.0213	0.6
0.8	94.35	0.01611	411.7	62.41	1041.7	62.41	1040.2	1102.6	0.1195	1.8773	1.9968	0.8
1.0	101.70	0.01614	333.6	69.74	1044.0	69.74	1036.0	1105.8	0.1327	1.8453	1.9779	1.0
1.2	107.88	0.01616	280.9	75.90	1046.0	75.90	1032.5	1108.4	0.1436	1.8190	1.9626	1.2
1.5	115.65	0.01619	227.7	83.65	1048.5	83.65	1028.0	1111.7	0.1571	1.7867	1.9438	1.5
2.0	126.04	0.01623	173.75	94.02	1051.8	94.02	1022.1	1116.1	0.1750	1.7448	1.9198	2.0
3.0	141.43	0.01630	118.72	109.38	1056.6	109.39	1013.1	1122.5	0.2009	1.6852	1.8861	3.0
4.0	152.93	0.01636	90.64	120.88	1060.2	120.89	1006.4	1127.3	0.2198	1.6426	1.8624	4.0
5.0	162.21	0.01641	73.53	130.15	1063.0	130.17	1000.9	1131.0	0.2349	1.6093	1.8441	5.0
6.0	170.03	0.01645	61.98	137.98	1065.4	138.00	996.2	1134.2	0.2474	1.5819	1.8292	6.0
7.0	176.82	0.01649	53.65	144.78	1067.4	144.80	992.1	1136.9	0.2581	1.5585	1.8167	7.0
8.0	182.84	0.01653	47.35	150.81	1069.2	150.84	988.4	1139.3	0.2675	1.5383	1.8058	8.0
9.0	188.26	0.01656	42.41	156.25	1070.8	156.27	985.1	1141.4	0.2760	1.5203	1.7963	9.0
10	193.19	0.01659	38.42	161.20	1072.2	161.23	982.1	1143.3	0.2836	1.5041	1.7877	10
14.696	211.99	0.01672	26.80	180.10	1077.6	180.15	970.4	1150.5	0.3121	1.4446	1.7567	14.696
15	213.03	0.01672	26.29	181.14	1077.9	181.19	969.7	1150.9	0.3137	1.4414	1.7551	15
20	227.96	0.01683	20.09	196.19	1082.0	196.26	960.1	1156.4	0.3358	1.3962	1.7320	20
25	240.08	0.01692	16.31	208.44	1085.3	208.52	952.2	1160.7	0.3535	1.3607	1.7142	25
30	250.34	0.01700	13.75	218.84	1088.0	218.93	945.4	1164.3	0.3682	1.3314	1.6996	30
35	259.30	0.01708	11.90	227.93	1090.3	228.04	939.3	1167.4	0.3809	1.3064	1.6873	35
40	267.26	0.01715	10.50	236.03	1092.3	236.16	933.8	1170.0	0.3921	1.2845	1.6767	40
45	274.46	0.01721	9.40	243.37	1094.0	243.51	928.8	1172.3	0.4022	1.2651	1.6673	45
50	281.03	0.01727	8.52	250.08	1095.6	250.24	924.2	1174.4	0.4113	1.2476	1.6589	50
55	287.10	0.01733	7.79	256.28	1097.0	256.46	919.9	1176.3	0.4196	1.2317	1.6513	55

(Continued)

TABLE AI.7B (CONTINUED)

Properties of Saturated Water (Liquid–Vapor): Pressure Table (English Units)

		Specific Volume ft³/lb		Internal Energy Blu/lb		Enthalpy Blu/lb			Entropy Blu/lb · °R			
Press lbf/in.²	Temperature (°F)	Sat. Liquid v_f	Sat. Vapor v_g	Sal. Liquid u_f	Sat. Vapor u_g	Sat. Liquid h_f	Evap. h_{fg}	Sat. Vapor h_g	Sat. Liquid s_f	Evap. s_{fg}	Sal. Vapor s_g	Press lbf/in.²
60	292.73	0.01738	7.177	262.1	1098.3	262.2	915.8	1178.0	0.4273	1.2170	1.6443	60
65	298.00	0.01743	6.647	267.5	1099.5	267.7	911.9	1179.6	0.4345	1.2035	1.6380	65
70	302.96	0.01748	6.209	272.6	1100.6	272.8	908.3	1181.0	0.4412	1.1909	1.6321	70
75	307.63	0.01752	5.818	277.4	1101.6	277.6	904.8	1182.4	0.4475	1.1790	1.6265	75
80	312.07	0.01757	5.474	282.0	1102.6	282.2	901.4	1183.6	0.4534	1.1679	1.6213	80
85	316.29	0.01761	5.170	286.3	1103.5	286.6	898.2	1184.8	0.4591	1.1574	1.6165	85
90	320.31	0.01766	4.898	290.5	1104.3	290.8	895.1	1185.9	0.4644	1.1475	1.6119	90
95	324.16	0.01770	4.654	294.5	1105.0	294.8	892.1	1186.9	0.4695	1.1380	1.6075	95
100	327.86	0.01774	4.434	298.3	1105.8	298.6	889.2	1187.8	0.4744	1.1290	1.6034	100
110	334.82	0.01781	4.051	305.5	1107.1	305.9	883.7	1189.6	0.4836	1.1122	1.5958	110
120	341.30	0.01789	3.730	312.3	1108.3	312.7	878.5	1191.1	0.4920	1.0966	1.5886	120
130	347.37	0.01796	3.457	318.6	1109.4	319.0	873.5	1192.5	0.4999	1.0822	1.5821	130
140	353.08	0.01802	3.221	324.6	1110.3	325.1	868.7	1193.8	0.5073	1.0688	1.5761	140
150	358.48	0.01809	3.016	330.2	1111.2	330.8	864.2	1194.9	0.5142	1.0562	1.5704	150
160	363.60	0.01815	2.836	335.6	1112.0	336.2	859.8	1196.0	0.5208	1.0443	1.5651	160
170	368.47	0.01821	2.676	340.8	1112.7	341.3	855.6	1196.9	0.5270	1.0330	1.5600	170
180	373.13	0.01827	2.553	345.7	1113.4	346.3	851.5	1197.8	0.5329	1.0223	1.5552	180
190	377.59	0.01833	2.405	350.4	1114.0	351.0	847.5	1198.6	0.5386	1.0122	1.5508	190
200	381.86	0.01839	2.289	354.9	1114.6	355.6	843.7	1199.3	0.5440	1.0025	1.5465	200
250	401.04	0.01865	1.845	375.4	1116.7	376.2	825.8	1202.1	0.5680	0.9594	1.5274	250
300	417.43	0.01890	1.544	393.0	1118.2	394.1	809.8	1203.9	0.5883	0.9232	1.5115	300
350	431.82	0.01912	1.327	408.7	1119.0	409.9	795.0	1204.9	0.6060	0.8917	1.4977	350
400	444.70	0.01934	1.162	422.8	1119.5	424.2	781.2	1205.5	0.6218	0.8638	1.4856	400
450	456.39	0.01955	1.033	435.7	1119.6	437.4	768.2	1205.6	0.6360	0.8385	1.4745	450

(Continued)

TABLE AI.7B (CONTINUED)

Properties of Saturated Water (Liquid–Vapor): Pressure Table (English Units)

		Specific Volume ft³/lb		Internal Energy Blu/lb		Enthalpy Blu/lb			Entropy Blu/lb · °R			
Press lbf/in.²	Temperature (°F)	Sat. Liquid v_f	Sat. Vapor v_g	Sal. Liquid u_f	Sat. Vapor u_g	Sat. Liquid h_f	Evap. h_{fg}	Sat. Vapor h_g	Sat. Liquid s_f	Evap. s_{fg}	Sal. Vapor s_g	Press lbf/in.²
500	467.13	0.01975	0.928	447.7	1119.4	449.5	755.8	1205.3	0.6490	0.8154	1.4644	500
550	477.07	0.01994	0.842	458.9	1119.1	460.9	743.9	1204,8	0.6611	0.7941	1.4451	550
600	486.33	0.02013	0.770	469.4	1118.6	471.7	732.4	1204.1	0.6723	0.7742	1.4464	600
700	503.23	0.02051	0.656	488.9	1117.0	491.5	710.5	1202.0	0.6927	0.7378	1.4305	700
800	518.36	0.02087	0.569	506.6	1115.0	509.7	689.6	1199.3	0.7110	0.7050	1.4160	800
900	532.12	0.02123	0.501	523.0	1112.6	526.6	669.5	1196.0	0.7277	0.6750	1.4027	900
1000	544.75	0.02159	0.446	538.4	1109.9	542.4	650.0	1192.4	0.7432	0.6471	1.3903	1000
1100	556.45	0.02195	0.401	552.9	1106.8	557.4	631.0	1188.3	0.7576	0.6209	1.3786	1100
1200	567.37	0.02232	0.362	566.7	1103.5	571.7	612.3	1183.9	0.7712	0.5961	1.3673	1200
1300	577.60	0,02269	0.330	579.9	1099.8	585.4	593.8	1179.2	0.7841	0.5724	1.3565	1300
1400	587.25	0.02307	0.302	592.7	1096.0	598.6	575.5	1174.1	0.7964	0.5497	1.3461	1400
1500	596.39	0.02346	0.277	605.0	1091.8	611.5	557.2	1168.7	0.8082	0.5276	1.3359	1500
1600	605.06	0.02386	0.255	616.9	1087.4	624.0	538.9	1162.9	0.8196	0.5062	1.3258	1600
1700	613.32	0.02428	0.236	628.6	1082.7	636.2	520.6	1156.9	0.8307	0.4852	1.3159	1700
1800	621.21	0.02472	0.218	640.0	1077.7	648.3	502.1	1150.4	0.8414	0.4645	1.3060	1800
1900	628.76	0.02517	0.203	651.3	1072.3	660.1	483.4	1143.5	0.8519	0.4441	1.2961	1900
2000	636.00	0.02565	0.188	662.4	1066.6	671.9	464.4	1136.3	0.8623	0.4238	1.2861	2000
2250	652.90	0.02698	0.157	689.9	1050.6	701.1	414.8	1115.9	0.8876	0.3728	1.2604	2250
2500	668.31	0.02860	0.131	717.7	1031.0	730.9	360.5	1091.4	0.9131	0.3196	1.2327	2500
2750	682.46	0.03077	0.107	747.3	1005.9	763.0	297.4	1060.4	0.9401	0.2604	1.2005	2750
3000	695.52	0.03431	0.084	783.4	968.8	802.5	213.0	1015.5	0.9732	0.1843	1.1575	3000
3203.6	705.44	0.05053	0.0505	872.6	872.6	902.5	0	902.5	1.0580	0	1.0580	3203.6

TABLE AI.8A

Properties of Superheated Water Vapor (SI Units)

T (°C)	v (m^3/kg)	u (kJ/kg)	h (kJ/kg)	s (kJ/kg · K)	v (m^3/kg)	u (kJ/kg)	h (kJ/kg)	s (kJ/kg · K)
	$p = 0.06$ bar $= 0.006$ MPa ($T_{sat} = 36.16$°C)				$p = 0.35$ bar $= 0.035$ MPa ($T_{sat} = 72.69$°C)			
Sat.	23.739	2425.0	2567.4	8.3304	4.526	2473.0	2631.4	7.7158
80	27.132	2487.3	2650.1	8.5804	4.625	2483.7	2645.6	7.7564
120	30.219	2544.7	2726.0	8.7840	5.163	2542.4	2723.1	7.9644
160	33.302	2602.7	2802.5	8.9693	5.696	2601.2	2800.6	8.1519
200	36.383	2661.4	2879.7	9.1398	6.228	2660.4	2878.4	8.3237
240	39.462	2721.0	2957.8	9.2982	6.758	2720,3	2956.8	K-4828
280	42.540	2781.5	3036.8	9.4464	7.287	2780.9	3036.0	8.6314
320	45.618	2843.0	3116.7	9.5859	7.815	2842.5	3116.1	8.7712
360	48.696	2905.5	3197.7	9.7180	8.344	2905.1	3197.1	8.9034
400	51.774	2969.0	3279.6	9.8435	8.872	2968.6	3279.2	9.0291
440	54.851	3033.5	3362.6	9.9633	9.400	3033.2	3362.2	9.1490
500	59.467	3132.3	3489.1	10.1336	10.192	3132.1	3488.8	9.3194
	$p = 0.70$ bar $= 0.07$ MPa ($T_{sat} = 89.95$°C)				$p = 1.0$ bar $= 0.10$ MPa ($T_{sat} = 99.63$°C)			
Sat.	2.365	2494.5	2660.0	7.4797	1.694	2506.1	2675.5	7.3594
100	2.434	2509.7	2680.0	7.5341	1.696	2506.7	2676.2	7.3614
120	2.571	2539.7	2719.6	7.6375	1.793	2537.3	2716.6	7.4668
160	2.841	2599.4	2798.2	7.8279	1.984	2597.8	2796.2	7.6597
200	3.108	2659.1	2876.7	8.0012	2.172	2658.1	2875.3	7.8343
240	3.374	2719.3	2955.5	8.1611	2.359	2718.5	2954.5	79949
280	3.640	2780.2	3035.0	8.3162	2.546	2779.6	3034.2	8.1445
320	3.905	2842.0	3115.3	8.4504	2.732	2841.5	3114.6	8 2849
360	4.170	2904.6	3196.5	8.5828	2.917	2904.2	3195.9	8.4175
400	4.434	2968.2	3278.6	8.7086	3.103	2967.9	3278.2	8.5435
440	4.698	3032.9	3361.8	8.8286	3.288	3032.6	3361.4	8.6636
500	5.095	3131.8	3488.5	8.9991	3.565	3131.6	3488.1	8.8342
	$p = 1.5$ bar $= 0.15$ MPa ($T_{sat} = 111.37$°C)				$p = 3.0$ bar $= 0.30$ MPa ($T_{sat} = 133.55$°C)			
Sat.	1.159	2519.7	2693.6	7.2233	0.606	2543.6	2725.3	6.9919
120	1.188	2533.3	2711.4	7.2693				
160	1.317	2595.2	2792.8	7.4665	0.651	2587.1	2782.3	7.1276
200	1.444	2656.2	2872.9	7.6433	0.716	2650.7	2865.5	7.3115
240	1.570	2717.2	2952.7	7.8052	0.781	2713.1	2947.3	7.4774
280	1.695	2778.6	3032.8	7.9555	0.844	2775.4	3028.6	7.6299
320	1.819	2840.6	3113.5	8.0964	0.907	2838.1	3110.1	7.7722
360	1.943	2903.5	3195.0	8.2293	0.969	2901.4	3192.2	7.9061
400	2.067	2967.3	3277.4	8.3555	1.032	2965.6	3275.0	8.0330
440	2.191	3032.1	3360.7	8.4757	1.094	3030.6	3358.7	8.1538
500	2.376	3131.2	3487.6	8,6466	1.187	3130.0	3486.0	8.3251
600	2.685	3301.7	3704.3	8.9101	1.341	33008	3703.2	8.5892

(Continued)

TABLE AI.8A (CONTINUED)

Properties of Superheated Water Vapor (SI Units)

T (°C)	v (m³/kg)	u (kJ/kg)	h (kJ/kg)	s (kJ/kg · K)	v (m³/kg)	u (kJ/kg)	h (kJ/kg)	s (kJ/kg · K)
	p = 5.0 bar = 0.50 MPa (T_{sat} = 151.86°C)				p = 7.0 bar = 0.70 MPa (T_{sat} = 164.97°C)			
Sat.	0.3749	2561.2	2748.7	6.8213	0.2729	2572.5	2763.5	6.7080
180	0.4045	2609.7	2812.0	6.9656	0.2847	2599.8	2799.1	6.7880
200	0.4249	2642.9	2855.4	7.0592	0.2999	2634.8	2844.8	6.8865
240	0.4646	2707.6	2939.9	7.2307	0.3292	2701.8	2932.2	7.0641
280	0.5034	2771.2	3022.9	7.3865	0.3574	2766.9	3017.1	7.2233
320	0.5416	2834.7	3105.6	7.5308	0.3852	2831.3	3100.9	7.3697
360	0.5796	2898.7	3188.4	7.6660	0.4126	2895.8	3184.7	7.5063
400	0.6173	2963.2	3271.9	7.7938	0.4397	2960.9	3268.7	7.6350
440	0.6548	3028.6	3356.0	7.9152	0.4667	3026.6	3353.3	7.7571
500	0.7109	3128.4	3483.9	8.0873	0.5070	3126.8	3481.7	7.9299
600	0.8041	3299.6	3701.7	8.3522	0.5738	3298.5	3700.2	8.1956
700	0.8969	3477.5	3925.9	8.5952	0.6403	3476.6	3924.8	8.4391
	p = 10.0 bar = 1.0 MPa (T_{sat} = 179.91°C)				p = 15.0 bar = 1.5 MPa (T_{sat} = 198.32°C)			
Sat.	0.1944	2583.6	2778.1	6.5865	0.1318	2594.5	2792.2	6.4448
200	0.2060	2621.9	2827.9	6.6940	0.1325	2598.1	2796.8	6.4546
240	0.2275	2692.9	2920.4	6.8817	0.1483	2676.9	2899.3	6.6628
280	0.2480	2760.2	3008.2	7.0465	0.1627	2748.6	2992.7	6.8381
320	0.2678	2826.1	3093.9	7.1962	0.1765	2817.1	3081.9	6.9938
360	0.2873	2891.6	3178.9	7.3349	0.1899	2884.4	3169.2	7.1363
400	0.3066	2957.3	3263.9	7.4651	0.2030	2951.3	3255.8	7.2690
440	0.3257	3023.6	3349.3	7.5883	0.2160	3018.5	3342.5	7.3940
500	0.3541	3124.4	3478.5	7.7622	0.2352	3120.3	3473.1	7.5698
540	0.3729	3192.6	3565.6	7.8720	0.2478	3189.1	3560.9	7.6805
600	0.4011	3296.8	3697.9	8.0290	0.2668	3293.9	3694.0	7.8385
640	0.4198	3367.4	3787.2	8.1290	0.2793	3364.8	3783.8	7.9391
	p = 20.0 bar = 2.0 MPa (T_{sat} = 212.42°C)				p = 30.0 bar = 3.0 MPa (T_{sat} = 233.90°C)			
Sat.	0.0996	2600.3	2799.5	6.3409	0.0667	2604.1	2804.2	6.1869
240	0.1085	2659.6	2876.5	6.4952	0.0682	2619.7	2824.3	6.2265
280	0.1200	2736.4	2976.4	6.6828	0.0771	2709.9	2941.3	6.4462
320	0.1308	2807.9	3069.5	6.8452	0.0850	2788.4	3043.4	6.6245
360	0.1411	2877.0	3159.3	6.9917	0.0923	2861.7	3138.7	6.7801
400	0.1512	2945.2	3247.6	7.1271	0.0994	2932.8	3230.9	6.9212
440	0.1611	3013.4	3335.5	7.2540	0.1062	3002.9	3321.5	7.0520
500	0.1757	3116.2	3467.6	7.4317	0.1162	3108.0	3456.5	7.2338
540	0.1853	3185.6	3556.1	7.5434	0.1227	3178.4	3546.6	7.3474
600	0.1996	3290.9	3690.1	7.7024	0.1324	3285.0	3682.3	7.5085
640	0.2091	3362.2	3780.4	7.8035	0.1388	3357.0	3773.5	7.6106
700	0.2232	3470.9	3917.4	7.9487	0.1484	3466.5	3911.7	7.7571

(Continued)

TABLE AI.8A (CONTINUED)

Properties of Superheated Water Vapor (SI Units)

T (°C)	v (m³/kg)	u (kJ/kg)	h (kJ/kg)	s (kJ/kg · K)	v (m³/kg)	u (kJ/kg)	h (kJ/kg)	s (kJ/kg · K)
	p = 40 bar = 4.0 MPa (T_{sat} = 250.4°C)				p = 60 bar = 6.0 MPa (T_{sat} = 275.64°C)			
Sat.	0.04978	2602.3	2801.4	6.0701	0.03244	2589.7	2784.3	5.8892
280	0.05546	2680.0	2901.8	6.2568	0.03317	2605.2	2804.2	5.9252
320	0.06199	2767.4	3015.4	6.4553	0.03876	2720.0	2952.6	6.1845
360	0.06788	2845.7	3117.2	6.6215	0.04331	2811.2	3071.1	6.3782
400	0.07341	2919.9	3213.6	6.7690	0.04739	2892.9	3177.2	6.5408
440	0.07872	2992.2	3307.1	6.9041	0.05122	2970.0	3277.3	6.6853
500	0.08643	3099.5	3445.3	7.0901	0.05665	3082.2	3422.2	6.8803
540	0.09145	3171.1	3536.9	7.2056	0.06015	3156.1	3517.0	6.9999
600	0.09885	3279.1	3674.4	7.3688	0.06525	3266.9	3658.4	7.1677
640	0.1037	3351.8	3766.6	7.4720	0.06859	3341.0	3752.6	7.2731
700	0.1110	3462.1	3905.9	7.6198	0.07352	3453.1	3894.1	7.4234
740	0.1157	3536.6	3999.6	7.7141	0.07677	3528.3	3989.2	7.5190
	p = 80 bar = 8.0 MPa (T_{sat} = 295.06°C)				p = 100 bar = 10.0 MPa (T_{sat} = 311.06°C)			
Sat.	0.02352	2569.8	2758.0	5.7432	0.01803	2544.4	2724.7	5.6141
320	0.02682	2662.7	2877.2	5.9489	0.01925	2588.8	2781.3	5.7103
360	0.03089	2772.7	3019.8	6.1819	0.02331	2729.1	2962.1	6.0060
400	0.03432	2863.8	3138.3	6.3634	0.02641	2832.4	3096.5	6.2120
440	0.03742	2946.7	3246.1	6.5190	0.02911	2922.1	3213.2	6.3805
480	0.04034	3025.7	3348.4	6.6586	0.03160	3005.4	3321.4	6.5282
520	0.04313	3102.7	3447.7	6.7871	0.03394	3085.6	3425.1	6.6622
560	0.04582	3178.7	3545.3	6.9072	0.03619	3164.1	3526.0	6.7864
600	0.04845	3254.4	3642.0	7.0206	0.03837	3241.7	3625.3	6.9029
640	0.05102	3330.1	3738.3	7.1283	0.04048	3318.9	3723.7	7.0131
700	0.05481	3443.9	3882.4	7.2812	0.04358	3434.7	3870.5	7.1687
740	0.05729	3520.4	3978.7	7.3782	0.04560	3512.1	3968.1	7.2670
	p = 120 bar = 12.0 MPa (T_{sat} = 324.75°C)				p = 140 bar = 14.0 MPa (T_{sat} = 336.75°C)			
Sat.	0.01426	2513.7	2684.9	5.4924	0.01149	2476.8	2637.6	5.3717
360	0.01811	2678.4	2895.7	5.8361	0.01422	2617.4	2816.5	5.6602
400	0.02108	2798.3	3051.3	6.0747	0.01722	2760.9	3001.9	5.9448
440	0.02355	2896.1	3178.7	6.2586	0.01954	2868.6	3142.2	6.1474
480	0.02576	2984.4	3293.5	6.4154	0.02157	2962.5	3264.5	6.3143
520	0.02781	3068.0	3401.8	6.5555	0.02343	3049.8	3377.8	6.4610
560	0.02977	3149.0	3506.2	6.6840	0.02517	3133.6	3486.0	6.5941
600	0.03164	3228.7	3608.3	6.8037	0.02683	3215.4	3591.1	6.7172
640	0.03345	3307.5	3709.0	6.9164	0.02843	3296.0	3694.1	6.8326
700	0.03610	3425.2	3858.4	7.0749	0.03075	3415.7	3846.2	6.9939
740	0.03781	3503.7	3957.4	7.1746	0.03225	3495.2	3946.7	7.0952

(Continued)

TABLE AI.8A (CONTINUED)

Properties of Superheated Water Vapor (SI Units)

T (°C)	v (m³/kg)	u (kJ/kg)	h (kJ/kg)	s (kJ/kg · K)	v (m³/kg)	u (kJ/kg)	h (kJ/kg)	s (kJ/kg · K)
	p = 160 bar = 16.0 MPa (T_{sat} = 347.44°C)				p = 180 bar = 18.0 MPa (T_{sat} = 357.06°C)			
Sat.	0.00931	2431.7	2580.6	5.2455	0.00749	2374.3	2509.1	5.1044
360	0.01105	2539.0	2715.8	5.4614	0.00809	2418.9	2564.5	5.1922
400	0.01426	2719.4	2947.6	5.8175	0.01190	2672.8	2887.0	5.6887
440	0.01652	2839.4	3103.7	6.0429	0.01414	2808.2	3062.8	5.9428
480	0.01842	2939.7	3234.4	6.2215	0.01596	2915.9	3203.2	6.1345
520	0.02013	3031.1	3353.3	6.3752	0.01757	3011.8	3378.0	6.2960
560	0.02172	3117.8	3465.4	6.5132	0.01904	3101.7	3444.4	6.4392
600	0.02323	3201.8	3573.5	6.6399	0.02042	3188.0	3555.6	6.5696
640	0.02467	3284.2	3678.9	6.7580	0.02174	3272.3	3663.6	6.6905
700	0.02674	3406.0	3833.9	6.9224	0.02362	3396.3	3821.5	6.8580
740	0.02808	3486.7	3935.9	7.0251	0.02483	3478.0	3925.0	6.9623
	p = 200 bar = 20.0 MPa (T_{sat} = 365.81°C)				p = 240 bar = 24.0 MPa			
Sat.	0.00583	2293.0	2409.7	4.9269				
400	0.00994	2619.3	2818.1	5.5540	0.00673	2477.8	2639.4	5.2393
440	0.01222	2774.9	3019.4	5.8450	0.00929	2700.6	2923.4	5.6506
480	0.01399	2891.2	3170.8	6.0518	0.01100	2838.3	3102.3	5.8950
520	0.01551	2992.0	3302.2	6.2218	0.01241	2950.5	3248.5	6.0842
560	0.01689	3085.2	3423.0	6.3705	0.01366	3051.1	3379.0	6.2448
600	0.01818	3174.0	3537.6	6.5048	0.01481	3145.2	3500.7	6.3875
640	0.01940	3260.2	3648.1	6.6286	0.01588	3235.5	3616.7	6.5174
700	0.02113	3386.4	3809.0	6.7993	0.01739	3366.4	3783.8	6.6947
740	0.02224	3469.3	3914.1	6.9052	0.01835	3451.7	3892.1	6.8038
800	0.02385	3592.7	4069.7	7.0544	0.01974	3578.0	4051.6	6.9567
	p = 280 bar = 28.0 MPa				p = 320 bar = 32.0 MPa			
400	0.00383	2223.5	2330.7	4.7494	0.00236	1980.4	2055.9	4.3239
440	0.00712	2613.2	2812.6	5.4494	0.00544	2509.0	2683.0	5.2327
480	0.00885	2780.8	3028.5	5.7446	0.00722	2718.1	2949.2	5.5968
520	0.01020	2906.8	3192.3	5.9566	0.00853	2860.7	3133.7	5.8357
560	0.01136	3015.7	3333.7	6.1307	0.00963	2979.0	3287.2	6.0246
600	0.01241	3115.6	3463.0	6.2823	0.01061	3085.3	3424.6	6.1858
640	0.01338	3210.3	3584.8	6.4187	0.01150	3184.5	3552.5	6.3290
700	0.01473	3346.1	3758.4	6.6029	0.01273	3325.4	3732.8	6.5203
740	0.01558	3433.9	3870.0	6.7153	0.01350	3415.9	3847.8	6.6361
800	0.01680	3563.1	4033.4	6.8720	0.01460	3548.0	4015.1	6.7966
900	0.01873	3774.3	4298.8	7.1084	0.01633	3762.7	4285.1	7.0372

TABLE AI.8B

Properties of Superheated Water Vapor (English Units)

T (°F)	v (ft³/lb)	u (Btu/lb)	h (Btu/lb)	s (Btu/lb·°R)	v (ft³/lb)	u (Btu/lb)	h (Btu/lb)	s (Btu/lb°R)
	$p = 1$ lbf/in.² ($T_{sat} = 101.7$°F)				$p = 1$ lbf/in.² ($T_{sat} = 162.2$°F)			
Sat.	333.6	1044.0	1105.8	1.9779	73.53	1063.0	1131.0	1.8441
150	362.5	1060.4	1127.5	2.0151				
200	392.5	1077.5	1150.1	2.0508	78.15	1076.0	1148.6	1.8715
250	422.4	1094.7	1172.8	2.0839	84.21	1093.8	1171.7	1.9052
300	452.3	1112.0	1195.7	2.1150	90.24	1111.3	1194.8	1.9367
400	511.9	1147.0	1241.8	2.1720	102.24	1146.6	1241.2	1.9941
500	571.5	1182.8	1288.5	2.2235	114.20	1182.5	1288.2	2.0458
600	631.1	1219.3	1336.1	2.2706	126.15	1219.1	1335.8	2.0930
700	690.7	1256.7	1384.5	2.3142	138.08	1256.5	1384.3	2.1367
800	750.3	1294.4	1433.7	2.3550	150.01	1294.7	1433.5	2.1775
900	809.9	1333.9	1483.8	2.3932	161.94	1333.8	1483.7	2 2158
1000	869.5	1373.9	1534.8	2.4294	173.86	1373.9	1534.7	2.2520
	$p = 10$ lbf/in.² ($T_{sat} = 193.2$°F)				$p = 14.7$ lbf/in.² ($T_{sat} = 212.0$°F)			
Sat.	38.42	1072.2	1143.3	1.7877	26.80	1077.6	1150.5	1.7567
200	38.85	1074.7	1146.6	1.7927				
250	41.95	1092.6	1170.2	1.8272	28.42	1091.5	1168.8	1.7832
300	44.99	1110.4	1193.7	1.8592	30.52	1109.6	1192.6	1.8157
400	51.03	1146.1	1240.5	1.9171	34.67	1145.6	1239.9	1.8741
500	57.04	1182.2	1287.7	1.9690	38.77	1181.8	1287.3	1.9263
600	63.03	1218.9	1335.5	2.0164	42.86	1218.6	1335.2	1.9737
700	69.01	1256.3	1384.0	2.0601	46.93	1256.1	1383.8	2.0175
800	74.98	1294.6	1433.3	2.1009	51.00	1294.4	1433.1	2.0584
900	80.95	1333.7	1483.5	2.1393	55.07	1333.6	1483.4	2.0967
1000	86.91	1373.8	1534.6	2.1755	59.13	1373.7	1534.5	2.1330
1100	92.88	1414.7	1586.6	2.2099	63.19	1414.6	1586.4	2.1674
	$p = 20$ lbf/in.² ($T_{sat} = 228.0$°F)				$p = 40$ lbf/in.² ($T_{sat} = 267.3$°F)			
Sat.	20.09	1082.0	1156.4	1.7320	10.50	1093.3	1170.0	1.6767
250	20.79	1090.3	1167.2	1.7475				
300	22.36	1108.7	1191.5	1.7805	11.04	1105.1	1186.8	1.6993
350	23.90	1126.9	1215.4	1.8110	11.84	1124.2	1211.8	1.7312
400	25.43	1145.1	1239.2	1.8395	12.62	1143.0	1236.4	1.7606
500	28.46	1181.5	1286.8	1.8919	14.16	1180.1	1284.9	1.8140
600	31.47	1218.4	1334.8	1.9395	15.69	1217.3	1333.4	1.8621
700	34.47	1255.9	1383.5	1.9834	17.20	1255.1	1382.4	1.9063
800	37.46	1294.3	1432.9	2.0243	18.70	1293.7	1432.1	1.9474
900	40.45	1333.5	1483.2	2.0627	20.20	1333.0	1482.5	1.9859
1000	43.44	1373.5	1534.3	2.0989	21.70	1373.1	1533.8	2.0223
1100	46.42	1414.5	1586.3	2.1334	23.20	1414.2	1585.9	2.0568

(Continued)

TABLE AI.8B (CONTINUED)

Properties of Superheated Water Vapor (English Units)

T (°F)	v (ft³/lb)	u (Btu/lb)	h (Btu/lb)	s (Btu/lb·°R)	v (ft³/lb)	u (Btu/lb)	h (Btu/lb)	s (Btu/lb°R)
	p = 60 lbf/in.² (T_{sat} = 292.7°F)				p = 80 lbf/in.² (T_{sat} = 312.1°F)			
Sat.	7.17	1098.3	1178.0	1.6444	5.47	1102.6	1183.6	1.6214
300	7.26	1101.3	1181.9	1.6496				
350	7.82	1121.4	1208.2	1.6830	5.80	1118.5	1204.3	1.6476
400	8.35	1140.8	1233.5	1.7134	6.22	1138.5	1230.6	1.6790
500	9.40	1178.6	1283.0	1.7678	7.02	1177.2	1281.1	1.7346
600	10.43	1216.3	1332.1	1.8165	7.79	1215.3	1330.7	1.7838
700	11.44	1254.4	1381.4	1.8609	8.56	1253.6	1380.3	1.8285
800	12.45	1293.0	1431.2	1.9022	9.32	1292.4	1430.4	1.8700
900	13.45	1332.5	1481.8	1.9408	10.08	1332.0	1481.2	1.9087
1000	14.45	1372.7	1533.2	1.9773	10.83	1372.3	1532.6	1.9453
1100	15.45	1413.8	1585.4	2.0119	11.58	1413.5	1584.9	1.9799
1200	16.45	1455.8	1638.5	2.0448	12.33	1455.5	1638.1	2.0130
	p = 100 lbf/in.² (T_{sat} = 327.8°F)				p = 120 lbf/in.² (T_{sat} = 341.3°F)			
Sat.	4.434	1105.8	1187.8	1.6034	3.730	1108.3	1191.1	1.5886
350	4.592	1115.4	1200.4	1.6191	3.783	1112.2	1196.2	1.5950
400	4.934	1136.2	1227.5	1.6517	4.079	1133.8	1224.4	1.6288
450	5.265	1156.2	1253.6	1.6812	4.360	1154.3	1251.2	1.6590
500	5.587	1175.7	1279.1	1.7085	4.633	1174.2	1277.1	1.6868
600	6.216	1214.2	1329.3	1.7582	5.164	1213.2	1327.8	1.7371
700	6.834	1252.8	1379.2	1.8033	5.682	1252.0	1378.2	1.7825
800	7.445	1291.8	1429.6	1.8449	6.195	1291.2	1428.7	1.8243
900	8.053	1331.5	1480.5	1.8838	6.703	1330.9	1479.8	1.8633
1000	8.657	1371.9	1532.1	1.9204	7.208	1371.5	1531.5	1.9000
1100	9.260	1413.1	1584.5	1.9551	7.711	1412.8	1584.0	1.9348
1200	9.861	1455.2	1637.7	1.9882	8.213	1454.9	1637.3	1.9679
	p = 140 lbf/in.² (T_{sat} = 353.1°F)				p = 160 lbf/in.² (T_{sat} = 363.6°F)			
Sat.	3.221	1110.3	1193.8	1.5761	2.836	1112.0	1196.0	1.5651
400	3.466	1131.4	1221.2	1.6088	3.007	1128.8	1217.8	1.5911
450	3.713	1152.4	1248.6	1.6399	3.228	1150.5	1246.1	1.6230
500	3.952	1172.7	1275.1	1.6682	3.440	1171.2	1273.0	1.6518
550	4.184	1192.5	1300.9	1.6945	3.646	1191.3	1299.2	1.6785
600	4.412	1212.1	1326.4	1.7191	3.848	1211.1	1325.0	1.7034
700	4.860	1251.2	1377.1	1.7648	4.243	1250.4	1376.0	1.7494
800	5.301	1290.5	1427.9	1.8068	4.631	1289.9	1427.0	1.7916
900	5.739	1330.4	1479.1	1.8459	5.015	1329.9	1478.4	1.8308
1000	6.173	1371.0	1531.0	1.8827	5.397	1370.6	1530.4	1.8677
1100	6.605	1412.4	1583.6	1.9176	5.776	1412.1	1583.1	1.9026
1200	7.036	1454.6	1636.9	1.9507	6.154	1454.3	1636.5	1.9358

(Continued)

TABLE AI.8B (CONTINUED)

Properties of Superheated Water Vapor (English Units)

T (°F)	v (ft³/lb)	u (Btu/lb)	h (Btu/lb)	s (Btu/lb·°R)	v (ft³/lb)	u (Btu/lb)	h (Btu/lb)	s (Btu/lb°R)
	$p = 180$ lbf/in.² ($T_{sat} = 373.1$°F)				$p = 200$ lbf/in.² ($T_{sat} = 381.8$°F)			
Sat.	2.533	1113.4	1197.8	1.5553	2.289	1114.6	1199.3	1.5464
400	2.648	1126.2	1214.4	1.5749	2.361	1123.5	1210.8	1.5600
450	2.850	1148.5	1243.4	1.6078	2.548	1146.4	1240.7	1.5938
500	3.042	1169.6	1270.9	1.6372	2.724	1168.0	1268.8	1.6239
550	3.228	1190.0	1297.5	1.6642	2.893	1188.7	1295.7	1.6512
600	3.409	1210.0	1323.5	1.6893	3.058	1208.9	1322.1	1.6767
700	3.763	1249.6	1374.9	1.7357	3.379	1248.8	1373.8	1.7234
800	4.110	1289.3	1426.2	1.7781	3.693	1288.6	1425.3	1.7660
900	4.453	1329.4	1477.7	1.8174	4.003	1328.9	1477.1	1.8055
1000	4.793	1370.2	1529.8	1.8545	4.310	1369.8	1529.3	1.8425
1100	5.131	1411.7	1582.6	1.8894	4.615	1411.4	1582.2	1.8776
1200	5.467	1454.0	1636.1	1.9227	4.918	1453.7	1635.7	1.9109
	$p = 250$ lbf/in.² ($T_{sat} = 401.0$°F)				$p = 300$ lbf/in.² ($T_{sat} = 417.4$°F)			
Sat.	1.845	1116.7	1202.1	1.5274	1.544	1118.2	1203.9	1.5115
450	2.002	1141.1	1233.7	1.5632	1.636	1135.4	1226.2	1.5365
500	2.150	1163.8	1263.3	1.5948	1.766	1159.5	1257.5	1.5701
550	2.290	1185.3	1291.3	1.6233	1.888	1181.9	1286.7	1.5997
600	2.426	1206.1	1318.3	1.6494	2.004	1203.2	1314.5	1.6266
700	2.688	1246.7	1371.1	1.6970	2.227	1244.0	1368.3	1.6751
800	2.943	1287.0	1423.2	1.7301	2.442	1285.4	1421.0	1.7187
900	3.193	1327.6	1475.3	1.7799	2.653	1326.3	1473.6	1.7589
1000	3.440	1368.7	1527.9	1.8172	2.860	1367.7	1526.5	1.7964
1100	3.685	1410.5	1581.0	1.8524	3.066	1409.6	1579.8	1.8317
1200	3.929	1453.0	1634.8	1.8858	3.270	1452.2	1633.8	1.8653
1300	4.172	1496.3	1689.3	1.9177	3.473	1495.6	1688.4	1.8973
	$p = 350$ lbf/in.² ($T_{sat} = 431.8$°F)				$p = 400$ lbf/in.² ($T_{sat} = 444.7$°F)			
Sat.	1.327	1119.0	1204.9	1.4978	1.162	1119.5	1205.5	1.4856
450	1.373	1129.2	1218.2	1.5125	1.175	1122.6	1209.5	1.4901
500	1.491	1154.9	1251.5	1.5482	1.284	1150.1	1245.2	1.5282
550	1.600	1178.3	1281.9	1.5790	1.383	1174.6	1277.0	1.5605
600	1.703	1200.3	1310.6	1.6068	1.476	1197.3	1306.6	1.5892
700	1.898	1242.5	1365.4	1.6562	1.650	1240.4	1362.5	1.6397
800	2.085	1283.8	1418.8	1.7004	1.816	1282.1	1416.6	1.6844
900	2.267	1325.0	1471.8	1.7409	1.978	1323.7	1470.1	1.7252
1000	2.446	1366.6	1525.0	1.7787	2.136	1365.5	1523.6	1.7632
1100	2.624	1408.7	1578.6	1.8142	2.292	1407.8	1577.4	1.7989
1200	2.799	1451.5	1632.8	1.8478	2.446	1450.7	1621.8	1.8327
1300	2.974	1495.0	1687.6	1.8799	2.599	1494.3	1686.8	1.8648

(Continued)

TABLE AI.8B (CONTINUED)

Properties of Superheated Water Vapor (English Units)

T (°F)	v (ft³/lb)	u (Btu/lb)	h (Btu/lb)	s (Btu/lb·°R)	v (ft³/lb)	u (Btu/lb)	h (Btu/lb)	s (Btu/lb°R)
	$p = 450$ lbf/in.² ($T_{sat} = 456.4$°F)				$p = 500$ lbf/in.² ($T_{sat} = 467.1$°F)			
Sat.	1.033	1119.6	1205.6	1.4746	0.928	1119.4	1205.3	1.4645
500	1.123	1145.1	1238.5	1.5097	0.992	1139.7	1231.5	1.4923
550	1.215	1170.7	1271.9	1.5436	1.079	1166.7	1266.6	1.5279
600	1.300	1194.3	1302.5	1.5732	1.158	1191.1	1298.3	1.5585
700	1.458	1238.2	1359.6	1.6248	1.304	1236.0	1356.7	1.6112
800	1.608	1280.5	1414.4	1.6701	1.441	1278.8	1412.1	1.6571
900	1.752	1322.4	1468.3	1.7113	1.572	1321.0	1466.5	1.6987
1000	1.894	1364.4	1522.2	1.7495	1.701	1363.3	1520.7	1.7371
1100	2.034	1406.9	1576.3	1.7853	1.827	1406.0	1575.1	1.7731
1200	2.172	1450.0	1630.8	1.8192	1.952	1449.2	1629.8	1.8072
1300	2.308	1493.7	1685.9	1.8515	2.075	1493.1	1685.1	1.8395
1400	2.444	1538.1	1741.7	1.8823	2.198	1537.6	1741.0	1.8704
	$p = 600$ lbf/in.² ($T_{sat} = 486.3$°F)				$p = 700$ lbf/in.² ($T_{sat} = 503.2$°F)			
Sat.	0.770	1118.6	1204.1	1.4464	0.656	1117.0	1202.0	1.4305
500	0.795	1128.0	1216.2	1.4592				
550	0.875	1158.2	1255.4	1.4990	0.728	1149.0	1243.2	1.4723
600	0.946	1184.5	1289.5	1.5320	0.793	1177.5	1280.2	1.5081
700	1.073	1231.5	1350.6	1.5872	0.907	1226.9	1344.4	1.5661
800	1.190	1275.4	1407.6	1.6343	1.011	1272.0	1402.9	1.6145
900	1.302	1318.4	1462.9	1.6766	1.109	1315.6	1459.3	1.6576
1000	1.411	1361.2	1517.8	1.7155	1.204	1358.9	1514.9	1.6970
1100	1.517	1404.2	1572.7	1.7519	1.296	1402.4	1570.2	1.7337
1200	1.622	1447.7	1627.8	1.7861	1.387	1446.2	1625.8	1.7682
1300	1.726	1491.7	1683.4	1.8186	1.476	1490.4	1681.7	1.8009
1400	1.829	1536.5	1739.5	1.8497	1.565	1535.3	1738.1	1.8321
	$p = 800$ lbf/in.² ($T_{sat} = 518.3$°F)				$p = 900$ lbf/in.² ($T_{sat} = 532.1$°F)			
Sat.	0.569	1115.0	1199.3	1.4160	0.501	1112.6	1196.0	1.4027
550	0.615	1138.8	1229.9	1.4469	0.527	1127.5	1215.2	1.4219
600	0.677	1170.1	1270.4	1.4861	0.587	1162.2	1260.0	1.4652
650	0.732	1197.2	1305.6	1.5186	0.639	1191.1	1297.5	1.4999
700	0.783	1222.1	1338.0	1.5471	0.686	1217.1	1331.4	1.5297
800	0.876	1268.5	1398.2	1.5969	0.772	1264.9	1393.4	1.5810
900	0.964	1312.9	1455.6	1.6408	0.851	1310.1	1451.9	1.6257
1000	1.048	1356.7	1511.9	1.6807	0.927	1354.5	1508.9	1.6662
1100	1.130	1400.5	1567.8	1.7178	1.001	1398.7	1565.4	1.7036
1200	1.210	1444.6	1623.8	1.7526	1.073	1443.0	1621.7	1.7386

(Continued)

TABLE AI.8B (CONTINUED)

Properties of Superheated Water Vapor (English Units)

T (°F)	v (ft³/lb)	u (Btu/lb)	h (Btu/lb)	s (Btu/lb·°R)	v (ft³/lb)	u (Btu/lb)	h (Btu/lb)	s (Btu/lb°R)
1300	1.289	1489.1	1680.0	1.7854	1.144	1487.8	1687.3	1.7717
1400	136.7	1534.2	1736.6	1.8167	1.214	1533.0	1735.1	1.8031
	$p = 1000$ lbf/in.² ($T_{sat} = 544.7$°F)				$p = 1200$ lbf/in.² ($T_{sat} = 567.4$°F)			
Sat.	0.446	1109.0	1192.4	1.3903	0.362	1103.5	1183.9	1.3673
600	0.514	1153.7	1248.8	1.4450	0.402	1134.4	1223.6	1.4054
650	0.564	1184.7	1289.1	1.4822	0.450	1170.9	1270.8	1.4490
700	0.608	1212.0	1324.6	1.5135	0.491	1201.3	1310.2	1.4837
800	0.688	1261.2	1388.5	1.5665	0.562	1253.7	1378.4	1.5402
900	0.761	1307.3	1448.1	1.6120	0.626	1301.5	1440.4	1.5876
1000	0.831	1352.2	1505.9	1.6530	0.685	1347.5	1499.7	1.6297
1100	0.898	1396.8	1562.9	1.6908	0.743	1393.0	1557.9	1.6682
1200	0.963	1441.5	1619.7	1.7261	0.798	1438.3	1615.5	1.7040
1300	1.027	1486.5	1676.5	1.7593	0.853	1483.8	1673.1	1.7377
1400	1.091	1531.9	1733.7	1.7909	0.906	1529.6	1730.7	t .7696
1600	1.215	1624.4	1849.3	1.8499	1.011	1622.6	1847.1	1.8290
	$p = 1400$ lbf/in.² ($T_{sat} = 587.2$ °F)				$p = 1600$ lbf/in.² = 605.1°F)			
Sat.	0.302	1096.0	1174.1	1.3461	0.255	1087.4	1162.9	1.3258
600	0.318	1110.9	1193.1	1.3641				
650	0.367	1155.5	1250.5	1.4171	0.303	1137.8	1227.4	1.3852
700	0.406	1189.6	1294.8	1.4562	0.342	1177.0	1278.1	1.4299
800	0.471	1245.8	1367.9	1.5168	0.403	1237.7	1357.0	1.4953
900	0.529	1295.6	1432.5	1.5661	0.466	1289.5	1424.4	1.5468
1000	0.582	1342.8	1493.5	1.6094	0.504	1338.0	1487.1	1.5913
1100	0.632	1389.1	1552.8	1.6487	0.549	1385.2	1547.7	1.6315
1200	0.681	1435.1	1611.4	1.6851	0.592	1431.8	1607.1	1.6684
1300	0.728	1481.1	1669.6	1.7192	0.634	1478.3	1666.1	1.7029
1400	0.774	1527.2	1727.8	1.7513	0.675	1524.9	1724.8	1.7354
1600	0.865	1620.8	1844.8	1.8111	0.755	1619.0	1842.6	1.7955
	$p = 1800$ lbf/in.² ($T_{sat} = 621.2$°F)				$p = 2000$ lbf/in.² ($T_{sat} = 636.0$°F)			
Sat.	0.218	1077.7	1150.4	1.3060	0.188	1066.6	1136.3	1.2861
650	0.251	1117.0	1200.4	1.3517	0.206	1091.1	1167.2	1.3141
700	0.291	1163.1	1259.9	1.4042	0.249	1147.7	1239.8	1.3752
750	0.322	1198.6	1305.9	1.4430	0.280	1187.3	1291.1	1.4216
800	0.350	1229.1	1345.7	1.4753	0.307	1220.1	1333.8	1.4562
900	0.399	1283.2	1416.1	1.5291	0.353	1276.8	1407.6	1.5126
1000	0.443	1333.1	1480.7	1.5749	0.395	1328.1	1474.1	1.5598
1100	0.484	1381.2	1542.5	1.6159	0.433	1377.2	1537.2	1.6017
1200	0.524	1428.5	1602.9	1.6534	0.469	1425.2	1598.6	1.6398
1300	0.561	1475.5	1662.5	1.6883	0.503	1472.7	1659.0	1.6751
MOO	0.598	1522.5	1721.8	1.7211	0.537	1520.2	1718.8	1.7082
1600	0.670	1617.2	1840.4	1.7817	0.602	1615.4	1838.2	1.7692

(Continued)

TABLE AI.8B (CONTINUED)

Properties of Superheated Water Vapor (English Units)

T (°F)	v (ft³/lb)	u (Btu/lb)	h (Btu/lb)	s (Btu/lb·°R)	v (ft³/lb)	u (Btu/lb)	h (Btu/lb)	s (Btu/lb°R)
	$p = 2500$ lbf/in.² ($T_{sat} = 668.3$°F)				$p = 3000$ lbf/in.² ($T_{sat} = 695.5$°F)			
Sat.	0.1306	1031.0	1091.4	1.2327	0.0840	968.8	1015.5	1.1575
700	0.1684	1098.7	1176.6	1.3073	0.0977	1003.9	1058.1	1.1944
750	0.2030	1155.2	1249.1	1.3686	0.1483	1114.7	1197.1	1.3122
800	0.2291	1195.7	1301.7	1.4112	0.1757	1167.6	1265.2	1.3675
900	0.2712	1259.9	1385.4	1.4752	0.2160	1241.8	1361.7	1.4414
1000	0.3069	1315.2	1457.2	1.5262	0.2485	1301.7	1439.6	1.4967
1100	0.3393	1366.8	1523.8	1.5704	0.2772	1356.2	1510.1	1.5434
1200	0.3696	1416.7	1587.7	1.6101	0.3086	1408.0	1576.6	1.5848
1300	0.3984	1465.7	1650.0	1.6465	0.3285	1458.5	1640.9	1.6224
1400	0.4261	1514.2	1711.3	1.6804	0.3524	1508.1	1703.7	1.6571
1500	0.4531	1562.5	1772.1	1.7123	0.3754	1557.3	1765.7	1.6896
1600	0.4795	1610.8	1832.6	1.7424	0.3978	1606.3	1827.1	1.7201
	$p = 3500$ lbf/in.²				$p = 4000$ lbf/in.²			
650	0.0249	663.5	679.7	0.8630	0.0245	657.7	675.8	0.8574
700	0.0306	759.5	779.3	0.9506	0.0287	742.1	763.4	0.9345
750	0.1046	1058.4	1126.1	1.2440	0.0633	960.7	1007.5	1.1395
800	0.1363	1134.7	1223.0	1.3226	0.1052	1095.0	1172.9	1.2740
900	0.1763	1222.4	1336.5	1.4096	0.1462	1201.5	1309.7	1.3789
1000	0.2066	1287.6	1421.4	1.4699	0.1752	1272.9	1402.6	1.4449
1100	0.2328	1345.2	1496.0	1.5193	0.1995	1333.9	1481.6	1.4973
1200	0.2566	1399.2	1565.3	1.5624	0.2213	1390.1	1553.9	1.5423
1300	0.2787	1451.1	1631.7	1.6012	0.2414	1443.7	1622.4	1.5823
1400	0.2997	1501.9	1696.1	1.6368	0.2603	1495.7	1688.4	1.6188
1500	0.3199	1552.0	1759.2	1.6699	0.2784	1546.7	1752.8	1.6526
1600	0.3395	1601.7	1831.6	1.7010	0.2959	1597.1	1816.1	1.6841
	$p = 4400$ lbf/in.²				$p = 4800$ lbf/in.²			
650	0.0242	653.6	673.3	0.8535	0.0237	649.8	671.0	0.8499
700	0.0278	732.7	755.3	0.9257	0.0271	725.1	749.1	0.9187
750	0.0415	870.8	904.6	1.0513	0.0352	832.6	863.9	1.0154
800	0.0844	1056.5	1125.3	1.2306	0.0668	1011.2	1070.5	1.1827
900	0.1270	1183.7	1287.1	1.3548	0.1109	1164.8	1263.4	1.3310
1000	0.1552	1260.8	1387.2	1.4260	0.1385	1248.3	1317.4	1.4078
1100	0.1784	1324.7	1469.9	1.4809	0.1608	1315.3	1458.1	1.4653
1200	0.1989	1382.8	1544.7	1.5274	0.1802	1375.4	1535.4	1.5133
1300	0.2176	1437.7	1614.9	1.5685	0.1979	1431.7	1607.4	1.5555
1400	0.2352	1490.7	1682.3	1.6057	0.2143	1485.7	1676.1	1.5934
1500	0.2520	1542.7	1747.6	1.6399	0.2300	1538.2	1742.5	1.6282
1600	0.2681	1593.4	1811.7	1.6718	0.2450	1589.8	1807.4	1.6605

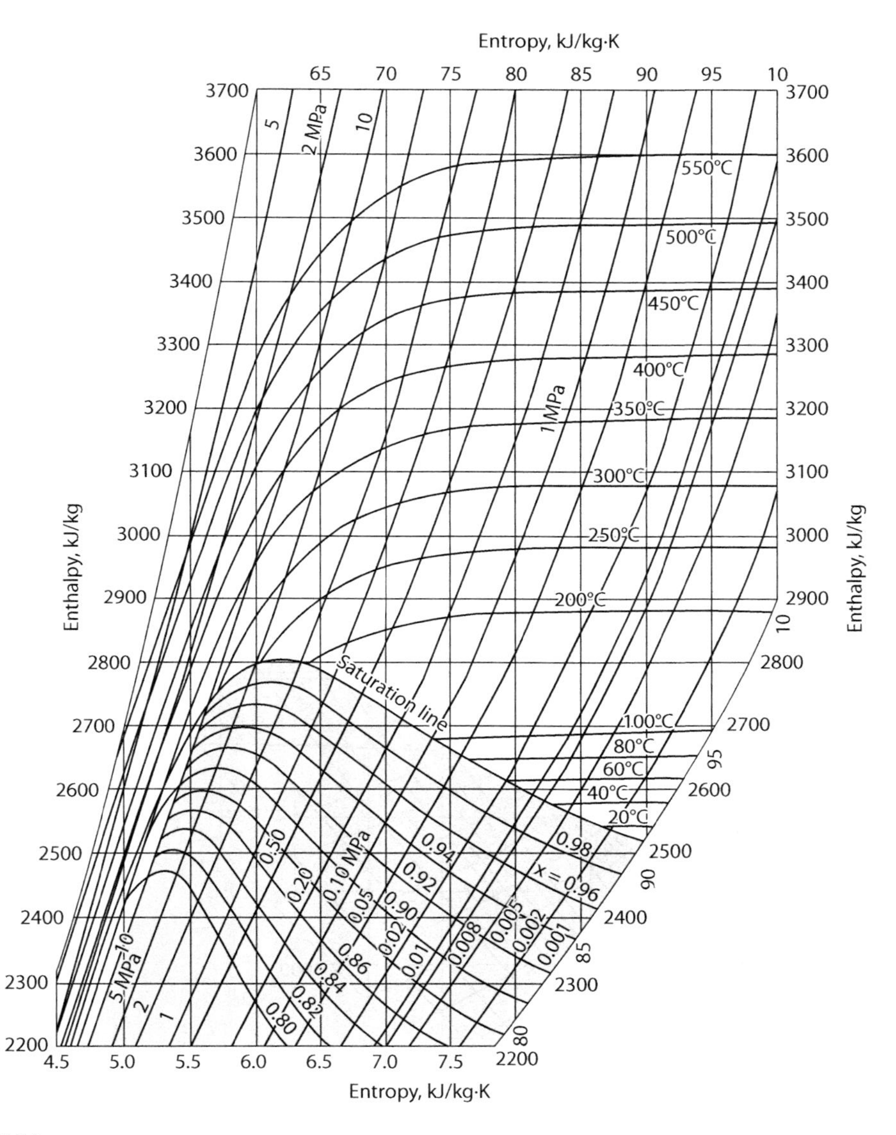

FIGURE AI.9A
Enthalpy-entropy diagram for water (SI units). (From Jones, J.B. and Hawkins, G.AI., *Engineering Thermodynamics*, Wiley, New York, 1986.)

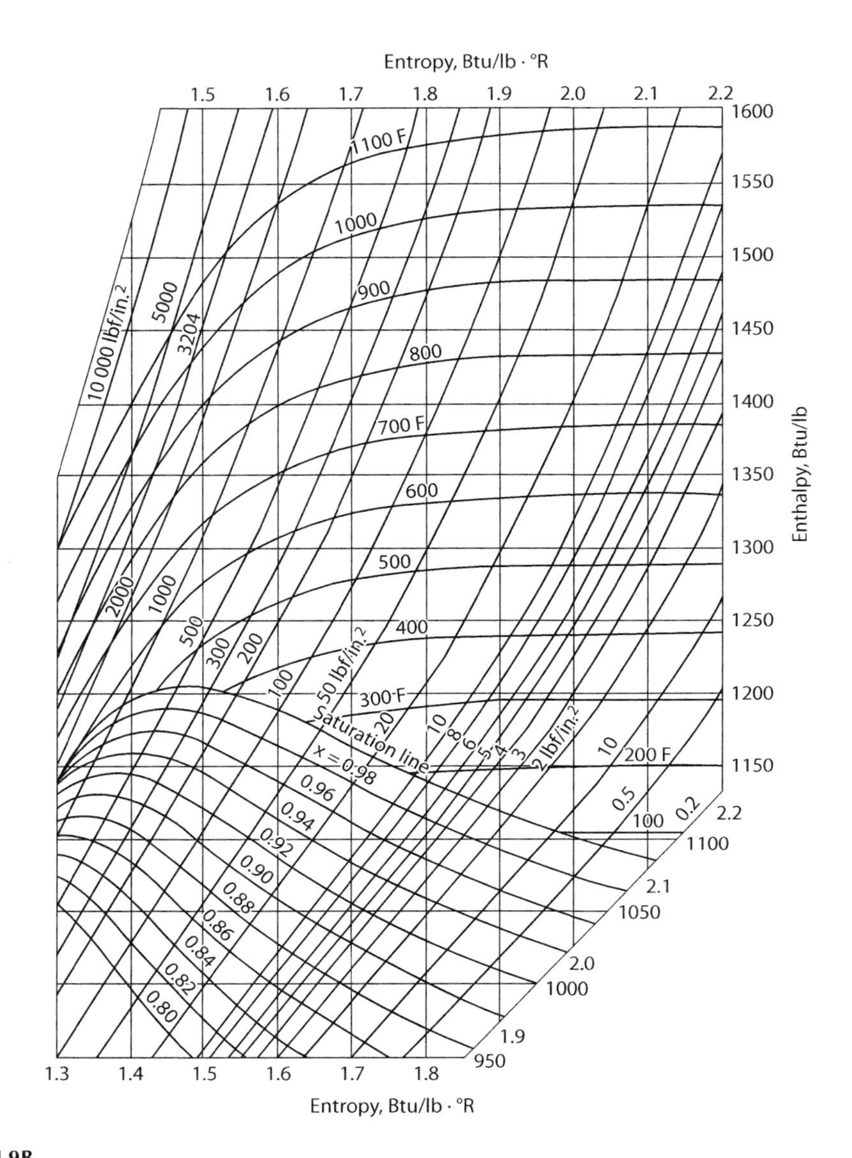

FIGURE AI.9B
Enthalpy-entropy diagram for water (English units). (From Jones, J.B. and Hawkins, G.AI., *Engineering Thermodynamics*, Wiley, New York, 1986.)

Appendix II: Some Formulas in Thermodynamics and Compressible Flows

The equation of state for ideal gases is given by

$$pv_s = RT \quad \text{or} \quad p = \rho RT \tag{AII.1}$$

For steady state and steady flows with a single inlet and exit, the energy equation or the first law of thermodynamics is written as

$$\dot{Q} + \dot{m}\left(h_1 + \frac{V_1^2}{2} + gz_1\right) = \dot{W}_s + \dot{m}\left(h_2 + \frac{V_2^2}{2} + gz_2\right) \tag{AII.2}$$

For adiabatic flows, the energy equation can be simplified as

$$h_1 + \frac{V_1^2}{2} + gz_1 = h_2 + \frac{V_2^2}{2} + gz_2 + E \tag{AII.3}$$

The energy given by Equation AII.3 can be further simplified for liquids and gases as follows:

$$\text{For liquids} \quad \frac{p_1}{\rho} + \frac{V_1^2}{2} + gz_1 = \frac{p_2}{\rho} + \frac{V_2^2}{2} + gz_2 + E \tag{AII.4a}$$

$$\text{For gases} \quad h_1 + \frac{V_1^2}{2} = h_2 + \frac{V_2^2}{2} + E \tag{AII.4b}$$

The equation of isentropic flows of ideal gases may be represented as

$$\frac{p}{\rho^k} = \text{const} \quad \text{or} \quad \frac{p_1}{p_2} = \left(\frac{\rho_1}{\rho_2}\right)^k \tag{AII.5a}$$

$$\text{or} \quad \frac{p}{T^{\frac{k}{k-1}}} = \text{const} \quad \text{or} \quad \frac{p_1}{p_2} = \left(\frac{T_1}{T_2}\right)^{\frac{k}{k-1}} \tag{AII.5b}$$

$$\text{or} \quad \frac{\rho}{T^{\frac{1}{k-1}}} = \text{const} \quad \text{or} \quad \frac{\rho_1}{\rho_2} = \left(\frac{T_1}{T_2}\right)^{\frac{1}{k-1}} \tag{AII.5c}$$

where k is the ratio of specific heats, i.e. $k = C_p/C_v$.

Although the specific heats are dependent only on temperature for ideal gases, they are treated as constant in the discussion on turbomachinery. In this context, they are defined as

$$C_p = \frac{dh}{dt} \Rightarrow \Delta h = C_p \Delta T \Rightarrow (h_2 - h_1) = C_p (T_2 - T_1) \tag{AII.6}$$

Since $C_p - C_v = R$, the specific heats, gas constant, and the ratio of specific heats are related by the following expressions:

$$C_p = \frac{kR}{k-1}; \qquad C_v = \frac{R}{k-1} \tag{AII.7}$$

The speed of sound in a fluid and Mach number are given by

$$c = \sqrt{\frac{kp}{\rho}} = \sqrt{kRT} \tag{AII.8}$$

$$M = \frac{V}{c} = \frac{V}{\sqrt{kRT}} \tag{AII.9}$$

For isentropic flows of ideal gases, the static and stagnation temperatures can be expressed in terms of the local Mach number as

$$\frac{T_0}{T} = 1 + \frac{k-1}{2} M^2 \tag{AII.10}$$

For isentropic flow between any two points,

$$\frac{T_2}{T_1} = \frac{1 + \frac{k-1}{2} M_2^2}{1 + \frac{k-1}{2} M_1^2} \tag{AII.11}$$

Using Equation AII.11 and Equations AII.5b and AII.5c, pressures and densities can be written in terms of local Mach numbers. In the absence of shock waves, the flows through turbomachines are considered to be isentropic.

Appendix III

Some Conversion Factors

Physical Quantity	Symbol	Conversion Factor
Area	A	1 $ft^2 = 0.0929\ m^2$
		1 $in^2 = 6.452 \times 10^4\ m^2$
Density	P	1 $lb_m/ft^3 = 16.018\ kg/m^3$
		1 $slug/ft^3 = 515.379\ kg/m^3$
Energy, heat	Q	1 Btu = 1055.1 J
		1 cal = 4.186 J
		$l(ft)(lb_f) = 1.3558$ J
		1 (hp)(h) $= 2.685 \times 10^6$J
Force	F	$lb_f = 4.448$ N
Heat flow rate	q	1 Btu/h = 0.2931 W
		1 Btu/s = 1055.1 W
Heat flux	q''	1 $Btu/(h)(ft^2) = 3.1525\ W/m^2$
Heat generation per unit volume	q_G	1 $Btu/(h)(ft^3) = 10.343\ W/m^3$
Heat transfer coefficient	h	1 $Btu/(h)(ft^2)(°F) = 5.678\ W/m^2\ K$
Length	L	1 ft = 0.3048 m
		1 in. = 2.54 cm = 0.0254 m
		1 mile = 1.6093 km = 1609.3 m
Mass	m	1 $lb_m = 0.4536$ kg
		1 slug = 14.594 kg
Mass flow rate	$\dot{m}$	1 $lb_m/h = 0.000126$ kg/s
		1 $lb_m/s = 0.4536$ kg/s
Power	W	1 hp = 745.7 W
		1 $(ft)(lb_r)/s = 1.3558$ W
		1 Btu/s = 1055.1 W
		1 Btu/h = 0.293 W
Pressure	P	1 $lb_f/in.^2 = 6894.8\ N/m^2$ (Pa)
		1 $lb_f/ft^2 = 47.88\ N/m^2$ (Pa)
		1 atm = 101,325 N/m^2 (Pa)
Specific energy	Q/m	1 $Btu/lb_m = 2326.1$ J/kg
Specific heat capacity	c	1 $Btu/(lb_m)(°F) = 4188$ J/kg K
Temperature	T	T(°R) = (9/5) T (K)
		T(°F) = [T(°C)](9/5) + 32
		T(°F) = [T(K) – 273.15](9/5) + 32
Thermal conductivity	k	1 Btu/(h)(ft)(°F) = 1.731 W/m K
Thermal diffusivity	α	1 $ft^2/s = 0.0929\ m^2/s$
		1 $fr/h = 2.581 \times 10^{-5}\ m^2/s$
Thermal resistance	R_t	1 (h)(°F)/Btu = 1.8958 K/W
Velocity	U	1 ft/s = 0.3048 m/s
		1 mph = 0.44703 m/s

(Continued)

(Continued)

Viscosity, dynamic	μ	1 $lb_m/(ft)(s) = 1.488\ N\ s/m^2$
		1 centipoise = $0.00100\ N\ s/m^2$
Viscosity, kinematic	ν	1 $ft^2/s = 0.0929\ m^2/s$
		1 $ft^2/h = 2.581 \times 10^{-5}\ m^2/s$
Volume	V	1 $ft^3 = 0.02832\ m^3$
		1 $in.^3 = 1.6387 \times 10^{-5}\ m^3$
		1 gal(U.S. liq.) = $0.003785\ m^3$

Index